Challenges and Research Trends of Renewable Energy Power System

Challenges and Research Trends of Renewable Energy Power System

Editor

Yogendra Arya

MDPI • Basel • Beijing • Wuhan • Barcelona • Belgrade • Manchester • Tokyo • Cluj • Tianjin

Editor
Yogendra Arya
J.C. Bose University of
Science & Technology, YMCA
Faridabad
India

Editorial Office
MDPI
St. Alban-Anlage 66
4052 Basel, Switzerland

This is a reprint of articles from the Special Issue published online in the open access journal *Energies* (ISSN 1996-1073) (available at: https://www.mdpi.com/journal/energies/special_issues/_Renewable_Power_System).

For citation purposes, cite each article independently as indicated on the article page online and as indicated below:

LastName, A.A.; LastName, B.B.; LastName, C.C. Article Title. *Journal Name* **Year**, *Volume Number*, Page Range.

ISBN 978-3-0365-7528-5 (Hbk)
ISBN 978-3-0365-7529-2 (PDF)

© 2023 by the authors. Articles in this book are Open Access and distributed under the Creative Commons Attribution (CC BY) license, which allows users to download, copy and build upon published articles, as long as the author and publisher are properly credited, which ensures maximum dissemination and a wider impact of our publications.
The book as a whole is distributed by MDPI under the terms and conditions of the Creative Commons license CC BY-NC-ND.

Contents

About the Editor vii

Preface to "Challenges and Research Trends of Renewable Energy Power System" ix

Sowmya Ramachandradurai, Narayanan Krishnan and Prabaharan Natarajan
Unintentional Passive Islanding Detection and Prevention Method with Reduced
Non-Detection Zones
Reprinted from: *Energies* **2022**, *15*, 3038, doi:10.3390/en15093038 1

Mohammed Ozayr Abdul Kader, Kayode Timothy Akindeji and Gulshan Sharma
A Novel Solution for Solving the Frequency Regulation Problem of Renewable Interlinked
Power System Using Fusion of AI
Reprinted from: *Energies* **2022**, *15*, 3376, doi:10.3390/en15093376 27

Anurag Gautam, Ibraheem, Gulshan Sharma, Pitshou N. Bokoro and Mohammad F. Ahmer
Available Transfer Capability Enhancement in Deregulated Power System through TLBO
Optimised TCSC
Reprinted from: *Energies* **2022**, *15*, 4448, doi:10.3390/en15124448 47

Milan Joshi, Gulshan Sharma, Pitshou N. Bokoro and Narayanan Krishnan
A Fuzzy-PSO-PID with UPFC-RFB Solution for an LFC of an Interlinked Hydro Power System
Reprinted from: *Energies* **2022**, *15*, 4847, doi:10.3390/en15134847 63

Hemant Ahuja, Arika Singh, Sachin Sharma, Gulshan Sharma and Pitshou N. Bokoro
Coordinated Control of Wind Energy Conversion System during Unsymmetrical Fault at Grid
Reprinted from: *Energies* **2022**, *15*, 4898, doi:10.3390/en15134898 81

**Sudhir Kumar Singh, Rajveer Singh, Haroon Ashfaq, Sanjeev Kumar Sharma,
Gulshan Sharma and Pitshou N. Bokoro**
Super-Twisting Algorithm-Based Virtual Synchronous Generator in Inverter Interfaced
Distributed Generation (IIDG)
Reprinted from: *Energies* **2022**, *15*, 5890, doi:10.3390/en15165890 97

Anubhav Kumar Pandey, Vinay Kumar Jadoun and Jayalakshmi N. Sabhahit
Real-Time Peak Valley Pricing Based Multi-Objective Optimal Scheduling of a Virtual Power
Plant Considering Renewable Resources
Reprinted from: *Energies* **2022**, *15*, 5970, doi:10.3390/en15165970 113

**Sivakavi Naga Venkata Bramareswara Rao, Venkata Pavan Kumar Yellapragada,
Kottala Padma, Darsy John Pradeep, Challa Pradeep Reddy, Mohammad Amir
and Shady S. Refaat**
Day-Ahead Load Demand Forecasting in Urban Community Cluster Microgrids Using Machine
Learning Methods
Reprinted from: *Energies* **2022**, *15*, 6124, doi:10.3390/en15176124 143

**Arunesh Kumar Singh, Tabish Tariq, Mohammad F. Ahmer, Gulshan Sharma,
Pitshou N. Bokoro and Thokozani Shongwe**
Intelligent Control of Irrigation Systems Using Fuzzy Logic Controller
Reprinted from: *Energies* **2022**, *15*, 7199, doi:10.3390/en15197199 169

Bhabasis Mohapatra, Binod Kumar Sahu, Swagat Pati, Mohit Bajaj, Vojtech Blazek, Lukas Prokop, Stanislav Misak, et al.
Real-Time Validation of a Novel IAOA Technique-Based Offset Hysteresis Band Current Controller for Grid-Tied Photovoltaic System
Reprinted from: *Energies* **2022**, *15*, 8790, doi:10.3390/en15238790 . . . . **189**

Kabulo Loji, Sachin Sharma, Nomhle Loji, Gulshan Sharma and Pitshou N. Bokoro
Operational Issues of Contemporary Distribution Systems: A Review on Recent and Emerging Concerns
Reprinted from: *Energies* **2023**, *16*, 1732, doi:10.3390/en16041732 . . . . **215**

About the Editor

Yogendra Arya

Yogendra Arya received his PhD in Electrical Engineering from Delhi Technological University, New Delhi, India, in 2018. He is currently working as an Associate Professor at J.C. Bose University of Science and Technology, YMCA, India. He has published 50+ research papers in reputed International journals. He received the "MSIT Best Faculty Award" in 2018 and "MSIT Certificate of Excellence in Research" in 2018 and 2019. He has been placed among "Top 2% of Researchers in the World" for 2019, 2020, and 2021 by Stanford University, USA. He is a regular reviewer of leading journals. He has Google Scholar citations of 2400+ and h-index of 29. He has received "Best Associate Editor Award-2021 and 2022" from *Journal of Electrical Engineering and Technology* (Springer). He is Associate Editor of a few SCIE journals. He is a Fellow of IETE, Senior Member of IEEE, AMIE(I) and Life Member of ISTE. His major areas of research interests include optimization techniques, fuzzy control, control system, AGC/LFC in smart grid, energy storage systems and renewable energy systems.

Preface to "Challenges and Research Trends of Renewable Energy Power System"

Renewable energy sources (RESs) like solar and wind integrated power systems are becoming increasingly important in addressing global energy challenges. However, there are still several challenges that need to be addressed to maximize the potential of RESs.

One of the main challenges is the intermittent nature of these energy sources, which means that they are not available at all times and also offers less or no inertia support to the system. This creates difficulties in matching energy supply and demand, and in maintaining the stability and reliability of the power grid. This challenge is being addressed through the development of energy storage technologies that can store surplus energy during periods of high production and release it during times of low production.

Another challenge is the need for new infrastructure and grid modernization to integrate RESs into the power grid. This requires significant investments in transmission lines, substations, and other equipments to connect RESs to the grid. In modern power systems, the liberalization of energy markets, revised incentives and technology upgrading policies are required to create competition among generation companies to the advantage of consumers.

Research in renewable energy is also focusing on increasing the efficiency and cost-effectiveness of renewable energy technologies, such as the development of more efficient solar cells and wind turbines. Overall, the challenges and research trends in renewable energy power systems are focused on increasing the efficiency, reliability, and cost-effectiveness of RESs, while also enhancing the integration of renewable energy into the existing power grids.

The chapters in this book demonstrate the importance of RESs penetrated power systems designing, working and properly monitoring in order to ensure continued good performance. Out of various timely topics few of the chapters are about the: Review of operational issues of contemporary distribution systems, Frequency regulation problem solution of renewable interlinked power system, Fuzzy-PSO-PID controller with UPFC-RFB for LFC of an interlinked hydropower system, Coordinated control of wind energy conversion system during unsymmetrical fault at the grid, Super-twisting algorithm-based virtual synchronous generator in inverter interfaced distributed generation, Real-time peak valley pricing based multi-objective optimal scheduling of a virtual power plant considering renewable resources, Day-ahead load demand forecasting in urban community cluster microgrids using machine learning, and Real-time validation technique-based offset hysteresis band current controller for grid-tied photovoltaic system.

Yogendra Arya
Editor

Unintentional Passive Islanding Detection and Prevention Method with Reduced Non-Detection Zones

Sowmya Ramachandradurai, Narayanan Krishnan * and Natarajan Prabaharan *

Department of Electrical and Electronics Engineering, SASTRA Deemed to be University, Thanjavur 613401, India; sowmya2621994@gmail.com
* Correspondence: narayanan.nnit@gmail.com (N.K.); prabaharan.nataraj@gmail.com (N.P.)

Abstract: Islanding detection and prevention are involved in tandem with the rise of large- and small-scale distribution grids. To detect islanded buses, either the voltage or the frequency variation has been considered in the literature. A modified passive islanding detection strategy that coordinates the V-F (voltage–frequency) index was developed to reduce the non-detection zones (NDZs), and an islanding operation is proposed in this article. Voltage and frequency were measured at each bus to check the violation limits by implementing the proposed strategy. The power mismatch was alleviated in the identified islands by installing a battery and a diesel generator, which prevented islanding events. The proposed strategy was studied on the three distinct IEEE radial bus distribution systems, namely, 33-, 69-, and 118-bus systems. The results obtained in the above-mentioned IEEE bus systems were promising when the proposed strategy was implemented. The results of the proposed strategy were compared with those of methods developed in the recent literature. As a result, the detection time and number of islanded buses are reduced.

Keywords: distributed generation (DG); backward and forward sweep (BFS) method; passive islanding detection; islanding detection time; islanding prevention; V-F (voltage–frequency) index; non-detection zone (NDZ)

1. Introduction

Nowadays, engineers incorporate renewable energy resources into distribution systems to bring down the usage of fossil fuels and to sustainably reduce greenhouse gas emissions. In addition, renewable energy resources are used to meet the power demand and to decrease the losses in distribution networks. On the other hand, the increasing penetration of DG imposes challenges in terms of islanding issues. The existing methods of passive islanding detection, such as artificial neural networks (ANNs) and intelligence-based and signal-transferring techniques, have more non-detection zones (NDZs), and the detection of islanding is much slower than with the proposed modified passive islanding detection strategy. The existing prevention methods are implemented with relays, which lead to fewer difficulties during the loading and generation mismatch than with the proposed prevention method.

1.1. Literature Review

In [1], the use of multiple distributed generation (DG) units in a radial system improved the voltage profile, increased flexibility, and reduced power losses in the system. The undeniable problem in the integration of distributed generation units is the islanding, even though there are numerous advantages [2]. In [3], as specified in IEEE 1547-2008, in a DG system, islanding was examined with a delay of a minimum of two seconds. Local methods and remote methods were used to detect islanding. The problem encountered by the local and remote methods was that of non-detection zone (NDZs); failure to identify islanding leads to non-detection zones (NDZs) [4].

In remote methods, communication-based islanding occurs where information is transferred between a utility and distributed generation units, thereby eliminating NDZs. The costs for the execution of these methodologies are high when compared to those of local islanding detection techniques, but identification of islanding by using this technique can be more effective and reliable than with other methods [5].

The electrical variables, such as frequency, voltage, current, etc., in a DG region are measured by using active methods and passive methods that are categorized as local methods [6]. A passive method sets threshold limits to detect islanding, and the power flow is maintained by managing the frequency and voltage. A protective relay is operated to isolate the DG connection when a bus system goes beyond the threshold limits [7]. Since passive methods have a minimal effect on the quality of power and reduce NDZs, they are considered to be the better methods. The different kinds of passive techniques for the detection of islanding include over-current (I), over-voltage (V), over frequency, under-current (I), under-voltage (V), under-frequency, etc. [8].

Artificial-intelligence-based techniques have been introduced to identify islanding in systems where passive detection methods are used. Artificial neural networks [9] and intelligence-based [10] and signal-transferring techniques [11] have been introduced to detect islanding events with passive detection techniques.

NDZs can be eliminated with active methods, but the system suffers from the deterioration of power quality; an active islanding detection strategy is briefly explained in [12]. In the case of the presence of multiple renewable energy resources or DG units, the operation of islanding is performed through the evaluation of the voltage for enormous power fluctuations at the foremost bus to disconnect various loads [13].

In order to prevent the occurrence of islanding, a logical regression classifier was used to accurately simulate DGs in a radial system in a MATLAB environment. This model contained numerous conditions of occurrence based on DG energy supplies, a low monitorable area, and multiple islanding states during operation [14]. Since the communication speeds of directional over-current relays and frequency relays are fast and their operation is reliable, they can be used in small-scale micro-grids [15]. Islanding can be prevented by under-voltage and under-current relays and over-voltage and over-current relays, which do not require communication-based preventive devices [16].

In a pole transformer, the nonlinear magnetizing property was developed for the prevention of islanding in a PV system [17,18]. Rotating-machine-based DGs were used for the prevention of islanding with reduced non-detection zones [19]. All of the methods mentioned above had various shortcomings in the detection and prevention of islanding, other than the passive islanding detection methods. The brief explanation of literature review is shown in Table 1.

1.2. Research Gap and Motivation

The work proposed here is a novel modified passive islanding detection strategy for detecting and preventing islanding with load-flow analysis using the backward and forward sweep (BFS) method [20], which is associated with passive islanding detection. In this work, the proposed strategy for the detection and prevention of islanding was validated in the presence of various DGs in a radial system. The V-F measurements were carried out for various IEEE radial bus distribution systems, such as 33-, 69- and 118-bus systems, with predefined threshold limits in order to identify the islanding in these distribution systems. A diesel generator and the battery were installed in the buses, which were islanded based upon their generation and load values, thus preventing their operation during the islanding of the system.

Table 1. Technical contributions of the proposed method with respect to existing methods.

Inference	Islanding Detection Techniques								
	Rate of Change of Power (ROCOP) [21]	Rate of Change of Frequency (ROCOF) [22]	Phase Jump Detection (PJD) [23]	Harmonic Distortion [24]	Over-/Under- Voltage and Frequency (OUV/OUF) [25]	Voltage Imbalance [24]	Voltage Variation Method [26]	Frequency Variation Method [26]	Proposed Method
NDZ	smaller than OUV/OUF	smaller than ROCOF	large than ROCOP	smaller than ROCOF	larger than PJD	smaller than voltage variation	larger than harmonic distortion	larger than voltage variation	reduced than other methods
Detection time	24–26 s	24 s	10–20 s	45 s	4–2 s	53 s	1.29 s	1.95 s	0.58 s
Number of islanded buses	more	more	more	more	more	more	more	more	less
Prevention	can prevent with a power control relay	can prevent with the rate of change of the frequency relay	can control with the slip-mode frequency-shift method	can control with a digital relay	can prevent with an under-/over- frequency/voltage relay	voltage relays are used	over-/under- voltage relays are used	rates of change of frequency relays are used	installing a battery or diesel generator prevents islanding

1.3. Contribution and Organization of the Paper

The main attributes proposed in this work are:

- An unintentional passive islanding detection method using the V-F index is proposed to detect islanding events.
- With the proposed strategy, the number of buses islanded and time taken for the detection of islanding are minimized when compared to those of the existing methods

This paper is organized as follows:

- Section 2 describes the model description of the DGs.
- The detection strategy for the proposed method is described in Section 3.
- The prevention strategy for the proposed method is described in Section 4.
- The results obtained for various bus systems and their discussions are provided in Section 5.
- Finally, in Section 6, the conclusion of this work is described.

2. Model Description

The DGs were installed for 33-, 69-, and 118-bus systems [27,28] by referring to existing papers. The modeling equations for different DGs from existing papers were discussed in order to identify how the DGs were modeled and installed in particular buses.

2.1. PV Modeling

Solar energy is converted into electric current through what is referred as the photovoltaic (PV) effect. The solar panels receive sunlight, which is an abundant and sustainable form of energy. The cost is high in the first stage of usage and is slowly reduced with increasing efficiency [29,30]. The power generation depends on the rating of the module, the atmospheric temperature, and the solar insolation. Solar insolation is indicated as follows [31]:

$$B_F(s) = s^{(\sigma-1)} \times \frac{\Gamma(\sigma + \mu)}{\Gamma(\sigma)\Gamma(\mu)} \times (1-s)^{(\mu-1)} \quad 0 \leq s \leq 1 \quad (1)$$

$$\mu = \frac{(\nu(1+\nu))}{\zeta^2} \times (1-\nu) - 1 \quad (2)$$

$$\sigma = \frac{\nu\mu}{1-\nu} \quad (3)$$

s, $B_F(s)$, and Γ are the solar insolation (kW/m²), beta distribution function (BDF), and gamma function of solar power, respectively. σ is measured as 0.999 and μ is measured as 0.055 from Equations (2) and (3).

The parameters of the BDF are given as $\Gamma(\sigma + \mu)$; ν and ζ are the mean and standard deviation, which are taken from actual data. The PV's power output $P_{out}(s)$ is obtained as follows [31]:

$$P_{out}(s) = N_p \times F_F \times V_{yi} \times I_{yi} \quad (4)$$

The I (current) and V (voltage) characteristics of a cell are evaluated as follows:

$$F_F(fill factor) = \frac{V_{MT} \times I_{MT}}{V_{opt} \times I_S} \quad (5)$$

$$V_{yi} = V_{opt} - K_{vc} \times T_{CYI} \quad (6)$$

$$I_{yi} = SI[I_S + K_{ic}(T_{CYI} - 25)] \quad (7)$$

$$T_{CYI} = T_{AT} + SI[(N_{OT} - 20)/0.8] \quad (8)$$

Coefficient of voltage (K_{vc}) = 14.40 mV/°C and the current (K_{ic}) is 1.22 mA/°C; N_{OT} is calculated as 43 °C, representing the optimal temperature of a PV cell; the open-circuit voltage (V_{opt}) is 21.98 V and the short-circuit current (I_S) is 5.32 A; the maximum power point tracking (MPPT) (I_{MT}) current is 4.76 A; and the maximum power point tracking (V_{MT}) voltage is 17.32 V. The solar insolation (SI), fill factor (F_F), the temperature at a specific cell (T_{CYI}), and the ambient temperature (T_{AT}) are calculated with 1 kV as the value of V_{yi} [31]. The above-mentioned values are substituted into Equations (4)–(8) to calculate the output power of the PV $P_{out}(s)$ [31].

2.2. Wind Modeling

A wind turbine produces power that is not constant and varies from second to second because the turbine's speed is not constant. Therefore, the units of the wind power generation profile are affected due to the intermittency of the wind speed. The modeling of the wind speed is framed by the Weibull probability distribution function (PDF), and the power output concerning speed is given below [31]:

$$P_{is}^W = \begin{cases} 0, & v_{ins}^c \geq v \quad or \quad v_{out}^{cp} \leq v \\ \frac{v - v_{ins}^c}{v_{rated} - v_{ins}^c} P_{i,rat}^W, & v_{ins}^c \leq v \leq v_{rated} \\ P_{i,rat}^W, & else \end{cases} \quad (9)$$

$P_{i,rat}^W$ = 0.5 MW (rated power), v_{ins}^c = 3 m/s (cut-in speed), P_{is}^W = 10 MW (generated power), v_{out}^{cp} = 25 m/s (cut-out speed), and v_{rated} = 13 m/s (rated speed of the turbine).

2.3. Hydro-Energy Modeling

The generation of hydro-energy requires a large area for installation and has low emissions. The conversion of a kinetic source into an electrical source of energy is executed in a hydro-energy system. This is considered as a dispatchable generation unit, and the hydropower output is calculated as [32]:

$$P = Y_{total} \times \varkappa \times g_a \times h_p \quad (10)$$

The output power is P, hydraulic efficiency (Y_{total}) is 75.1%, water density ($\varkappa$) is 1000 kg/m^3, pressure head (h_p) is 2.25 m, and (g_a) = 9.81 m/s^2 (acceleration due to gravity).

3. Proposed Passive Method for Islanding Detection

The method proposed here is a passive method in which an islanded bus is identified in the presence of DG units in a distribution system with different ratings. The power flow is executed by using the BFS algorithm, and the DG is placed according to the results obtained from this algorithm. The islanding is detected by considering the frequency and voltage values.

The voltage index was used to detect islanding in [13]. Beginning with a load disconnection strategy, the generalization of the foremost bus was initiated with the line current (IG_{rms}) at each DG bus. The sampling frequency under consideration was 1.66 kHz, and the time was 2 s. The product of the sampling frequency and time provided 3333 samples per cycle with a 60 Hz base frequency [13].

$$K_{V1} = |v_{ba} - v_{sy}| \times aav \quad (11)$$

$$K_0 = N \times \Delta v_o \times V_{av} \quad (12)$$

The values determined by the equations given above [13] are K_{V1} = 0.35, which is the voltage index, K_0 = 0.17, which is the threshold index, and 'N' = 3333 samples/cycle. The average voltage V_{av} is calculated with respect to the change in the system frequency, which is 0.12 p.u. The average accumulative voltage (aav) is the mean voltage variation/cycle with respect to time [13].

In [33], the islanding was determined in different phases; in phase I, islanding was suspected due to changes in voltage or frequency. In the subsequent phase, islanding was detected with a change in voltage (reactive power) and change in frequency (active power).

The proposed strategy is expressed mathematically as

$$X_1 = \sum_{1}^{3353} V_{mag} \sin N_{sa} \omega t = \sum_{1}^{3353} v_{mag} \sin 2\pi f N_{sa} t \tag{13}$$

Since $\omega = 2\pi f$,

$$X_2 = \sum_{1}^{3353} (1/N_{sa} \times abs(X_1)) \tag{14}$$

$$V_{ope(new)} = V_{ba} - V_{ope} \tag{15}$$

$$\Delta V_o = V_{ba} - V_{ope(new)} \times X_2 \tag{16}$$

$$\Delta f = 60 - f_d \tag{17}$$

$$K_{constant} = \int_0^{n_b} (\Delta f \times 2\pi)/60 \tag{18}$$

$$f_d = 1 + f \times [P_{load(n_b)}/P_{actual\ load(n_b)}] \times (V_{ba}/V_{ope})] \tag{19}$$

$$\Delta c_s = (\Delta f/60) \times n_b \tag{20}$$

$$K_1 = X_2/V_{ba} \text{ and } K_2 = K_{constant} \times \Delta V_o \tag{21}$$

Table 2 shows the values of specific parameters of single buses used in the proposed V-F index method for 33-, 69-, and 118-bus systems, which were evaluated with Equations (13)–(21). The K_1 and K_2 values were evaluated using the frequency, base voltage, change in voltage, number of samples, voltage magnitude, and frequency variation at an individual bus with respect to the nominal frequency.

Table 2. Ratings of the proposed method.

Variables	Ratings (Bus System)		
	33	69	118
K_1—V-F index and K_2—threshold limits represented in Equation (21)	0.09684 and 0.10504	0.09884 and 0.1504	0.099924 and 0.2432
X_1 and X_2—The phase voltage related to the time, frequency, and voltage represented in Equations (13) and (14)	1.5925 and 1.226	1.7925 and 1.326	1.6251 and 0.2432
$P_{load(n_b)}$—The real power value of one bus with respect to the next bus used in Equation (19)	7.2 kW	8.7 kW	9.3 kW
$P_{actual\ load(n_b)}$—The actual load used in Equation (19)	20 kW	22 kW	23 kW
N_{sa}—The number of samples used in Equations (13) and (14)	3353	3353	3353
$V_{ope(new)}$—The new operating voltage calculated using Equation (15)	11.66 kV	12.1 kV	9.2 kV
V_{ope}—The system operating voltage used in Equation (19)	0.995 kV	0.998 kV	0.996 kV

Table 2. *Cont.*

Variables	Ratings (Bus System)		
	33	69	118
V_{ba}—The base voltage used in Equations (15), (16), (19), and (21)	12.66 kV	12.66 kV	11 kV
Δf—The violated frequency represented in Equation (17)	0.36 Hz	0.38 Hz	0.39 Hz
ΔV_o—The change in voltage represented in Equation (16)	1.226 kV	1.226 kV	1.120 kV
f_d—The determined frequency represented in Equation (19)	59.64 Hz	59.65 Hz	59.61 Hz

The frequency variation at an individual bus with respect to the nominal frequency is evaluated with Equation (18). Followed by the number of samples, the phase voltage in the network is associated with the time, frequency, and magnitude of the voltage, which are calculated by using X_1 and X_2 from Equations (13) and (14). The number of samples (3353) is calculated with the product of the sampling frequency (1.676 kHz) and the time of 2 s. For each bus in the considered distribution systems, the above values should be calculated. The loads of all of the buses vary for each hour, according to which the samples are also changed for each bus and frequency value that is measured in the radial distribution network. An islanded bus is detected with Equation (21). If $K_1 > K_2$, islanding has occurred. K_1 and K_2 are unique and, hence, are calculated at each step for individual buses.

4. Proposed Strategy for Islanding Prevention

A battery or diesel generator is installed in an islanded bus based on the load and generation. If the generation is less than the load, a diesel generator is installed to overcome the power deficiency. If the generation is greater than the load at the islanded bus, a battery is installed to absorb the excess power. So, installing a battery or diesel generator at an islanded bus provides a power balance. Once the power is balanced, islanding in an appropriate bus can be prevented with stable voltage values.

The diesel generator is modeled as

$$P_{Gen} < P_{Load}, |P_{Gen} - P_{Load}| = P_{Diesel}. \tag{22}$$

The battery is modeled as

$$P_{Gen} > P_{Load}, |P_{Gen} - P_{Load}| = P_{B(ch)} \tag{23}$$

or

$$P_j + (X_{ij} \times Q_j) - (\Delta V \times V_i)/R_{ij} = P_{B(dch)} \tag{24}$$

$$\Delta V = (V_{i-1} - V_i) \tag{25}$$

For PV power alone, $P_{Gen} = P_{PV}$; for wind power alone, $P_{Gen} = P_{Wind}$; for hydropower alone, $P_{Gen} = P_{Hydro}$. For the combination of PV–hydro, $P_{Gen} = P_{PV} + P_{Hydro}$; for the combination of PV–wind, $P_{Gen} = P_{PV} + P_{Wind}$; for the combination of wind–hydro, $P_{Gen} = P_{Wind} + P_{Hydro}$. $P_{PV}, P_{Wind}, P_{Hydro}$, and P_{Diesel} are the power generation by the photovoltaic, wind, hydro, and diesel generation units. P_{Gen} is the power generation and P_{Load} is the load at the islanded bus. $P_{B(ch)}$ is the charging battery power and $P_{B(dch)}$ is the discharging battery power. ΔV represents changes in voltage, and X_{ij} and R_{ij} are the resistance and reactance of the bus. P_j is the real power, and Q_j is the reactive power. V_{i-1} is the voltage magnitude of the previous iteration; V_i is the voltage magnitude of a particular bus of the current iteration.

The size of the diesel generator is fixed by finding the difference between the generation and load values of an islanded bus. If the generation value is 200 kW and the load is 400 kW, the difference is 200 kW. Here, the load is greater than the generation, so the diesel generator is given the rating of 200 kW to balance the load and generation at the islanded bus. Likewise, the size of the battery is fixed by finding the difference between the load and generation values of an islanded bus. If the generation value is 800 kW and the load is 200 kW, the difference is 600 kW. Here, the generation is greater than the load, so the battery is given the rating of 600 kW to absorb the generated and to balance the load and generation at the islanded bus.

The operations of detection and prevention are presented in Figure 1 and can be explained as follows:

- Initially, the load-flow method is carried out. By using Equation (21), the values of K_1 and K_2 are calculated. When K_1 is greater than K_2, islanding occurs; otherwise, islanding does not occur in a particular bus, and the buses in the network are treated as islanded buses when there are ±2% changes in the frequency and voltage values with respect to the nominal frequency and voltage.
- After the occurrence of islanding, the condition between the load and generation is checked. If the load is less than the generation, a battery is installed to prevent islanding, and if the generation is less than the load, a diesel generator is installed to prevent islanding.
- Here, the diesel generators are installed as dispatchable DG units in order to support the network in such conditions where non-dispatchable DG units do not provide the power.
- Since a battery is a storage device, it can charge or discharge power, and it cannot generate power by itself. However, a diesel generator can generate power and support the grid as well. Hence, a diesel generator is also used to prevent islanding.
- After the placement, the system is rejoined with the grid, and the islanding detection method is performed as described in Section 3. Now, no buses are islanded due to the power balance, and the result of $K_1 < K_2$ is provided.

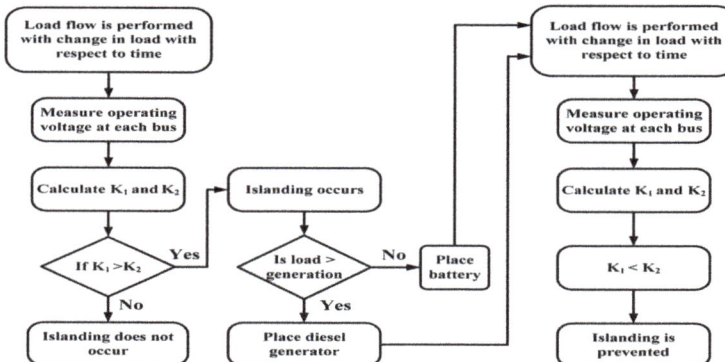

Figure 1. Flowchart for passive islanding detection.

5. Results and Discussion

The BFS load-flow method gave the results by which changes in the V-F index were determined through the placement of several DG units by referring to existing papers on all distribution systems, and it was simulated in MATLAB [34]. A number of DGs were used to enhance the stability of the system and reduce the losses in the system. The loads were varied linearly with respect to time to detect islanding for all DG combinations. The loads in all DG combinations were not varied until 0.5 s, and after 0.5 s, the loads were steadily increased in steps of 10% of the system's base load. While increasing the load, there was a distortion in voltage and frequency values for different time durations for the various DG combinations in all of the bus systems.

5.1. Islanding Detection and Prevention for the 33-Bus System

Islanding detection with deviations in voltage (v) and frequency (f) was performed with PV, hydro, and wind placement. The 33-bus system with DG units under consideration is shown in Figure 2. PV units were placed at buses 14, 24, and 29, and these buses had ratings of 691, 986.1, and 1277.3 kW, respectively [35]; the wind unit had ratings of 722.56 kW at the 14th bus and 813.89 kW at the 30th bus [27]. At the 13th bus, 24th bus, and 30th bus. a hydro units were placed with ratings of 537.8, 1058.9, and 967.7 kW, respectively [35].

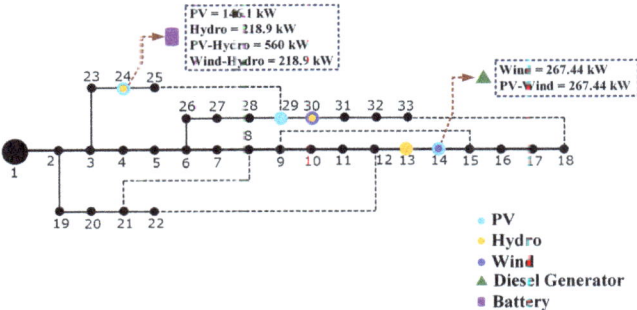

Figure 2. The structure of the 33-bus system.

Table 3 shows the results of exiting techniques with the proposed strategy for different DG units. The time of detection was measured with the changes in voltage and frequency at islanded buses. The loads in all DG combinations were not varied until 0.5 s, and after 0.5 s, the loads were constantly increased in steps of 10% of the system's base load. While increasing the load, there were variations in voltage and frequency at 1.02 s (PV), 0.75 s (wind), 0.60 s (hydro), 0.75 s (PV–hydro), 0.58 s (PV–wind), and 0.75 s (wind–hydro).

In the presence of PV, hydro, the combination of PV and hydro, and the combination of hydro and wind, the method using the V-F index identified islanding at the 24th bus. Using the (proposed) V-F index method, islanding was noticed at 117.6% of the system's base load for PV, 108.6% of the system's base load for hydro, 110.3% of the system's base load for DG with PV and hydro together, and 110.22% of the system's base load for DG with hydro and wind together. Using existing methods [13,33], islanding was noticed at 122.1% and 123.3% of the system's base load for PV, 110.3% and 114.2% of the system's base load for hydro, 123.23% and 112.7% of the system's base load for DG with PV and hydro together, and 117.3% and 119.2% of the system's base load for DG with hydro and wind together. In the presence of wind and the combination of PV and wind, the proposed method identified islanding at the 14th bus. Using the proposed (V-F index) method, islanding was determined at 119.6% of the system's base load for wind and 111.22% of the system's base load for DG with PV and wind together. However, when using existing methods [13,33], islanding was determined at 121.1% and 125.3% of the system's base load for wind and 121.23% and 120.7% of the system's base load for DG with PV and wind together.

In the proposed strategy, when the PV unit was placed, islanding was detected at the 24th bus at 1.02 s, and the detection time using voltage variation [33] was 1.29 s; that when using frequency variation [13] was 1.95 s. Islanding occurred at the 24th bus because the load was greater than the generation. In the 24th bus, the load was 840 kW and the generation was 986.1 kW. When placing a battery with a rating of 146.1 kW at bus 24, islanding was prevented. By using the proposed strategy, accuracy was achieved in detection; the time taken for the detection of islanding was also reduced by 0.27 s compared to voltage variation [33] and by 0.93 s compared to frequency variation [13].

Table 3. Results of islanding detection and prevention for the 33-bus system.

| Type of DG | DG Buses | Novel Passive Islanding Detection ||||| Islanding Prevention at the Identified Bus for the Proposed V-F Index Method ||
| | | Voltage Variation [33] || Frequency Variation [13] || Proposed V-F Index Method || | |
		Islanded Buses	Detection Time [s]	Islanded Buses	Detection Time [s]	Islanded Bus	Detection Time [s]	Diesel Generator (kW)	Battery (kW)
PV	14, 24, 29	24, 29	1.29	14, 24, 29	1.95	24	1.02	-	146.1
Wind	14, 30	14, 30	0.99	14, 30	0.98	14	0.75	267.44	-
Hydro	13, 24, 30	13, 30	0.82	24, 30	0.65	24	0.60	-	218.9
PV–Hydro	13, 14, 24, 29, 30	14, 24, 30	1.054	13, 29	1.032	24	0.75	-	560
PV–Wind	14, 24, 29, 30	14, 24, 30	0.62	14, 30	0.88	14	0.58	267.44	-
Wind–Hydro	31, 14, 24, 30	14, 24, 30	1.98	14, 30	1.99	24	0.75	-	218.9

When a wind unit was placed, the detection time using voltage variation [33] was 0.99 s and that using frequency variation [13] was 0.98 s. When the proposed strategy was used, islanding was detected at the 14th bus in 0.75 s. The load at the 14th bus was 990 kW and the generation at the 14th bus was 722.56 kW. Thus, the islanding occurred because the load was higher than the generation. To prevent this, a diesel generator with a rating of 267.44 kW was placed at that particular bus. The detection time was reduced by 0.24 s when compared to voltage variation [33] and by 0.23 s when compared to frequency variation [13].

When a hydro unit was placed, islanding occurred in 0.60 s at the 24th bus in the proposed strategy; it occurred with the generation of 1058.9 kW and load of 840 kW. The islanding was detected with the voltage variation [33] in 0.82 s, and it was detected with the frequency variation [13] in 0.65 s. At the 24th bus, the generation was 1058.9 kW and the load was 840 kW, which resulted in islanding. Here, the generation was greater than the load, so placing a battery with a rating of 218.9 kW prevented islanding. The time taken for the detection of islanding was reduced by 0.22 s when compared with the voltage variation [33] and by 0.05 s when compared with the frequency variation [13].

When the PV–hydro combination was used, islanding was detected in 0.75 s at the 24th bus by the proposed strategy. The time taken for the detection of islanding for the photovoltaic–wind combination was 0.58 s at bus 14. The generation and load at the 24th bus were 2045 and 1485 kW, respectively, for the PV–hydro combination, and at the 14th bus, they were 1413.56 and 1146.12 kW, respectively, for the PV–wind combination, which led to islanding. For the PV–hydro combination, the voltage variation method [33] detected the islanded buses in 1.054 s, and the method of frequency variation [13] detected the islanded buses in 1.032 s. For the PV–wind combination, the voltage variation method [33] detected the islanded buses in 0.62 s, and the frequency variation method [13] detected the islanded buses in 0.88 s. Placing a battery with a rating of 560 kW at the 24th bus prevented islanding with the PV–hydro combination, and placing a diesel generator with a rating of 267.44 kW at the 14th bus prevented islanding with the PV–wind combination.

For the PV–hydro combination, the detection time was reduced by 0.304 s when compared with that of voltage variation [33] and by 0.252 s when compared with that of frequency variation [13]. For the PV–wind combination, the detection time was reduced by 0.04 s when compared with that of voltage variation [33] and by 0.3 s when compared with that of frequency variation [13].

In the placement of the wind–hydro combination, islanding was detected in 0.75 s at the 24th bus with the proposed strategy. Here, at the 24th bus, the generation was 1058.9 kW and the load was 840 kW, which led to islanding. The voltage variation method [33] detected the islanded buses in 1.98 s, and the frequency variation method [13] detected the islanded buses in 0.75 s. By placing a battery with a rating of 218.9 kW at the 24th bus, islanding was prevented by maintaining the frequency and voltage. By placing a battery with a rating of 218.9 kW at the 24th bus, islanding was prevented. The detection time of the proposed strategy was reduced by 1.23 s when compared with that of voltage variation [33] and by 1.24 s when compared with that of frequency variation [13].

The NDZs were reduced by the proposed method in the presence of PV units, hydro units, the combination of PV–hydro, and the combination of wind–hydro in comparison with the results of existing methods. This is because the number of buses connected to the 24th bus (the islanded bus) was only two buses, unlike in existing methods. Likewise, in the presence of wind units and the combination of PV–wind, the NDZs were reduced by the proposed method. This is because the number of buses connected to the 14th bus was only five, unlike in existing methods. The proposed V-F index method is also suitable for low-power mismatches. In the 33-bus system, the islanded bus was identified at the load of 146 kW.

With the V-F index method, the voltage and frequency are continuously monitored (passive islanding), and K_1 and K_2 are calculated for different cases from Equation (21). The occurrence of islanding was mostly at the 24th bus because K_1 was greater than K_2.

For PV alone, the K_2 and K_1 values were calculated as 0.8532 and 0.98652. For hydro, the K_2 and K_1 values were 0.9235 and 0.93568. For PV–hydro and wind–hydro, the K_2 and K_1 values were 0.95362 and 0.9632 or 0.99386 and 0.9982, respectively. Islanding was detected at the 24th bus because K_1 was greater than K_2 for all of the cases. This was because of the greater generation in the bus, leading to variations in the frequency limits. The frequency changes led to voltage variations because the changes in the voltage (Δv) varied with the frequency violations, as shown in Equation (16). Figures 3 and 4 show the voltage values and frequency values of all of the buses. Buses in which the values deviated from the fixed values of voltage (0.99 to 1.05 p.u.) and fixed values of frequency (±2% of the rated frequency) were considered islanded buses.

Figure 5 has ones and zeros, which represent the non-islanded and islanded buses respectively, for various DG combinations, which were obtained using Equation (21).

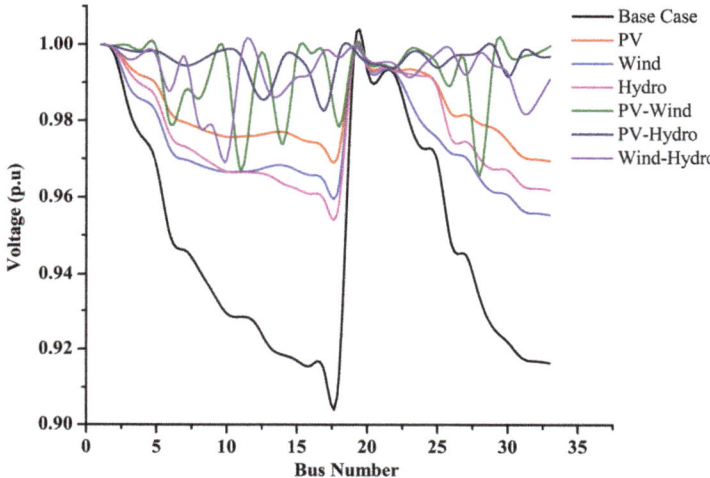

Figure 3. Voltage values of the 33-bus system.

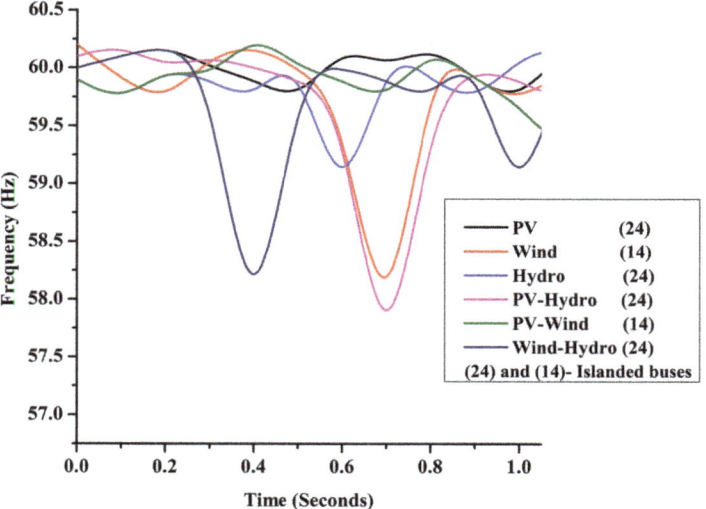

Figure 4. Frequency values of the 33-bus system.

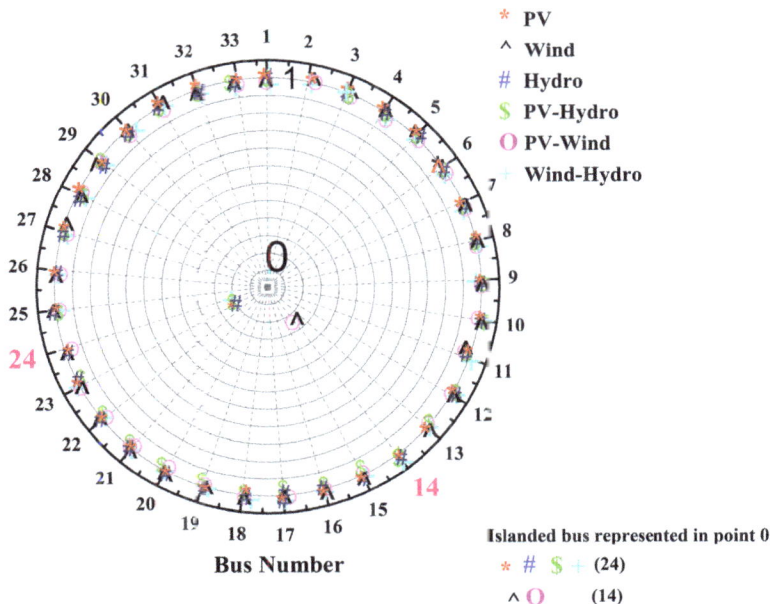

Figure 5. Detection of islanding.

Quantitative Analysis of the 33-Bus System

Reliability indices were calculated based on customer interruption (islanded buses). To ensure the system's performance, reliability indices were measured, as shown in Table 4.

Table 4. Reliability evaluation of the 33-bus system after reinforcement.

Cases	Voltage Variation [33]					Frequency Variation [13]					Proposed V-F Index Method				
	ENS	AENS	SAIDI	SAIFI	ASAI	ENS	AENS	SAIDI	SAIFI	ASAI	ENS	AENS	SAIDI	SAIFI	ASAI
PV	29,701.11	671.10	0.37	3.95	1.99	30,211.1	669.02	0.39	3.66	1.72	16,710.1	357.11	0.25	0.27	1.51
Wind	28,522.03	662.12	0.34	3.72	1.83	27,101.2	621.01	0.35	3.52	1.69	15,021.3	351.15	0.21	0.24	1.45
Hydro	26,210.01	572.10	0.31	2.57	1.79	26,101.2	602.1	0.32	3.37	1.63	13,101.6	272.16	0.19	0.21	1.32
PV–Hydro	24,311.11	525.12	0.29	2.42	1.71	24,001.7	597.4	0.29	3.02	1.58	9347.02	202.10	0.15	0.19	1.29
PV–Wind	21,010.12	511.03	0.25	2.25	1.45	21,295.2	531.8	0.25	2.92	1.52	7142.07	189.02	0.12	0.13	1.21
Wind–Hydro	17,821.51	502.12	0.21	1.99	1.41	15,150.5	521.9	0.19	2.82	1.47	5215.01	169.70	0.09	0.08	1.16

After installing the battery or diesel generator, the reliability values are improved by around 30% to 97% for various reliability indices, and the values of K_1 (V-F index values) are less than the values of K_2 (threshold limits).

5.2. Islanding Detection and Prevention for the 69-Bus System

Islanding detection with the deviation in voltage (v) and frequency (f) was performed with the placement of PV, hydro, and wind units. The 69-bus system with the DG units under consideration is shown in Figure 6. At buses 11, 18, and 61, PV units were placed, and these buses had ratings of 501.2, 482.2, and 1770.4 kW, respectively [35]. A wind unit with a rating of 409.6 kW was placed at the 18th bus, and one with a rating of 1338.73 kW was placed at the 61st bus [27]. At the 11th bus, 21st bus, and 61st bus, hydro units were placed with ratings of 707.1, 256, and 1875.2 kW, respectively [35].

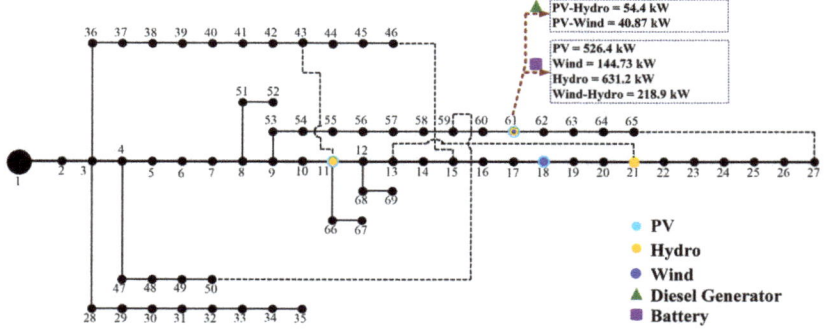

Figure 6. The structure of the 69-bus system.

Table 5 shows the results of the exiting techniques with the proposed strategy for different DG units. The time taken for the detection of islanding was measured with the variation in voltage (v) and frequency (f) at the islanded buses. The loads in all DG combinations were not varied until 0.5 s, and after 0.5 s, the loads were constantly increased in steps of 10% of the system's base load. While increasing the load, there was a deviation in the voltage and frequency at 1.35 s (PV), 1.55 s (wind), 0.62 s (hydro), 0.65 s (PV–hydro), 0.57 s (PV–wind) and 0.57 s (wind–hydro).

In the presence of PV, hydro, wind, the combination of PV and hydro, the combination of PV and wind, and the combination of hydro and wind, the method of the V-F index identified islanding at the 61st bus. Using the (proposed) V-F index method, islanding was noticed at 119.6% of the system's base load for PV, 110.6% of the system's base load for hydro, 112.4% of the system's base load for wind, 111.3% of the system's base load for DG with PV and hydro together, 110.26% of the system's base load for DG with PV and wind together, and 111.25% of the system's base load for DG with hydro and wind together. However, when using existing methods [13,33], the islanding is determined at 125.1% and 129.3% of the system's base load for PV, 112.3% and 116.2% of the system's base load for hydro, 127.3% and 112.2% of the system's base load for wind, 125.3% and 112.7% of the system's base load for DG with PV and hydro together, 117.3% and 119.2% of the system's base load for DG with PV and wind together, and 115.3% and 113.2% of the system's base load for DG with hydro and wind together.

Table 5. Results of islanding detection and prevention for the 69-bus system.

Type of DG	DG Buses	Novel Passive Islanding Detection						Islanding Prevention at the Identified Bus for the Proposed V-F Index Method	
		Voltage Variation [33]		Frequency variation [13]		Proposed V-F Index Method		Diesel Generator (kW)	Battery (kW)
		Islanded Buses	Detection Time [s]	Islanded Buses	Detection Time [s]	Islanded Bus	Detection Time [s]		
PV	11, 18, 61	11, 61	1.56	18, 61	1.72	61	1.35	-	526.4
Wind	18, 61	18, 61	1.92	18, 61	1.92	61	1.55	-	144.73
Hydro	11, 21, 61	21, 61	1.75	11, 61	1.56	61	0.62	-	631.2
PV–Hydro	11, 18, 21, 61	18, 61	0.96	21, 61	0.75	61	0.65	54.4	-
PV–Wind	11, 18, 61	11, 61	0.92	11, 61	0.66	61	0.57	40.87	-
Wind–Hydro	11, 18, 21, 61	18, 61	0.75	21, 61	0.63	61	0.57	-	218.9

In the proposed strategy, when a PV unit was placed, the islanding was detected at the 61st bus in 1.35 s. The detection time when using voltage variation [33] was 1.56 s and that when using frequency variation [13] was 1.72 s. The islanding occurred at the 61st bus because the load was less than the generation. There was a 1244 kW load, and the generation was 1770.4 kW. Placing a battery with a rating of 526.4 kW at the 61st bus prevented islanding. By using the proposed strategy, accuracy was achieved in the detection; in addition, the time taken for the detection of islanding was reduced by 0.21 s compared to that of the voltage variation [33] and 0.37 s compared to that of the frequency variation [13].

With the presence of wind, hydro, PV–hydro, PV–wind, and wind–hydro, islanding occurred at the 61st bus according to the proposed strategy. The voltage variation method [33] detected islanded buses in 1.92 s for wind, 1.75 s for hydro, 0.96 s for PV–hydro, 0.92 s for PV–wind, and 0.75 s for wind–hydro. The frequency variation method [13] detected islanded buses in 1.92 s for wind, 1.56 s for hydro, 0.75 s for PV–hydro, 0.66 s for PV–wind, and 0.63 s for wind–hydro. Islanding was detected by the proposed strategy at the 61st bus in 1.55 s for wind alone, 0.62 s for hydro alone, 0.65 s for PV–hydro, 0.57 s for PV–wind, and 0.57 s for wind–hydro.

Islanding occurred at the 61st bus due to a mismatch between the generation and load. Placement of wind units provided a load of 1194 kW and generation of 1338.73 kW, which led to islanding. The hydro placement led to islanding because the generation was greater than the load, with a load of 1244 kW and generation of 1875.2 kW. For the wind–hydro placement, islanding occurred at the same bus because the generation was greater than the load, with a load of 2995.03 kW and generation of 3213.93 kW. Placing a battery with a rating of 144.73, 631.2, or 218.9 kW at the 61st bus for wind, hydro, or wind–hydro, respectively, prevented islanding. By using the proposed strategy, accuracy was achieved in detection; in addition, the time taken for the detection of islanding for wind, hydro, and wind–hydro was reduced by 0.37, 1.13, and 0.18 s, respectively, in comparison with that of the voltage variation [33]. The time time taken for the detection of detection of islanding for wind, hydro, and wind–hydro was reduced by 0.37, 0.94, and 0.06 s, respectively, in comparison with that of the frequency variation [13].

For the PV–hydro and PV–wind placements, islanding occurred at the 61st bus because the load was greater than the generation, with loads of 3700 and 3150 kW and generation of 3645.6 and 3109.13 kW. Placing a diesel generator with a rating of 54.4 or 40.87 kW prevents islanding at the 61st bus. The detection times when using voltage variation [33] were 0.96 and 0.92 s, and those found when using frequency variation [13] were 0.75 and 0.66 s. The proposed strategy detected islanding in 0.65 and 0.57 s. By using the proposed strategy, accuracy was achieved in detection; in addition, the time taken for the detection of islanding for PV–hydro and PV–wind was reduced by 0.31 and 0.35 s in comparison with that of voltage variation [33]. The time taken for the detection of islanding for PV–hydro and PV–wind was reduced by 0.1 and 0.09 s in comparison with that of frequency variation [13].

The NDZs were reduced by the proposed method for all cases of DG units in comparison with the existing methods. This was because the number of buses connected to the 61st bus (the islanded bus) was only five, unlike in the existing methods. The proposed V-F index method was also suitable for low-power mismatches. In the 69-bus system, the islanded bus was identified at the load of 40 kW.

In the V-F index method, the voltage and frequency are continuously monitored (passive islanding), and K_1 and K_2 are calculated for different cases by using Equation (21). The occurrence of islanding was mostly at the 61st bus, as the K_2 and K_1 values for PV were 0.9532 and 0.97652, the K_2 and K_1 values for wind were 0.9235 and 0.93568, the K_2 and K_1 values for hydro were 0.9335 and 0.96568, the K_2 and K_1 values for PV–hydro were 0.95362 and 0.97386, the K_2 and K_1 values for PV–wind were 0.9432 and 0.9682, and the K_2 and K_1 values for wind–hydro were 0.9632 and 0.9782, respectively. Islanding was detected at the 61st bus for the above cases because K_1 was greater than K_2. This was due to greater generation or load in the bus, leading to a violation of the frequency limits. Frequency violations led to voltage variations because the change in the voltage (Δv) varied with the frequency violations, as shown in Equation (16). Figures 7 and 8 show the voltage values and frequency values of all of the buses. Buses in which the values deviated from the fixed values of voltage (0.99 to 1.05 p.u.) and fixed values of frequency ($\pm 2\%$ of the rated frequency) were considered as islanded buses. Figure 9 has ones and zeros that represent non-islanded buses and islanded buses, respectively, for various DG combinations, which were obtained by using Equation (21).

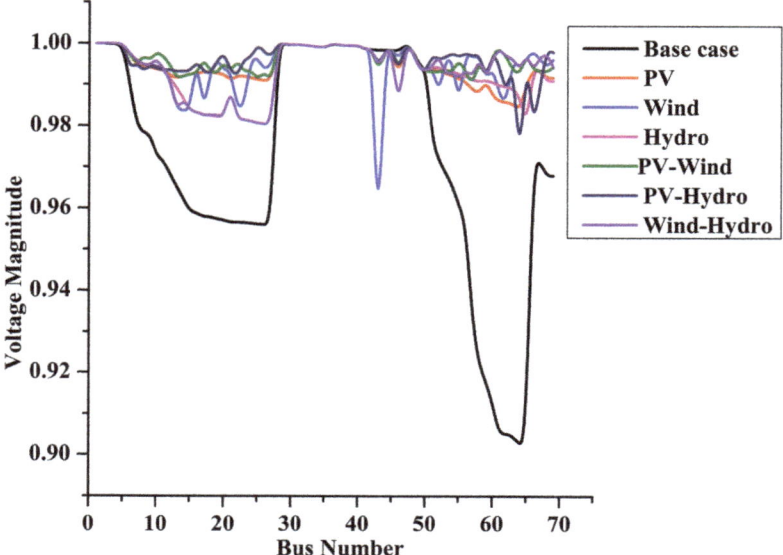

Figure 7. Voltage values of the 69-bus system.

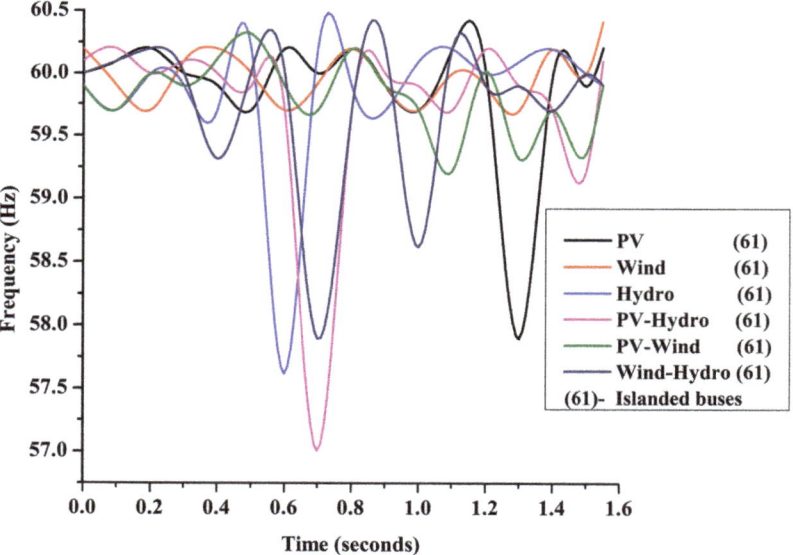

Figure 8. Frequency values of the 69-bus system.

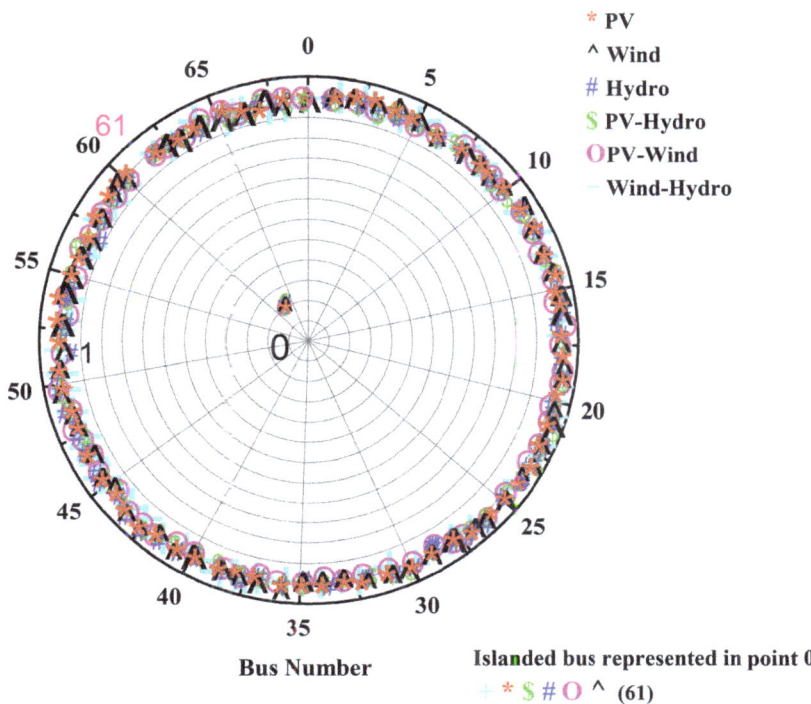

Figure 9. Detection of islanding.

Quantitative Analysis of the 69-Bus System

Reliability indices were calculated based on customer interruption (islanded buses). To ensure the system's performance, reliability indices were measured, as shown in Table 6.

Table 6. Reliability evaluation of the 69-bus system after reinforcement.

Cases	Voltage Variation [33]					Frequency Variation [13]					Proposed V-F Index Method				
	ENS	AENS	SAIDI	SAIFI	ASAI	ENS	AENS	SAIDI	SAIFI	ASAI	ENS	AENS	SAIDI	SAIFI	ASAI
PV	28,741.9	597.36	0.35	3.42	1.72	27,881.5	599.5	0.27	3.35	1.84	16,757.14	249.64	0.20	0.16	1.35
Wind	27,422.03	594.33	0.32	3.39	1.70	26,122.5	596.3	0.25	3.29	1.75	16,527.3	225.97	0.15	0.13	1.32
Hydro	26,122.13	570.21	0.31	2.57	1.68	24,325.1	570.3	0.22	2.99	1.71	7002.11	200.677	0.12	0.11	1.27
PV–Hydro	24,012.01	555.01	0.29	2.46	1.65	23,210.1	477.2	0.20	2.95	1.67	6950.21	179.544	0.10	0.07	1.24
PV–Wind	20,911.10	549.01	0.25	2.32	1.59	20,195.2	460.2	0.17	2.75	1.62	4709.12	168.422	0.07	0.05	1.14
Wind–Hydro	19,810.51	519.23	0.21	1.97	1.55	20,050.5	435.2	0.15	2.32	1.45	3790.02	150.011	0.05	0.02	1.12

After installing the battery or diesel generator, the reliability values are improved by around 20% to 99% for the various reliability indices, and the values of K_1 (V-F index values) are less than the values of K_2 (threshold limits).

5.3. Islanding Detection and Prevention for the 118-Bus System

Islanding detection with the deviation in voltage (v) and frequency (f) was performed with the placement of PV, hydro, and wind units. The 118-bus system with DG under consideration is shown in Figure 10. At buses 20, 30, 47, 73, 80, 90, and 110, PV units i with ratings of 2.0856, 3.3381, 2.1249, 2.794, 2.0369, 2.6069, and 3.1877 MW, respectively, were placed [36]. Wind units were placed at the second bus with a rating of 2.10000 MW, at the fifth bus with a rating of 1.70000 MW, at the 12th bus with a rating of 1.65000 MW, at the

44th bus with a rating of 2.00000 MW, at the 53rd bus with a rating of 1.55000 MW, at the 82nd bus with a rating of 1.85000 MW, and at the 86th bus with a rating of 1.95000 MW [28]. Hydro units were placed at various buses, such as at the 20th bus with a rating of 2.0187 MW, the 39th bus with a rating of 3.2905 MW, the 47th bus with a rating of 2.0615 MW, the 74th bus with a rating of 2.4092 MW, the 85th bus with a rating of 1.7437 MW, the 90th bus with a rating of 2.5473 MW, and the 110th bus with a rating of 3.1775 MW [36].

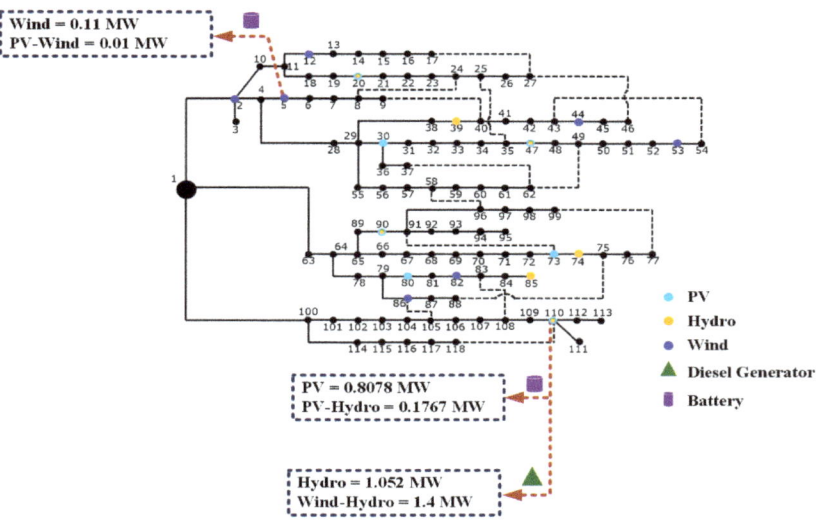

Figure 10. The structure of the 118-bus system.

Table 7 shows the results of the exiting techniques with the proposed strategy for different DG units. The time taken for the detection of islanding was measured with the variations in voltage (v) and frequency (f) at the islanded buses. The loads in all DG combinations were not varied until 0.5 s, and after 0.5 s, the loads were steadily increased in steps of 10% of the system's base load. While increasing the load, there were deviations in the voltage and frequency values at 1.25 s for PV, 1.30 s for wind, 1.50 s for hydro, 1.79 s for the combination of PV–hydro, 0.55 s for the combination of PV–wind, and 0.93 s for the combination of wind–hydro.

Table 7. Results of islanding detection and prevention for the 118-bus system.

Type of DG	DG Buses	Novel Passive Islanding Detection				Proposed V-F Index Method		Islanding Prevention at the Identified Bus for the Proposed V-F Index Method	
		Voltage Variation [33]		Frequency Variation [13]				Diesel Generator (kW)	Battery (kW)
		Islanded Buses	Detection Time [s]	Islanded Buses	Detection Time [s]	Islanded Bus	Detection Time [s]		
PV	20, 39, 47, 73, 80, 90, 110	47, 80	1.39	80, 110	1.59	110	1.25	-	0.8078
Wind	5, 82, 86	5, 82, 86	1.75	5, 86	1.32	5	1.30	-	0.11
Hydro	39, 47, 110	39, 47, 110	1.55	39, 110	1.55	110	1.50	1.052	-
PV–Hydro	20, 80, 90, 110	80, 110	1.92	80, 110	1.99	110	1.79	-	0.1767
PV–Wind	5, 39, 44, 47, 82	5, 39, 82	0.59	5, 82	0.75	5	0.55	-	0.01
Wind–Hydro	74, 82, 86, 110	86, 110	1.99	86, 110	0.99	110	0.93	1.4	-

In the presence of PV and hydropower generators, the method of the V-F index identified islanding at the 110th bus. Using the (proposed) V-F index method, the islanding was noticed at 114.9% of the base load for PV and 120.3% of the system's base load for hydro. However, with the existing methods [13,33], the islanding was detected at 120.9% of the system's base load and 123.5% of the system's base load for PV and 128.3% of the system's base load and 114.7% of the system's base load for hydro. In the presence of wind and DG with hydro and wind together, the proposed method detected islanding at the fifth bus. With the proposed method, the islanding was detected at 112.5% of the base load for wind and 113.6% of the system's base load for DG with hydro and wind together. However, with the existing methods [13,33], the islanding was detected at 128.3% of the base load and 114.7% of the system's base load for wind and 130% of the system's base load and 124.3% of the base load for the combination of hydro and wind. In the presence of DG with PV and hydro together and DG with hydro and wind together, the proposed method detected islanding at the 110th bus. With the proposed method, the islanding was detected at 128.7% of the system's base load for DG with PV and hydro together and 111.3% of the system's base load for DG with hydro and wind together. However, with the existing methods [13,33], the islanding was detected at 129.5% of the system's base load and 130% of the system's base load for DG with PV and hydro together and at 112% of the system's base load and 113.4% of the system's base load for DG with hydro and wind together.

With the proposed strategy, when the PV unit was placed, the islanding was detected at the 110th bus in 1.25 s. The voltage variation method [33] detected the islanded bus in 1.39 s, and the frequency variation method [13] detected the islanded bus in 1.59 s. Islanding occurred at this bus because the generation was less than the load; the 110th bus had a load of 3.9955 MW, and the generation was 3.1877 MW. Islanding was prevented by installing a diesel generator with a rating of 0.8078 MW at the 110th bus. By using the proposed strategy, accuracy was achieved in the detection; in addition, the time taken for the detection of islanding was reduced by 0.14 s in comparison with that of voltage variation [33] and by 0.3 s in comparison with that of frequency variation [13].

With the proposed strategy, when the wind unit was placed, the islanding was detected at the fifth bus in 1.30 s. The voltage variation method [33] detected the islanded bus in 1.75 s, and the frequency variation method [13] detected the islanded bus in 1.32 s. Islanding occurred at this bus because the load was less than the generation. The fifth bus had a load of 1.59 MW and the generation was 1.700 MW, which led to islanding. A battery with a rating of 0.11 MW is installed to prevent islanding. By using the proposed strategy, accuracy was achieved in the detection; in addition, the time taken for the detection of islanding was reduced by 0.45 s in comparison with that of voltage variation [33] and by 0.02 s in comparison with that of frequency variation [13].

With the proposed strategy, when the hydro unit was placed, the islanding occurred at the 110th bus in 1.50 s. The voltage variation method [33] detected the islanded bus in 1.55 s, and the frequency variation method [13] detected the islanded bus in 1.55 s. Islanding occurred at this bus because the generation was less than the load. The 110th bus had the generation of 3.1775 MW and a load of 4.230 MW, which led to islanding. Installing a diesel generator with a rating of 1.0525 MW prevented islanding at this bus. By using the proposed strategy, accuracy was achieved in the detection and prevention; in addition, the time taken for the detection of islanding was reduced by 0.05 s in comparison with that of voltage variation [33] and by 0.05 s in comparison with that of frequency variation [13].

When PV–hydro and wind–hydro were placed, islanding occurred at bus 110. The islanding detection times for the combinations of PV–hydro and wind–hydro were 1.79 and 0.93 s at the 110th bus. At the 110th bus, the generation was 6.3652 MW and the load was 6.18850 MW for the PV–hydro combination, and the generation was 3.1775 MW and the load was 4.58 MW for the wind–hydro combination. The voltage variation method [33] detected the islanded bus in 1.92 s and the frequency variation method [13] detected the islanded bus in 1.99 s for the combination of PV–hydro. For the wind–hydro combination, the voltage variation method [33] detected the islanded bus in 1.99 s and the frequency variation method [13] detected the islanded bus in 0.99 s. By using the proposed strategy, accuracy was achieved in the detection; in addition, the time taken for the detection of islanding was reduced by 0.13 and 1.06 s in comparison with that of voltage variation [33] for the PV–hydro combination and wind–hydro combination, and the time taken for the detection of islanding was reduced by 0.2 and 0.06 s in comparison with that of frequency variation [13] for the PV–hydro combination and wind–hydro combination. Installing a battery with a rating of 0.1767 MW at the 110th bus prevented islanding for the PV–hydro combination. Installing a diesel generator with a rating of 1.4 MW prevented islanding at the 110th bus for the wind–hydro combination.

When the PV–wind combination was placed, islanding occurred in 0.93 s at the fifth bus with the proposed strategy. Bus 5 had the generation of 1.700 MW and a load of 1.69 MW, which led to islanding. The method of voltage variation [33] detected the islanded bus in 1.99 s, and the frequency variation method [13] detected the islanded bus in 0.99 s. Installing a battery with a rating of 0.01 MW at the fifth bus prevented islanding By using the proposed strategy, accuracy was achieved in the detection; in addition, the time taken for the detection of islanding was reduced by 0.04 s in comparison with that of voltage variation [33] and by 0.2 s in comparison with that of frequency variation [13].

The NDZs were reduced by the proposed method in the presence of PV, hydro, the combination of PV–hydro, and the combination of wind–hydro in comparison with those of existing methods. This was because the number of buses connected to the 110th bus (the islanded bus) was only four, unlike in the existing methods. Likewise, in the presence of wind and the combination of PV–wind, the NDZs were reduced by the proposed method This was because the number of buses connected to the fifth bus was only five, unlike in the existing methods. The proposed V-F index method is also suitable for low-power mismatches. In the 118-bus system, the islanded bus was identified at a load of 10 kW.

In the V-F index method, the voltage and frequency are continuously monitored (passive islanding), and K_1 and K_2 are calculated for different cases by using Equation (21) The occurrence of islanding was mostly in the 110th bus, as the K_2 and K_1 values for PV were 0.9632 and 0.98652; for hydro, the K_2 and K_1 values were 0.9635 and 0.98568; for PV–hydro, the K_2 and K_1 values were 0.97362 and 0.99386; for wind–hydro, the K_2 and K_1 values were 0.9535 and 0.9782. Islanding was mostly detected at the 110th bus for the above cases because K_1 was greater than K_2. This was due to the greater generation or load in the bus, leading to the violation of the frequency limits. Frequency violations led to voltage variations because the changes in the voltage (Δv) varied with the frequency violations, as shown in Equation (16). Figures 11 and 12 show the voltage values and frequency values of all of the buses. Buses in which the values deviated from the fixed values of voltage (0.99 to 1.05 p.u.) and fixed values of frequency ($\pm 2\%$ of the rated frequency) were considered as islanded buses. Figure 13 has ones and zeros that represent the non-islanded and islanded buses, respectively, for various DG combinations, which were obtained by using Equation (21).

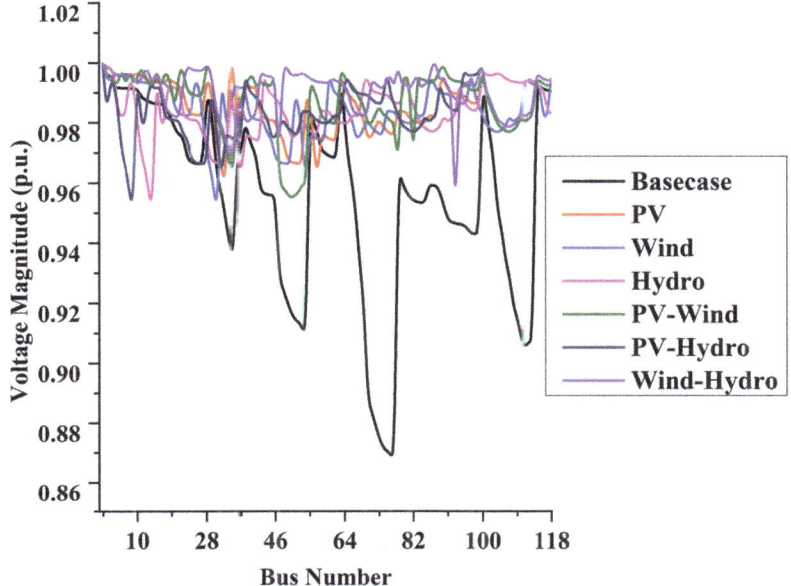

Figure 11. Voltage values of the 118-bus system.

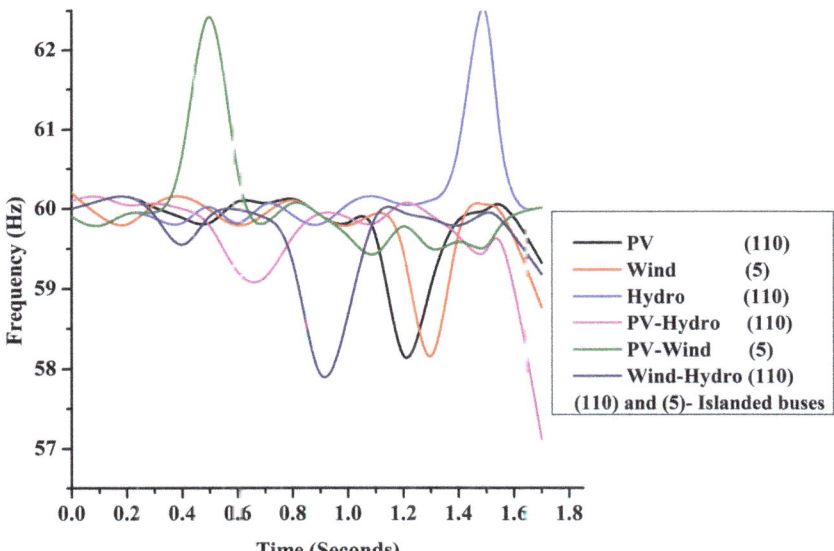

Figure 12. Frequency values of the 118-bus system.

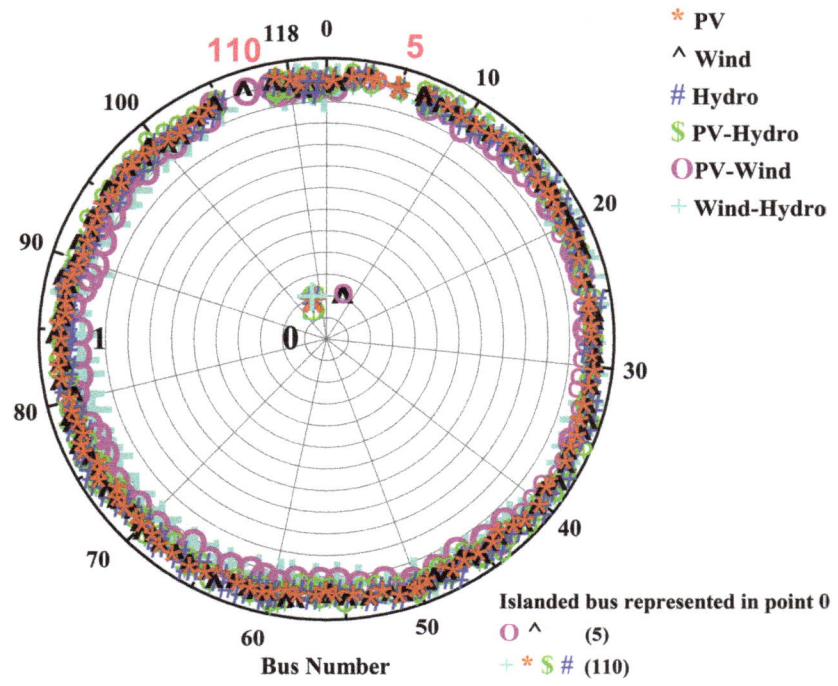

Figure 13. Detection of islanding.

Quantitative Analysis of the 118-Bus System

Reliability indices were calculated based on customer interruption (islanded buses). To ensure the system's performance, reliability indices were measured, as shown in Table 8.

Table 8. Reliability evaluation of the 118-bus system after reinforcement.

Cases	Voltage Variation [33]					Frequency Variation [13]					Proposed V-F Index Method				
	ENS	AENS	SAIDI	SAIFI	ASAI	ENS	AENS	SAIDI	SAIFI	ASAI	ENS	AENS	SAIDI	SAIFI	ASAI
PV	30,541.03	695.21	0.39	3.57	1.97	28,881.3	670.32	0.31	3.47	1.89	19,957.14	269.52	0.23	0.19	1.35
Wind	29,422.03	624.33	0.35	3.49	1.70	27,122.3	656.3	0.29	3.37	1.85	17,527.3	245.97	0.21	0.15	1.33
Hydro	26,122.13	590.21	0.32	2.27	1.68	26,325.1	590.3	0.26	3.31	1.80	8952.11	230.677	0.20	0.13	1.29
PV–Hydro	25,312.01	575.01	0.30	2.21	1.65	25,210.1	577.2	0.23	2.99	1.75	6545.21	189.544	0.17	0.10	1.25
PV–Wind	20,911.10	569.01	0.29	2.19	1.59	21,195.2	540.2	0.22	2.87	1.72	5950.12	178.422	0.12	0.07	1.19
Wind–Hydro	17,810.51	549.23	0.25	1.99	1.55	20,150.5	495.2	0.19	2.52	1.50	4990.02	155.011	0.09	0.04	1.11

After installing the battery or diesel generator, the reliability values are improved by around 21% to 98% for the various reliability indices, and the values of K_1 (V-F index values) are less than the values of K_2 (threshold limits).

The separate measurement of the deviations of frequency and voltage led to NDZs in the existing methods. The changes in frequency and voltage were simultaneously measured to reduce the NDZs in the proposed method. In addition, the proposed method helped in determining the exact islanded bus with various DG types. The effect of islanding was reduced by installing either a diesel generator or a battery unit, depending upon the power balance in the identified islands. The research findings from the three different bus systems can be summarized as follows:

- In the 33-bus system, the vulnerable bus identified was bus 24 because of the value of K_1 (V-F index values) was greater than that of K_2 (threshold limits) for PV, hydro, and PV–hydro and wind–hydro combinations. With the proposed V-F index method, the detection time for impending islanding was 23.37% faster than that of the voltage variation method and 62.62% faster than that of the frequency variation method in the presence of PV. The proposed method was 30.98% faster than the voltage variation method and 28% faster than the frequency variation method in the presence of hydro. The proposed method was 33.33% faster than the voltage variation method and 31.46% faster than the frequency variation method in the presence of PV–hydro. The proposed method was 90% faster than the voltage variation method and 90.51% faster than the frequency variation method in the presence of wind–hydro. Likewise, bus 14 was the vulnerable bus for wind and the PV–wind combination. With the proposed V-F index method, the detection time for impending islanding was 27.58% faster than that of the voltage variation method and 26.58% faster than that of the frequency variation method in the presence of wind. The proposed method was 6.66% faster than the voltage variation method and 41.09% faster than the frequency variation method in the presence of PV–wind. The generation was greater than the load for PV, hydro, and the PV–hydro and wind–hydro combinations. So, a battery was installed for power balance at the 24th bus. The generation was less than the load for wind and the PV–wind combination. So, a diesel generator was installed for power balance at the 14th bus for wind and the PV–wind combination.
- In the 69-bus system, the vulnerable bus identified was bus 61 because of the value of K_1 (V-F index values) was greater than that of K_2 (threshold limits) for PV, wind, hydro, PV–hydro, PV–wind, and wind–hydro. With the proposed V-F index method, the detection time for impending islanding was 14.43% faster than that of the voltage variation method and 24.10% faster than that of the frequency variation method in the presence of PV. The proposed method was 21.32% faster than the voltage variation method and 21.32% faster than the frequency variation method in the presence of wind. The proposed method was 95.35% faster than the voltage variation method and 86.23% faster than the frequency variation method in the presence of hydro. The proposed method was 38.50% faster than the voltage variation method and 14.28% faster than the frequency variation method in the presence of PV–hydro. The proposed method was 46.97% faster than the voltage variation method and 14.63% faster than the frequency variation method in the presence of PV–wind. The proposed method was 27.77% faster than the voltage variation method and 10% faster than the frequency variation method in the presence of wind–hydro. The generation was greater than the load for PV, wind, hydro, and the wind–hydro combination. So, a battery was installed for power balance at the 61st bus. The generation was less than the load for PV–hydro and the combination of PV–wind. So, a diesel generator was installed for power balance at the 61st bus for PV–hydro and the combination of PV–wind.
- In the 118-bus system, the vulnerable bus identified was bus 110 because the value of K_1 (V-F index values) was greater than that of K_2 (threshold limits) for PV, hydro, PV–hydro, and wind–hydro. With the proposed V-F index method, the detection time for impending islanding was 10.60% faster than that of the voltage variation method and 23.94% faster than that of the frequency variation method in the presence of PV. The proposed method was 3.27% faster than the voltage variation method and 3.27% faster than the frequency variation method in the presence of hydro. The proposed method was 7% faster than the voltage variation method and 10.58% faster than the frequency variation method in the presence of PV–hydro. The proposed method was 72.67% faster than the voltage variation method and 6.2% faster than the frequency variation method in the presence of wind–hydro. Likewise, bus 5 was the vulnerable bus for wind and the PV–wind combination. With the proposed V-F index method, the detection time for impending islanding was 27.58% faster than that of the voltage variation method and 1.52% faster than that of the frequency variation method in the

presence of wind. The proposed method was 7% faster than the voltage variation method and 30.76% faster than the frequency variation method in the presence of PV–wind. The generation was greater than the load for PV, wind, PV–hydro, and PV–wind. So, a battery was installed for power balance at the 110th bus for PV and the combination of PV–hydro, as well as at the fifth bus for wind and the combination of PV–wind. For hydro and the combination of wind–hydro, the generation was less than the load. So, a diesel generator was installed for power balance at the 110th bus for hydro and the combination of wind–hydro.

6. Conclusions

A modified method for the passive detection of islanding was proposed in this work, along with a prevention strategy that uses dispatchable and non-dispatchable DG units. Identification of islanded buses is performed by increasing the load in steps of 10% of the base load. The placement of a diesel generator or battery depends on the power balance in the identified island. The proposed islanding detection time is fast, and the number of islanded buses is less than that of the existing methods. The proposed method detects islanding for even smaller load variations than with the existing methods. The NDZs are reduced through the simultaneous measurement of voltage and frequency variations. The effectiveness of the proposed strategy was tested on IEEE 33-, 69-, and 118-bus systems.

In the 33-bus system, the generation was greater than the load for PV, hydro, and the PV–hydro and wind–hydro combinations. So, a battery was installed at the 24th bus. The generation was less than the load for wind and the PV–wind combination. So, a diesel generator was installed at the 14th bus. In the 69-bus system, the generation was greater than the load for PV, wind, hydro, and the wind–hydro combination. So, a battery was installed at the 61st bus. The generation was less than the load for the PV–hydro and PV–wind combinations. So, a diesel generator was installed at the 61st bus. In the 118-bus system, the generation was greater than the load for PV, wind, and the PV–hydro and PV–wind combinations. So, a battery was installed at the 110th bus for PV and the combination of PV–hydro and at the fifth bus for wind and the combination of PV–wind. The generation was less than the load for hydro and the combination of wind–hydro. So, a diesel generator was installed at the 110th bus for hydro and the combination of wind–hydro. In the future, various corrective methods can be implemented to reduce power imbalances. The main disadvantage is that passive methods of islanding detection require the inverter to be slightly out of time with the grid. This requires further research and investigation.

Author Contributions: Conceptualization, S.R.; Formal analysis, S.R.; Methodology, S.R.; Supervision, N.K. and N.P.; Writing—original draft, S.R.; Writing—review & editing, N.K. and N.P. All authors have read and agreed to the published version of the manuscript.

Funding: The authors have not received any funding for this work.

Institutional Review Board Statement: Not applicable.

Informed Consent Statement: Not applicable.

Data Availability Statement: Not applicable.

Conflicts of Interest: The authors have no conflict of interest to declare.

References

1. Selim, A.; Kamel, S.; Mohamed, A.A.; Elattar, E.E. Optimal Allocation of Multiple Types of Distributed Generations in Radial Distribution Systems Using a Hybrid Technique. *Sustainability* **2021**, *13*, 6644. [CrossRef]
2. Ji, X.; Zhang, X.; Zhang, Y.; Yin, Z.; Yang, M.; Han, X. Three-Phase Symmetric Distribution Network Fast Dynamic Reconfiguration Based on Timing-Constrained Hierarchical Clustering Algorithm. *Symmetry* **2021**, *13*, 1479. [CrossRef]
3. Shahid, M.U.; Alquthami, T.; Siddique, A.; Munir, H.M.; Abbas, S.; Abbas, Z. RES Based Islanded DC Microgrid with Enhanced Electrical Network Islanding Detection. *Energies* **2021**, *14*, 8432. [CrossRef]

7. Cebollero, J.A.; Cañete, D.; Martín-Arroyo, S.; García-Gracia, M.; Leite, H. A Survey of Islanding Detection Methods for Microgrids and Assessment of Non-Detection Zones in Comparison with Grid Codes. *Energies* **2022**, *15*, 460. [CrossRef]
8. Abokhalil, A.G.; Awan, A.B.; Al-Qawasmi, A.R. Comparative Study of Passive and Active Islanding Detection Methods for PV Grid-Connected Systems. *Sustainability* **2018**, *10*, 1798. [CrossRef]
9. Mohamed, A.A.; Kamel, S.; Selim, A.; Khurshaid, T.; Rhee, S.B. Developing a Hybrid Approach Based on Analytical and Metaheuristic Optimization Algorithms for the Optimization of Renewable DG Allocation Considering Various Types of Loads. *Sustainability* **2021**, *13*, 4447. [CrossRef]
10. Lopez, J.R.; Ibarra, L.; Ponce, P.; Molina, A. A Decentralized Passive Islanding Detection Method Based on the Variations of Estimated Droop Characteristics. *Energies* **2021**, *14*, 7759. [CrossRef]
11. Bukhari, S.B.A.; Mehmood, K.K.; Wadood, A.; Park, H. Intelligent Islanding Detection of Microgrids Using Long Short-Term Memory Networks. *Energies* **2021**, *14*, 5762. [CrossRef]
12. Montoya, O.D.; Arias-Londoño, A.; Grisales-Noreña, L.F.; Barrios, J.Á.; Chamorro, H.R. Optimal Demand Reconfiguration in Three-Phase Distribution Grids Using an MI-Convex Model. *Symmetry* **2021**, *13*, 1124. [CrossRef]
13. Bakhshi-Jafarabadi, R.; Sadeh, J.; Rakhshani, E.; Popov, M. High power quality maximum power point tracking-based islanding detection method for grid-connected photovoltaic systems. *Int. J. Electr. Power Energy Syst.* **2021**, *131*, 107103. [CrossRef]
14. Hashemi, F.; Ghadimi, N.; Sobhani, B. Islanding detection for inverter-based DG coupled with using an adaptive neuro-fuzzy inference system. *Int. J. Electr. Powe Energy Syst.* **2013**, *45*, 443–455. [CrossRef]
15. Ray, P.K.; Mohanty, S.R.; Kishor, N. Disturbance detection in grid-connected distributed generation system using wavelet and S-transform. *Electr. Power Syst. Res.* **2011**, *81*, 805–819. [CrossRef]
16. Abd-Elkader, A.G.; Saleh, S.M.; Eiteba, M.B.M. A passive islanding detection strategy for multi-distributed generations. *Int. J. Electr. Power Energy Syst.* **2018**, *99*, 146–155. [CrossRef]
17. Liu, S.; Ji, F.; Li, Y.; Xiang, J. Islanding detection method based on system identification. *IET Power Electron.* **2016**, *9*, 2095–2102. [CrossRef]
18. Ahmadzadeh-Shooshtari, B.; Hamedani Golshan, M.E.; Sadeghkhani, I. A combined method to efficiently adjust frequency-based anti-islanding relays of synchronous distributed generation. *Int. Trans. Electr. Energy Syst.* **2015**, *25*, 3042–3059. [CrossRef]
19. Zamani, M.A.; Sidhu, T.S.; Yazdani, A. A Protection Strategy and Microprocessor-Based Relay for Low-Voltage Microgrids. *IEEE Trans. Power Deliv.* **2011**, *26*, 1873–1883. [CrossRef]
20. Ustun, T.S.; Ozansoy, C.; Zayegh, A. Modeling of a Centralized Microgrid Protection System and Distributed Energy Resources According to IEC 61850-7-420. *Power Syst. IEEE Trans.* **2012**, *27*, 1560–1567. [CrossRef]
21. Sebastián, R. Battery energy storage for increasing stability and reliability of an isolated Wind Diesel power system. *IET Renew. Power Gener.* **2017**, *11*, 296–303. [CrossRef]
22. Abd-Elkader, A.G.; Saleh, S.M. Zero non-detection zone assessment for anti-islanding protection in rotating machines based distributed generation system. *Int. J. Energy Res.* **2021**, *45*, 521–540. [CrossRef]
23. Jabari, F.; Sohrabi, F.; Pourghasem, P.; Mohammadi-Ivatloo, B. Backward-forward sweep based power flow algorithm in distribution systems. In *Optimization of Power System Problems*; Springer: Berlin/Heidelberg, Germany, 2020; pp. 365–382.
24. Redfern, M.; Usta, O.; Fielding, G. Protection against loss of utility grid supply for a dispersed storage and generation unit. *IEEE Trans. Power Deliv.* **1993**, *8*, 948–954. [CrossRef]
25. Freitas, W.; Xu, W.; Affonso, C.M.; Huang, Z. Comparative analysis between ROCOF and vector surge relays for distributed generation applications. *IEEE Trans. Power Deliv.* **2005**, *20*, 1315–1324. [CrossRef]
26. IEA-PVPS. *Evaluation of Islanding Detection Methods for Photovoltaic Utility-Interactive Power Systems*; Report IEA PVPS T5-09; IEA-PVPS: Paris, France, 2002.
27. Jang, S.I.; Kim, K.H. An islanding detection method for distributed generations using voltage unbalance and total harmonic distortion of current. *IEEE Trans. Power Deliv.* **2004**, *19*, 745–752. [CrossRef]
28. De Mango, F.; Liserre, M.; Dell'Aquila, A. Overview of anti-islanding algorithms for pv systems. Part ii: Activemethods. In Proceedings of the 2006 12th International Power Electronics and Motion Control Conference, Portorož, Slovenia, 30 August–1 September 2006; pp. 1884–1889.
29. Ramachandradurai, S.; Krishnan, N. Islanding-based reliability enhancement and power loss minimization by network reconfiguration using PSO with multiple DGs. *Int. Trans. Electr Energy Syst.* **2021**, *31*, e12987. [CrossRef]
30. Pottukkannan, B.; Suvarchala, K.; Muthukannan, P.; Thangaraj, Y. Optimal Allocation of Renewable Distributed Generation in Radial Distribution Network. *J. Adv. Res. Dyn. Control Syst.* **2018**, *10*, 11.
31. Fang, X.; Li, F.; Wei, Y.; Azim, R.; Xu, Y. Reactive power planning under high penetration of wind energy using Benders decomposition. *IET Gener. Transm. Distrib.* **2015**, *9*, 1835–1844. [CrossRef]
32. Sahu, B.K. A study on global solar PV energy developments and policies with special focus on the top ten solar PV power producing countries. *Renew. Sustain. Energy Rev.* **2015**, *43*, 621–634. [CrossRef]
33. Toledo, O.M.; Oliveira Filho, D.; Diniz, A.S.A.C. Distributed photovoltaic generation and energy storage systems: A review. *Renew. Sustain. Energy Rev.* **2010**, *14*, 506–511. [CrossRef]
34. Soroudi, A.; Aien, M.; Ehsan, M. A probabilistic modeling of photo voltaic modules and wind power generation impact on distribution networks. *IEEE Syst. J.* **2011**, *6*, 254–259. [CrossRef]

32. Bhandari, B.; Poudel, S.R.; Lee, K.T.; Ahn, S.H. Mathematical modeling of hybrid renewable energy system: A review on small hydro-solar-wind power generation. *Int. J. Precis. Eng. Manuf. Green Technol.* **2014**, *1*, 157–173. [CrossRef]
33. Narayanan, K.; Siddiqui, S.A.; Fozdar, M. Hybrid islanding detection method and priority-based load shedding for distribution networks in the presence of DG units. *IET Gener. Transm. Distrib.* **2017**, *11*, 586–595.
34. Mathworks. MATLAB. 2019. Available online: https://ww2.mathworks.cn/en/products/new_products/release2019a.html (accessed on 1 March 2022).
35. Prakash, D.B.; Lakshminarayana, C. Multiple DG placements in distribution system for power loss reduction using PSO Algorithm. *Procedia Technol.* **2016**, *25*, 785–792. [CrossRef]
36. Jamil Mahfoud, R.; Sun, Y.; Faisal Alkayem, N.; Haes Alhelou, H.; Siano, P.; Shafie-khah, M. A novel combined evolutionary algorithm for optimal planning of distributed generators in radial distribution systems. *Appl. Sci.* **2019**, *9*, 3394. [CrossRef]

Article

A Novel Solution for Solving the Frequency Regulation Problem of Renewable Interlinked Power System Using Fusion of AI

Mohammed Ozayr Abdul Kader [1], Kayode Timothy Akindeji [1] and Gulshan Sharma [2,*]

[1] Department of Electrical Power Engineering, Durban University of Technology, Durban 4001, South Africa; ozayr.ak@gmail.com (M.O.A.K.); timothya@dut.ac.za (K.T.A.)
[2] Department of Electrical Engineering Technology, University of Johannesburg, Johannesburg 2006, South Africa
* Correspondence: gulshans@uj.ac.za

Abstract: The requirement for clean energy has increased drastically over the years due to the emission of CO_2 and the degrading of the environment by introducing Renewable Energy Systems (RES) into the existing power grid. While these systems are a positive change, they come at a cost, with some issues relating to the stability of the grid and feasibility. Hence, this research paper closely investigates the modeling and interlinking of photovoltaic (PV)-based solar power and Double-Fed Induction Generator (DFIG)-based wind turbines with the conventional power systems. RES has been known to contribute to a highly non-linear system and complexity. To return the power systems to their original state after a load disturbance, a novel control technique based on the fractional-order Type-2 Fuzzy logic system, well developed via particle swarm optimization (PSO), has been utilized for solving the frequency control problem of a renewable interlinked power system. The efficacy of the proposed technique is validated for various possible operating conditions and the system results are compared with some of the recent methods with and without including non-linearity, and the performance of the controllers is superimposed on frequency/time graphs for ease of understanding to show the benefits of the proposed research work.

Keywords: Renewable Energy Systems (RES); solar PV; wind power; Double-Fed Induction Generators (DFIG); Area Control Error (ACE); Fuzzy Type-2 (FT2); Fractional Order PID (FOPID)

1. Introduction

In recent years, the implementation of clean energy into the power system has been the focus of the modern era. Multiple applications, such as hydro, wind, and solar power, are being interconnected with the existing power system to help curb some of the negative environmental effects. These methods are highly sought after but have negative effects when increasing the system frequency disturbances and financial costs [1]. Renewable energy has been known for its highly non-linear contribution to the grid. For instance, PV-based solar energy makes use of the unlimited light energy from the sun called irradiance. The energy output is dependent on the amount of irradiance that can be produced. This contributes to a highly variable system due to the fluctuations of sunlight that are available in a particular area. Further, wind energy has been shown to have a similar effect, with the wind becoming highly unpredictable with its constant wind speeds. Even though this is the case, the temperature of the world has been rising every year, which makes solar energy a good alternative for energy generation. Similarly, studies have shown that wind levels are bound to increase in the future.

DFIG wind-based energy schemes have been shown to assist with system stability after a fault has occurred but rely on the type of control method option. One of the phenomena experienced by series-compensated wind power systems is sub-synchronous resonance. Dynamic reactive power reference signals at the point of common coupling are required to ensure that the reactive power supply is upheld [2,3]. DFIG shares properties of inductive and synchronous generators, which further contribute to the power system stability. Without interfering with voltage control, the damping of power variations in the power system can be improved through effective control. One of the control methods that researchers use is diode rectifiers for wind power systems, which contributes to stability error reduction while the output active power increases. With active and reactive power playing an important role, information gathered from scholars states that wind turbines with varying wind speed over time decrease voltage fluctuations [4–7]. The governors of thermal power units do not have the necessary measures to reduce frequency deviations, with its stagnant response and lack of control.

The PV-based solar system incorporates an inverter within its system, which reduces system inertia and makes the system more susceptible to disturbances. While disturbances pose a problem to the power grid, the efficiency of wind and solar systems is very low and they are proposed, depending on the application, to be used as a coupled generation system. Maximum Power Point Tracking (MPPT) control for the renewable system is an effective way to maximize the low efficiency given while being connected to the power grid [8–15]. This system inertia was responsible, in conventional power systems, for suppressing the small frequency excursions in the wake of unexpected load alterations.

One of the frequency control methods used is fractional-order PID (FOPID). This type of controller has been shown to positively abolish steady-state error, transient disturbance reduction, system non-linearities, and uncertainties. With its multiple parameters that are required, which are difficult to determine, manual tuning or algorithms are used to decipher the appropriate gains. The system's robustness and dynamic characteristics improve to a certain extent.

Performance criteria are used to prove this by calculating the area of the control. One such method is used is Integral Time Absolute Error (ITAE), which can be tabulated to evaluate the system with ease of understanding [16–19]. It is important that, if fluctuations occur, the system must return to its nominal value. In a two-area power system, these controllers are required in both areas to maintain the power interchange at scheduled values, as well as to minimize the frequency deviations for unexpected load alterations. PID controllers have the disadvantage of noise occurring in the derivative area of the equation. They also are linear and symmetric, which makes the performance of the controller vary. Therefore, the additional parameters for FOPID are input into the formula to help mitigate these [20].

Further enhancement and optimization, such as artificial intelligence (AI) techniques, i.e., fuzzy logic (FL), have been integrated and tested successfully for applications that require control. An extension and advancement to conventional FL is the Fuzzy Type-2 system, which includes an NT-type reduction process that converts type-2 to type-1 and then obtains the final defuzzification result. The Fuzzy Type-2 system is much more suitable for highly non-linear systems such as renewable interlinked power systems. Fuzzy Type-2 has been shown to have better performance than Fuzzy Type-1 and PI controller methods in multi-area power systems.

Optimization techniques applied, such as particle swarm optimization, are proven to positively affect results, further enhancing them. Artificial intelligence, used for optimization, has been shown to utilize multiple methods structured from the behavioral patterns of living organisms to successfully solve uncertainties in the power system. The tuning of adaptive parameters by performing an input delay has also been illustrated to produce positive results for the load frequency of a two-area power system.

By utilizing a PID controller with a Type-2 Fuzzy system, the output scaling factor has increased system performance and maintained stability, with an input delay and uncertainties. By adjusting or including an increased number of rules for the interference mechanism, the controller's performance can be increased in fuzzy logic systems, according to some researchers. The control algorithm has been shown to be highly effective and contributes to a significant performance increase regarding settling times, overshoots, and other performance criteria [21–26].

Given the above discussion, this work sets out to achieve the following:

- To model a two-area interlinked power system with coal-based generation and with an equal capacity for each area. These systems' areas are connected via an AC tie-line.
- To model the solar PV model in the transfer function domain and to interlink it in the coal-based power system. To model the wind system with DFIG and to use this model to assist in frequency excursion of the power system.
- To interlink the concept of Fuzzy Type-2 with fractional-order PID and result in a Fuzzy Type-2 FOPID for a renewable integrating power system. The proposed design is tested for various cases, including and without including non-linearity, and the application results are shown graphically to demonstrate the benefits of the proposed research work.

This research paper is divided into five sections. Section 1 is the introduction, with a literature review of the problem. Section 2 discusses the model of the conventional system with PV and DFIG. The concept and model of Fuzzy Type-2 with FOPID are formulated in Section 3. Section 4 shows the detailed explanation of results, with conclusions in Section 5.

2. Modeling of Renewable Interlinked Power System

2.1. Interlinked Power System with Wind and Solar Energy Systems

An interlinked thermal power system with a solar farm and wind farm connected to each area is shown in Figure 1, with an electrical representation in Figure 2. These areas are interconnected via an AC tie-line. Both areas have a thermal power system, which consists of a governor, turbine, and generator model that has the relevant parameters for gains and time constants. The non-linearities present, such as Governor Dead Band (GDB) and Generation Rate Constant (GRC), have been included to consider the ramp rate constraints and upper to lower bound constraints. The signal from both areas connects to the synchronizing coefficient via the tie-line and provides the relevant inputs and outputs for the rest of the interconnection power system. The interconnected system integrates with primary control action, which is well known as the speed governor mechanism, and the proposed Fuzzy Type-2 FOPID. This controller uses multiple parameters for fine-tuning to reduce the uncertainties and disturbances in the system. The change in power demand is present in both Area 1 and Area 2 for analysis of the controller performance to bring the system back to steady-state condition, i.e., frequency and tie-power deviations.

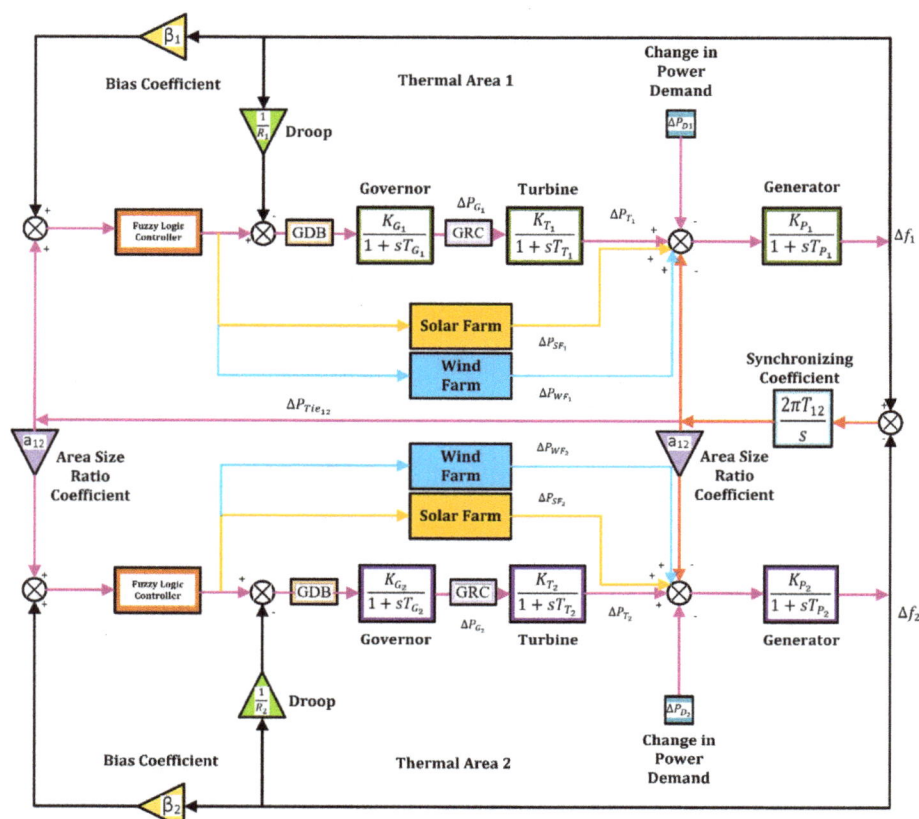

Figure 1. System model.

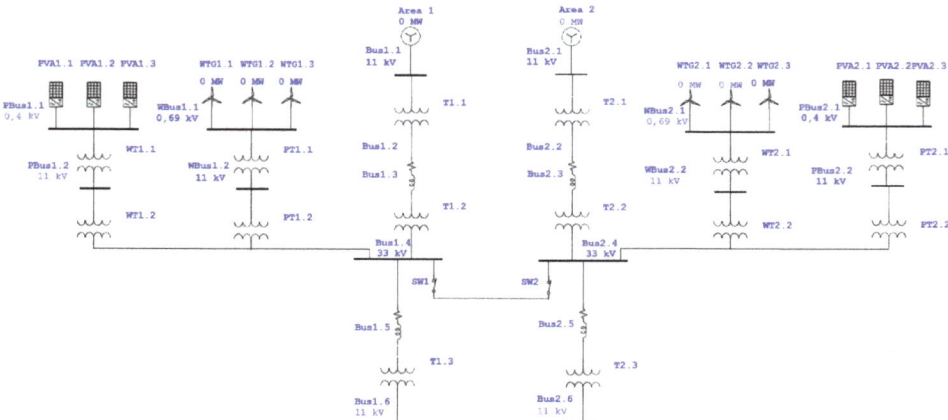

Figure 2. Single-line electrical diagram of interconnected system using ETAP software.

2.2. Solar PV System

A PV system consists of the PV panel, Maximum Power Point Tracking (MPPT), Sinusoidal Pulse Width Modulation Inverter (SPWM), and filter. A solar cell panel is made up of semiconductors that have been doped to contribute to the deficit of free electrons on the reverse side and surplus on the front side. This panel works with the photovoltaic process, where photons are absorbed in the solar cells. The solar cell can produce an output voltage of 0.3–0.6 V. This is dependent on the temperature and irradiance [8–10]. The solar equivalent circuit is shown in Figure 3. An equation for the solar panel can be derived from Equations (1) and (2) and is shown below:

$$I = I_1 - I_0 \left(e^{\frac{(V-IR_S)}{AkT}} - 1 \right) - \frac{V - IR_S}{R_{SH}} \quad (1)$$

where

I = Output current of PV array;
I_1 = Array current generated by the incident sunlight;
I_0 = Reverse saturation current of the PV array;
V = Output voltage of the PV array;
R_S = Equivalent series resistance of the array;
R_{SH} = Equivalent parallel resistance of the array;
A = Diode quality factor (ranging 0–2);
k = Boltzmann constant (1.380649 × 10^{-23} m^2 kg s^{-2} K^{-1});
T = Temperature (°C or K).

$$I_1 = \left(\frac{\lambda}{1000} \right) [I_{SC} + k_1(T - 25)] \quad (2)$$

where

λ = Irradiance (0–2500 W/m^2);
I_{SC} = Short-circuit current.

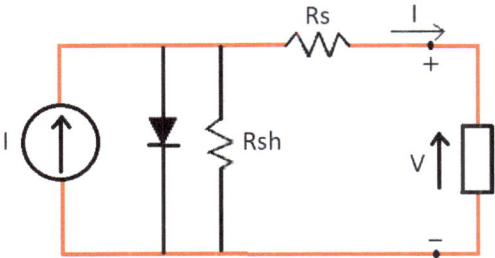

Figure 3. Solar cell equivalent circuit.

To harness the maximum power from a solar panel, a method called Maximum Power Point Tracking (MPPT) is utilized to provide input voltage regulation and improve efficiency. The voltage can be regulated through a booster DC/DC converter to deliver maximum power to the load. This type of converter can provide a higher output voltage than input voltage, which is discovered with the duty cycle of the gate pulse to the MOSFET switch [10]. Boost converters have two modes, which are the ON and OFF states, as shown in Equations (3) and (4), respectively. The ON and OFF state operation is shown in Figures 4 and 5.

$$ON\ STATE \left\{ L\frac{di_1}{dt} = V_{PV},\ C\frac{dV_0}{dt} + \frac{V_0}{R} = 0 \right. \quad (3)$$

where

L = Inductance (H);
C = Capacitance (F);
R = Resistance (Ω);
V_0 = Output voltage (V);
V_{PV} = Photovoltaic voltage (V).

$$\text{OFF STATE} \left\{ L\frac{di_2}{dt} + V_0 = V_{PV}, \; i_2 - C\frac{dV_0}{dt} + \frac{V_0}{R} = 0 \right. \tag{4}$$

where

i_2 = Current in the inductor (A).

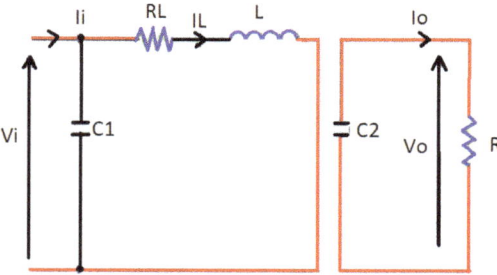

Figure 4. ON state operation.

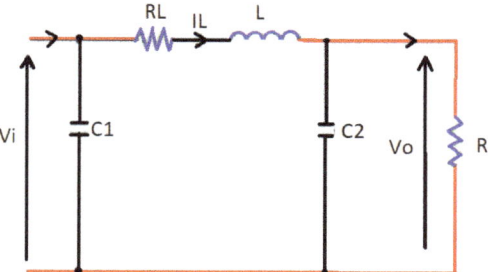

Figure 5. OFF state operation.

For the conversion of DC to AC power, an inverter is utilized in this application. The Sinusoidal Pulse Width Modulation inverter is used to maintain a constant voltage. Finally, a filter is used to remove disturbances within the output power signal. From these components, a transfer function equation is derived, which can be seen in [8], and finally, the structure of the equation can be seen in Figure 6.

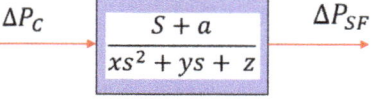

Photovoltaic System
Transfer Function

Figure 6. Model of photovoltaic panel transfer function.

2.3. Wind Turbine System

DFIG can contribute to frequency regulation, but the frequency changes of the wind turbine are ignored previously due to the separate inertia. The operations are controlled via electronic controllers for communication between the grids. Power reserve control through speed and pitch control can assist with frequency control in the power system. The DFIG releases kinetic energy to support system inertia due to the additional inertia control loop that is frequency-sensitive to the system. Governor setting and system inertial response are researched for the frequency control of DFIG. Extracting the kinetic energy of the turbine blades from DFIG-based wind turbines contributes to the reduction of the rotor speed, which responds to the deviation in frequency to improve the frequency of the power system [3–5]. As the only form of tracking of non-conventional machine equivalent controllers, the inertial control adds a signal to the power reference output in Equation (5) according to [5]. The frequency behind a high-pass filter is represented as Δf, the constant weighting the frequency deviation derivative is K_{df} and the frequency deviation is K_{pf}. When the frequency transient is over, the equivalent non-conventional machine moves back to the optimal speeds. By forcing the speed to track the desired speed reference, a power reference in Equation (6) is devised. The PI controller is utilized with design constants of K_P and K_I. This controller is used for fast recovery speeds and transient speed variations. For non-conventional generators, from Equations (5) and (6), the total active power reference for non-conventional generators is given in Equation (7). In a short period, frequency transients generally occur. A slow PI controller is provided by p_ω^*. There are no dynamics in the power reference p_f^* and non-conventional total power injection if the power p_{NC} is regulated by high-speed power electronics. The equation can be seen in (8). The injected power before the frequency transient is shown as p_{NC}^0. The inertial control affects the power system. The system damping is provided by K_{pf} and system inertia is regulated by K_{df}. The non-conventional generating machine contributes to system inertia, and in conventional inertial control, the system inertia converts to H^*, as shown in Equation (9). The modified inertial control for a DFIG is given in Equation (10). We use a washout filter for the change in frequency having a time constant T_ω. The reference point is given in Equation (11). The frequency change measured where the wind turbine connects to the network is ΔX_2, and R is the speed regulation. Using the stored kinetic energy, the change in frequency during load disturbance is detected by the DFIG. The proposed controller provides fast active power injection control. During any disturbance, active power is injected by the wind turbine, which is ΔP_{NC}. By maintaining the reference rotor speed where maximum output power is obtained, the power injected by a wind turbine is differentiated with $\Delta P_{NC,ref}$. The wind turbine's mechanical power is shown in Equation (12). The model of the DFIG system is shown in Figure 7. The wind model is made up of frequency measurement, a washout filter, droop, a speed controller, mechanical inertia, and finally the wind turbine.

$$P_f^* = -K_{df}\frac{d\Delta f}{dt} - K_{pf}\Delta f \tag{5}$$

$$P_\omega^* = K_P(\omega_e^* - \omega_e) + K_I \int (\omega_e^* - \omega_e)dt \tag{6}$$

$$P_{f\omega}^* = \left[-K_{df}\frac{d\Delta f}{dt} - K_{pf}\Delta f\right] - \left[K_P(\omega_e^* - \omega_e) + K_I \int (\omega_e^* - \omega_e)dt\right] \tag{7}$$

$$P_{NC} = \left[-K_{df}\frac{d\Delta f}{dt} - K_{pf}\Delta f\right] - P_{NC}^0 \tag{8}$$

$$(2H + K_{df})\frac{d\Delta f}{dt} = P_f - D\Delta f = P_G + P_{NC} + P_T - P_L - \left(K_{pf} + D\right)\Delta f \tag{9}$$

$$\frac{2H}{f}\frac{d\Delta f}{dt} = P_f - D\Delta f = P_G + P_{NC} + P_T - P_L - D\Delta f \tag{10}$$

$$P_f^* = \frac{1}{R}(\Delta X_2) \tag{11}$$

$$P_{mech} = \left(\frac{\frac{1}{2}(\rho Ar)}{S_n} C_{p.opt}\right)\omega_s^3 \tag{12}$$

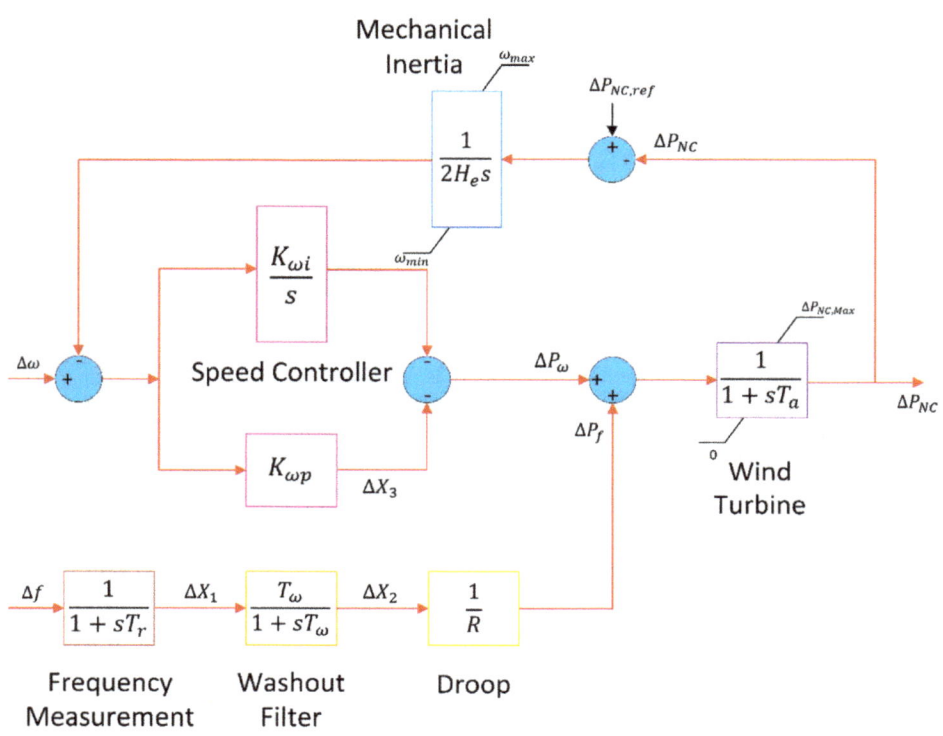

Figure 7. Model of DFIG-based wind turbines with inertia control.

3. Modeling of Fuzzy Type-2 FOPID Controller

The controller consists of a FOPID together with fuzzy action due to its positive advantages in solving disturbances and stability applications. The tuning of FOPID consists of five parameters, which are K_P, K_I, K_D, λ, and μ. These parameters are highly variable according to the output and are complex; they must be tuned via well-known Particle Swarm Optimization (PSO). The chosen variables are selected to provide the best possible outcome for the controller. Further, various optimization techniques can be utilized but are not guaranteed to provide the optimal outputs. Through fractional calculus, more adjustable parameters can be provided, which assist with tuning the controllers. The flexibility, stability, and control effect are improved with FOPID, which acts as a filter for an infinite dimension. The FOPID has a memory function that is related to the entire history in the fractional differentiation. The far and close errors have small and larger response factors respectively. The future and present information is influenced through this. Therefore, this provides good applications for boiler–turbine systems. The arrangement of the controller can be seen below in Figure 8.

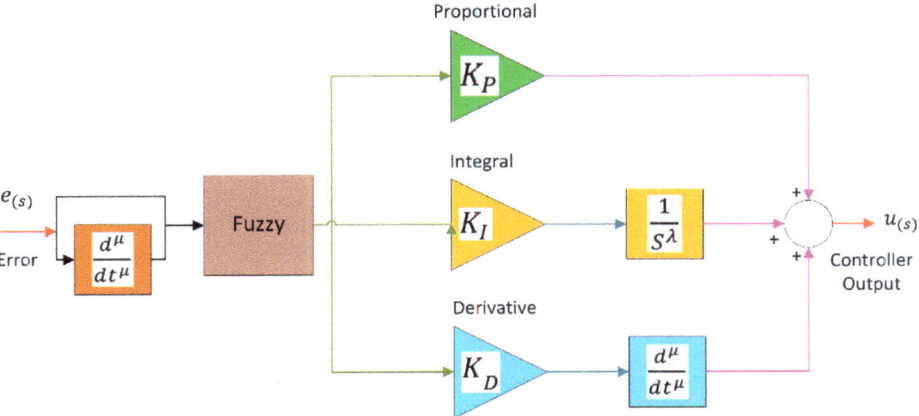

Figure 8. Fractional-order fuzzy logic controller.

The Fuzzy Type-2 controller consists of five elements. These elements are the fuzzifier, fuzzy interference, fuzzy rules, type reducer, and defuzzifier. Each process plays an important role in the output of data. The input of data is fuzzified through the introduction of membership functions for ease of understanding and classification. The data are changed to a fuzzified input by the use of fuzzy applications into membership functions to establish a rule strength. The Mamdani fuzzy interference system is used due to its advantages of intuitiveness, widespread acceptance, and interpretable rule base. Combining the rule strength and the output membership function to find the consequence of the rule, Mamdani FIS is used. The structure of a Fuzzy Type-2 system is similar to that of a Fuzzy Type-1 system, with the only difference being the type reduction process function, which allows the controller to better handle system uncertainties because it can model them and minimize their effect. If all uncertainties disappear, the Type-2 Fuzzy sets convert to Type-1, which thereafter leads to the final defuzzification result (Figure 9).

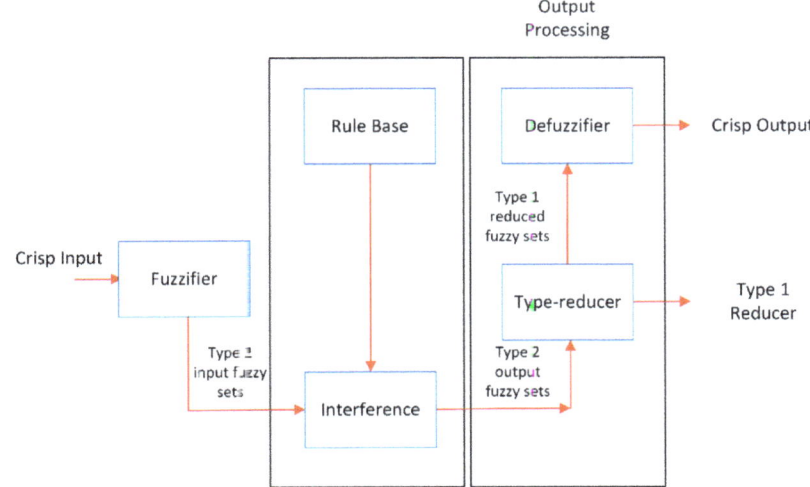

Figure 9. Fuzzy Type-2 logic system block diagram [26].

For this research, the Nie–Tan reduction method (NT) was utilized with no iterative process, which improves the type reduction efficiency. A Fuzzy Type-2 system has more design degrees of freedom than a Fuzzy Type-1 system, because Type-2 has more parameters than Type-1. As random uncertainties flow through a system and their effects can be evaluated using the mean and the variance, linguistic and random uncertainties flow through a Fuzzy Type-2 system, and their effects can be evaluated using the defuzzified output and the type-reduced output of the system. Often used in intervals, the variance provides a measure of dispersion about the mean. The defuzzified output, which provides a measure of dispersion, is the interpretation of the type-reduced output. The type-reduced set also increases as linguistic or random uncertainties increase, because the variance increases as the random uncertainty increases. A Fuzzy Type-1 system is comparable to a probabilistic system through the first moment, whereas a Fuzzy Type-2 system is comparable through the first and second moments. The rules are based on the individual's application of the information data. The fuzzy yield is made up of 49 laws from seven triangular membership functions on information and yield data. The logic statements "if" and "then" are used to determine the yield at this point. These rules have been utilized for FT2-FOPID, which can be seen in Tables 1 and 2. The output distribution is defuzzified to produce crisp outputs. The membership functions are generally used from negative one to positive one and the design membership functions for error and error deviation are shown in Figure 10.

Table 1. Fuzzy Type-2 FOPID rule base.

ACE/dACE	NB	NM	NS	ZE	PS	PM	PB
NB	NB	NB	NB	NB	NM	NS	ZE
NM	NB	NB	NB	NM	NS	ZE	PS
NS	NB	NB	NM	NS	ZE	PS	PM
ZE	NL	NM	NS	ZE	PS	PM	PB
PS	NM	NS	ZE	PS	PM	PB	PB
PM	NS	ZE	PS	PM	PB	PB	PB
PB	ZE	PS	PM	PB	PB	PB	PB

Table 2. Rule base statements.

Rule	Statement
1	If ACE is A and dACE is A, then dACE is NB
2	If ACE is B and dACE is A, then dACE is NM
3	If ACE is C and dACE is A, then dACE is NS
4	If ACE is D and dACE is A, then dACE is ZE
5	If ACE is E and dACE is A, then dACE is PS
6	If ACE is F and dACE is A, then dACE is PM
7	If ACE is G and dACE is A, then dACE is PB

The rules are based on a Boolean system of true or false statements to provide valuable flexibility for reasoning, thereby considering the inaccuracies and uncertainties in the system. In a fuzzy logic system, there is no absolute true or false, but it is partially true or false. The rule base contains if and then conditions to govern the decision-making system, which is very important for the output results. The rules are input into the interference system, which matches the current fuzzy inputs with each rule statement, which then produces the required outputs to perform control actions. This helps to remove uncertainties and disturbance to an acceptable level. Some of the rule base statements are shown in Table 2, where the two inputs are checked against the interference system consisting of the set of rules and the output is given to the type reducer.

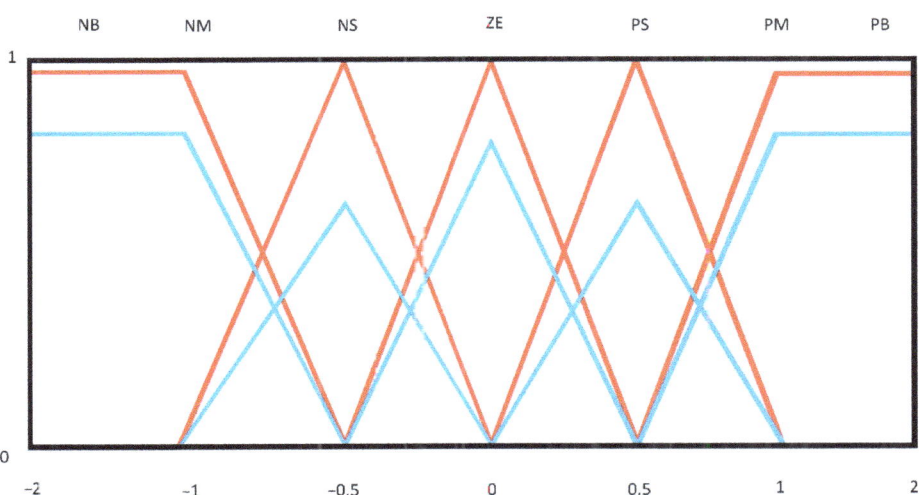

Figure 10. Fuzzy Type-2 primary membership function of error and error deviation.

The input values of a Fuzzy Type-2 system have membership functions ranging from upper membership functions to lower membership functions. This provides two fuzzy values for each Type-2 membership function. With the rules discussed previously, the fuzzy operator is applied to the fuzzified values of the membership functions. The minimum and maximum output value for the fuzzy set of each rule is the result of the fuzzy operator to the fuzzy values of upper and lower membership functions.

4. Simulation and Analysis of Results

The research work displayed shows the analysis of the thermal power system integrated with PV and wind-based power generation in each area, which is connected via an AC tie-line that contributes to the role of balancing supply and demand loads. The frequency of the designed interconnected system is analyzed and studied for the behavior of the signal. The investigation is meant to demonstrate the integration of clean energy through renewable energy sources effectively within existing thermal power systems while assisting with variable load changes within the grid. Each control area action is limited by using the GDB and GRC non-linearities as it makes the action of secondary controllers more practical and realizable.

For this work, the controller FT2-FOPID is showcased. While the Fuzzy Type-2 system can be coupled with either PI, PD or PID, the FOPID has been proven to provide favorable results in control problems due to the two additional freedom adjustable parameters. FOPID and FT1-FOPID have been shown to have difficulty in dealing with the uncertainty of systems; therefore, FOPID with Type-2 Fuzzy was applied. The controllers aim to restore the frequency and power deviations over tie-lines to their original state within the least amount of time while producing less settling time, low overshoot, and no oscillations. The different types of controller configurations are used for comparison of the outputs. The ACE and dACE are inputs of the fuzzy system.

The output of the fuzzy system is composed of seven areas: NB, NM, NS, ZE, PS, PM, and PB. These are used within the triangular uncertainty member function class for ease of understanding. The reduced rule base with non-linear membership functions for Fuzzy Type-2 is shown in Table 1 and Figure 10. The output of the fuzzy logic is defuzzified as Type-1 reduced sets, which produce real values from crisp values. The input of the FOPID is derived from the output signal of the fuzzy logic system. The parameter gains for the

FOPID, i.e., K_P, K_I, K_D, λ, and μ, are calculated through PSO with 50 iterations to produce the best results.

The simulated results are quantitatively given using the performance index Integral of Time Absolute Error (ITAE) and Integral Absolute Error (IAE). These indices can calculate the area of the error, which assists with higher accuracy for the analysis of the controller performance, especially in graphical representations. The comparison of performance was carried out using the output values given in Tables 3–6.

Table 3. ITAE results obtained for various controllers for demand change of 1% in Area 1.

Controllers	ITAE
PID with no RES	0.9433
PID with RES	2.269
FOPID with RES	0.02066
FT1-FOPID with RES	0.01362
FT2-FOPID with RES	0.009286

Table 4. IAE results obtained for various controllers for demand change of 1% in Area 1.

Controllers	IAE
PID with no RES	0.05726
PID with RES	0.08157
FOPID with RES	0.007249
FT1-FOPID with RES	0.001953
FT2-FOPID with RES	0.001161

Table 5. ITAE results obtained for various controllers for demand change of 1% in Area 1 and 2% in Area 2.

Controllers	ITAE
PID with no RES	2.093
PID with RES	4.057
FOPID with RES	0.06176
FT1-FOPID with RES	0.0401
FT2-FOPID with RES	0.02749

Table 6. IAE results obtained for various controllers for demand change of 1% in Area 1 and 2% in Area 2.

Controllers	IAE
PID with no RES	0.1139
PID with RES	0.1638
FOPID with RES	0.02205
FT1-FOPID with RES	0.005757
FT2-FOPID with RES	0.00347

The integrated power system is simulated using a 1% load demand change in Area 1 for analysis purposes. The results of all the areas can be seen and are arranged in a way that is easy to analyze in Figures 11–13, especially with the ITAE and IAE values. It can be easily seen that the penetration of RES affects the system negatively by providing high oscillations and making the system extremely non-linear in all three graphs. When comparing the results in Area 1, it can be seen that the FOPID has better results than traditional PID controllers, even though there are slight oscillations present. PID with RES has displayed higher overshoots than the rest of the controller configurations, with an ITAE performance of 2.269. PID has been shown to produce multiple larger oscillations in all three depictions and cannot overcome the fluctuations from the high penetration of renewable energy systems. Therefore, this type of controller is not suitable for these applications. In Area 1 and Area 2, and for tie-line deviations, the fuzzy integrated systems have shown good results, with the lowest settling time ranging from 0 to 5 s and returning to steady state, with little to no oscillations and minimum overshoot of these controllers. The controllers, including fuzzy, do not exceed 0.015, as displayed in the stand-alone FOPID and PID controllers for the area of error using ITAE. The stand-alone FOPID controller does return to the initial state but takes a long time, which is around 8–10 s, with an error of 0.02066 using ITAE. The FT2-FOPID has been shown to have the best results using IAE, with the smallest error of 0.001161 Therefore, the response time and performance of the FT2-FOPID are far superior to the traditional frequency controllers and Fuzzy Type-1 logic systems.

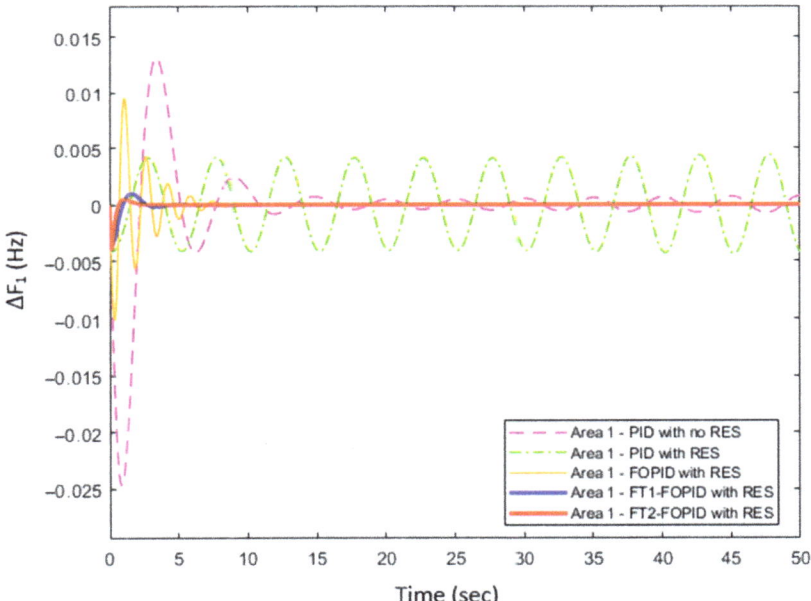

Figure 11. Results for 1% load alteration in Area 1.

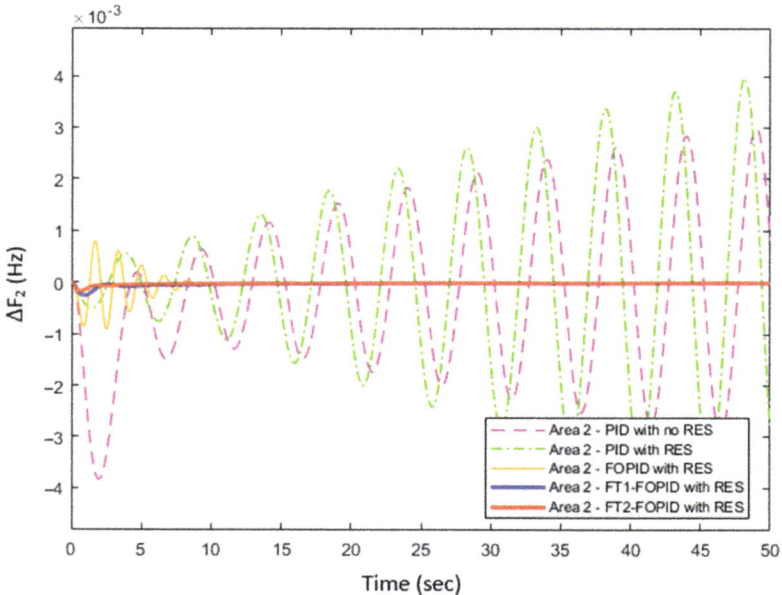

Figure 12. Results for 1% load alteration in Area 2.

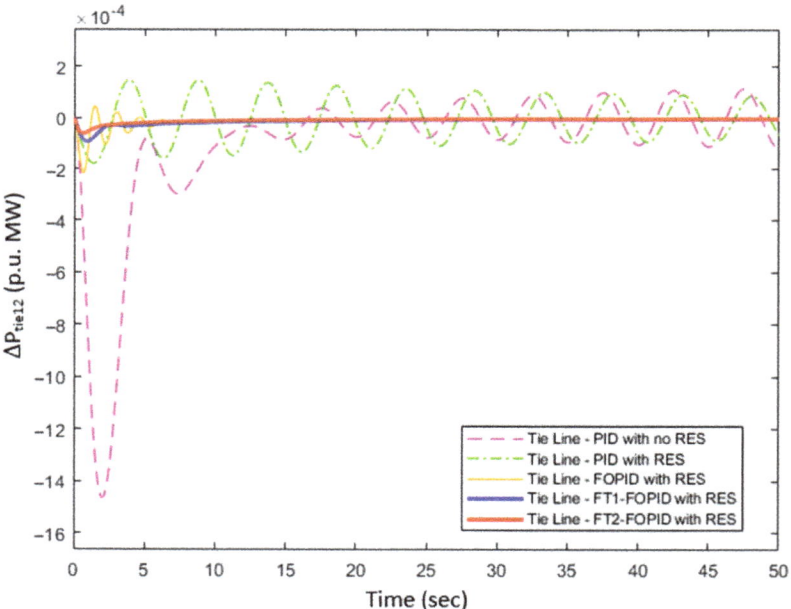

Figure 13. Results for 1% load alteration in tie-line.

Figures 14–16 are graphically presented to show the load disturbance of both areas in the power system using 1% and 2%, respectively. The results displayed are similar to the previous depictions showing load disturbance in a single area. It is still evident that the inclusion of the RES in both areas contributes to a highly non-linear system, which negatively impacts the power system, as shown in Tables 2 and 4 for PID with RES. However, comparing the tie-line representations in Figures 13 and 16, Figure 16 showcases fewer oscillations for the inclusion of RES compared to no RES. The system balances itself with the load demands and supply for the areas that contribute to this. However, it still has oscillations present and does not return to a steady state. The FOPID has displayed better performance than the PID controller with the inclusion of the additional fractional-order parameters. Proof of this is shown in Table 4; for comparison, the error is simulated as 0.06176 for FOPID with RES and 4.057 for PID with RES. This can be confirmed with the second performance criterion in Table 6, with 0.02205 for FOPID with RES and 0.1638 for PID with RES. The controller brings the system to normal conditions with a lower wavelength and quicker response time. There are still minimal oscillations present but much less than the PID controller. The overshoot in Area 2 is shown to be very high due to the 2% load disturbance present, which contributes to the negative results displayed. In the graphs presented, the settling time has an average value of 5–10 s for FOPID. The integration of fuzzy logic in the system has drastically improved the performance of the controllers with the additional iterations of the logical system processes. With the fastest response time, least overshoot, best settling time, and no oscillations, the fuzzy logic controllers have the best performance with the lowest error for the two areas and the tie-line region.

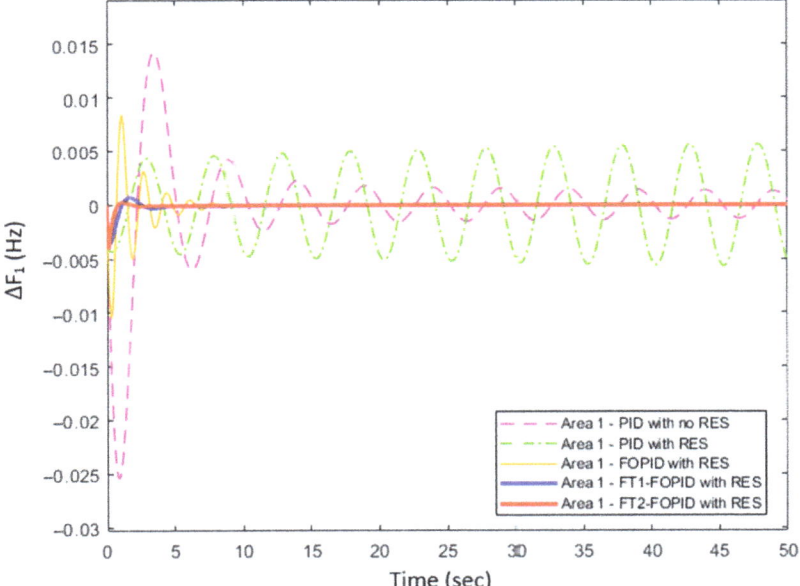

Figure 14. Results for 1% and 2% load alteration in Area 1.

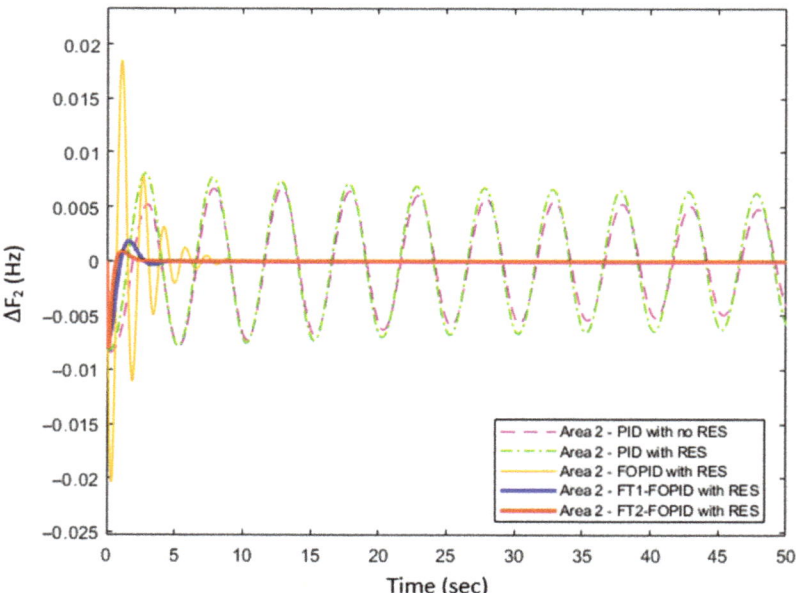

Figure 15. Results for 1% and 2% load alteration in Area 2.

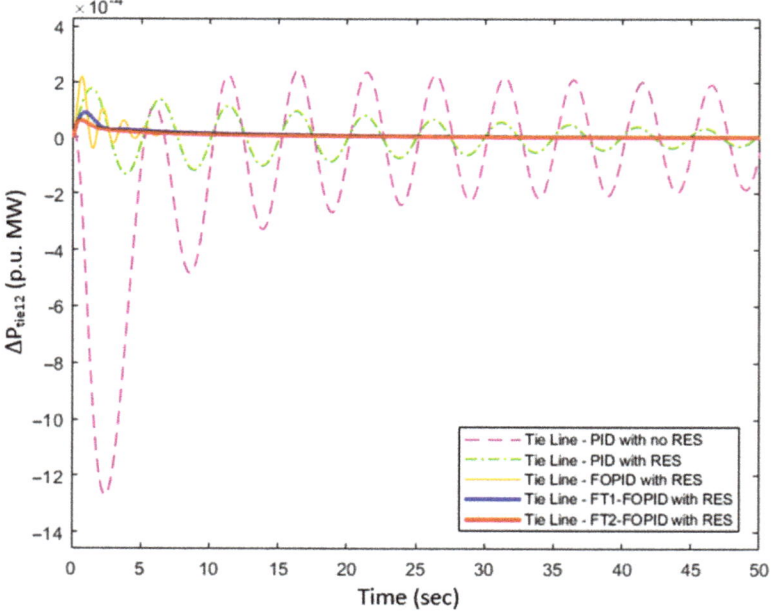

Figure 16. Results for 1% and 2% load alteration in Tie Line.

The Fuzzy Type-2 system has superior performance to Type-1, which is evident in the depictions and the ITAE/IAE error values. With the NT-type reduction in Type-2, the results are further fine-tuned to provide optimal performance. The wind and solar power have made the system highly non-linear due to the changing of the wind rotor speeds and the irradiance from the solar panels, shading of panels, and also inverters. The individual RES does have components and mechanisms in place to help curb these disturbances, but does not fully reduce them to zero. Therefore, a secondary controller is required for additional assistance to mitigate these disturbances. Even though RES provides clean and renewable energy to the system, the higher the RES capacity, the greater the control required for interconnected systems to work together. From all the results displayed, the FT2-FOPID has the greatest performance compared to the rest of the controllers, with a slightly better edge than the fuzzy Type-1, which does not exceed ±0.005 Hz. The error between the Type-1 and Type-2 Fuzzy systems has a difference of approximately 0.02 for a change in demand for Area 1 and Area 2. The results illustrated on the above graph and the ITAE/IAE values clearly show that RES creates disturbances when coupled with the power system. With the introduction of PID, the results had some control but did not come to steady-state conditions. FOPID with RES has been shown to provide better results than PID with or without RES; therefore, FOPID was used as the coupling controller for the fuzzy logic systems. The FT2-FOPID has proven itself to have lower overshoot and almost non-existent oscillations present. Even with the penetration of RES, the Fuzzy Type-2 controller can handle the non-linearity that could harm the power system. The applications of artificial intelligence can assist with control methods and solve unprecedented problems. With more processes being introduced, the controller can obtain better outputs and help the power system to overcome the disturbances as soon as possible.

5. Conclusions

This paper presents an attempt to model and integrate renewable energy sources such as wind and solar power within a thermal power system. The thermal power system is interconnected via an AC tie-line to other areas having thermal power generation and resulting in an interconnected system. This system is controlled with the assistance of a Type-2 Fuzzy logic controller together with FOPID. Multiple configurations of controllers have been simulated, compared, and analyzed to produce the most efficient output. The controller's main objective is to highlight the best performance in the overshoot, oscillations, and settling time of the frequency and power interchange over tie-lines while experiencing a sudden change in load demand. The addition of DFIG-based wind turbines assists with the stability of the power system, while the PV-based solar system introduces fluctuations in the system due to its inverter, contributing to the slight disturbances even after the signal has been filtered. The system when interconnected with clean energy systems can still become stable through the introduction of auxiliary control methods utilizing Type-2 Fuzzy logic with FOPID, well developed via PSO. The results guarantee that the proposed design is well suited for a renewable interlinked power system in comparison to results obtained via other control techniques. Further, the non-linearities' inclusion has shown a negative impact on all controllers' output; nonetheless, Type-2 Fuzzy logic with FOPID is effective in providing acceptable results for the power system. As an increase in renewable energy generation is becoming highly desirable, this research paper can assist as the foundation for the gradually increasing penetration of clean energy within various countries' fossil-fuel-driven power systems.

Author Contributions: All authors envisioned the study and participated in the development of the concept and collection of information. Introduction, M.O.A.K.; Modeling, M.O.A.K.; Simulation, M.O.A.K.; Analysis, M.O.A.K. and G.S.; Conclusions, M.O.A.K. and G.S.; Investigation, M.O.A.K. and G.S.; Writing—original draft preparation, M.O.A.K.; Writing—review and editing, M.O.A.K. and G.S.; Visualization, M.O.A.K. and G.S.; Supervision, M.O.A.K., K.T.A. and G.S. All authors have read and agreed to the published version of the manuscript.

Funding: Not applicable.

Institutional Review Board Statement: Not applicable.

Informed Consent Statement: Not applicable.

Data Availability Statement: Not applicable.

Conflicts of Interest: The authors declare no conflict of interest.

References

1. Thopil, M.; Bansal, R.; Zhang, L.; Sharma, G. A review of grid connected distributed generation using renewable energy sources in South Africa. *Energy Strategy Rev.* **2018**, *21*, 88–97. [CrossRef]
2. Gopalan, J.; Mampilly, B. A Versatile Statcom for SSR Mitigation and Dynamic Reactive Power Management in DFIG Based Wind Farm Connected to Series Compensated Line. In Proceedings of the 2020 IEEE International Power and Renewable Energy Conference, Karunagappally, India, 30 October–1 November 2020.
3. Sharma, G.; Niazi, K.R. Recurrent ANN based AGC of a two-area power system with DFIG based wind turbines considering asynchronous tie-lines. In Proceedings of the 2014 International Conference on Advances in Engineering & Technology Research (ICAETR—2014), Unnao, India, 1–2 August 2014.
4. Li, J.; Wang, X.; Lyu, J.; Zong, H.; Xue, T.; Fang, Z.; Cai, X. Stability Analysis of Wind Farm Connected to Hybrid HVDC Converter. In Proceedings of the 2020 IEEE 9th International Power Electronics and Motion Control Conference (IPEMC2020-ECCE Asia), Nanjing, China, 29 November–2 December 2020.
5. Niazi, K.; Sharma, G. Study on Dynamic Participation of Wind Turbines in Automatic Generation Control of Power Systems. *Electr. Power Compon. Syst.* **2014**, *43*, 44–55. [CrossRef]
6. Huang, W.; Zhang, W. Research on Distributed wind Power Reactive Voltage Coordinated Control Strategy Connected to Distribution Network. In Proceedings of the 2021 4th International Conference on Energy, Electrical and Power Engineering (CEEPE), Chongqing, China, 23–25 April 2021.
7. Mahdi, A.; Tang, W.; Wu, Q. Derivation of a complete transfer function for a wind turbine generator system by experiments. In Proceedings of the 2011 IEEE Power Engineering and Automation Conference, Wuhan, China, 8–9 September 2011.
8. Estrice, M.; Sharma, G.; Akindeji, K.; Davidson, I. Frequency Regulation Studies of Interconnected PV Thermal Power System. In Proceedings of the 2020 International SAUPEC/RobMech/PRASA Conference, Cape Town, South Africa, 29–31 January 2020.
9. Estrice, M.; Sharma, G.; Akindeji, K.; Davidson, I. Application of AI for Frequency Normalization of Solar PV-Thermal Electrical Power System. In Proceedings of the 2020 International Conference on Artificial Intelligence, Big Data, Computing and Data Communication Systems (icABCD), Durban, South Africa, 6–7 August 2020.
10. Tomy, F.; Prakash, R. Load frequency control of a two area hybrid system consisting of a grid-connected PV system and thermal generator. *Int. J. Res. Eng. Technol.* **2014**, *3*, 573–580.
11. Cao, F.; Qiu, J.; Jing, Z. Performance simulation and distribution strategy of solar and wind coupled power generation systems in Northwest China. In Proceedings of the 2020 12th IEEE PES Asia-Pacific Power and Energy Engineering Conference (APPEEC), Nanjing, China, 20–23 September 2020.
12. Xing, C.; Xi, X.; He, X.; Xiang, C.; Li, S.; Xu, Z. Research on Maximum Power Point Tracking Control of Wind-Solar Hybrid Generation System. In Proceedings of the 2021 China International Conference on Electricity Distribution (CICED), Shanghai, China, 7–9 April 2021.
13. Abd-Elazim, S.; Ali, E. Load frequency controller design of a two-area system composing of PV grid and thermal generator via firefly algorithm. *Neural Comput. Appl.* **2016**, *30*, 607–616. [CrossRef]
14. Anuradha, A.Y.; Sinha, S. Solar-Wind Based Hybrid Energy System: Modeling and Simulation. In Proceedings of the 2021 4th International Conference on Recent Developments in Control, Automation & Power Engineering (RDCAPE), Noida, India, 7–8 October 2021.
15. Zhang, X.; Spencer, J.; Guerrero, J. Small-Signal Modeling of Digitally Controlled Grid-Connected Inverters with LCL Filters. *IEEE Trans. Ind. Electron.* **2013**, *60*, 3752–3765. [CrossRef]
16. Mohamed, R.; Boudy, B.; Gabbar, H. Fractional PID Controller Tuning Using Krill Herd for Renewable Power Systems Control. In Proceedings of the 2021 IEEE 9th International Conference on Smart Energy Grid Engineering (SEGE), Oshawa, ON, Canada, 11–13 August 2021.
17. Sondhi, S.; Hote, Y. Fractional order PID controller for load frequency control. *Energy Convers. Manag.* **2014**, *85*, 343–353. [CrossRef]
18. Shouran, M.; Alsseid, A. Cascade of Fractional Order PID based PSO Algorithm for LFC in Two-Area Power System. In Proceedings of the 2021 3rd International Conference on Electronics Representation and Algorithm (ICERA), Yogyakarta, Indonesia, 29–30 July 2021.
19. Li, P.; Zheng, M.; Zhong, D.; Zheng, Y.; Zhang, G. Design of Fractional Order Controller for Microgrid Based on Model Analysis. In Proceedings of the 2021 40th Chinese Control Conference (CCC), Shanghai, China, 26–28 July 2021.
20. Shi, J. A Fractional Order General Type-2 Fuzzy PID Controller Design Algorithm. *IEEE Access* **2020**, *8*, 52151–52172. [CrossRef]
21. Shakibjoo, A.; Moradzadeh, M.; Din, S.; Mohammadzadeh, A.; Mosavi, A.; Vandevelde, L. Optimized Type-2 Fuzzy Frequency Control for Multi-Area Power Systems. *IEEE Access* **2022**, *10*, 6989–7002. [CrossRef]
22. Joshi, M.; Sharma, G.; Davidson, I. Load Frequency Control of Hydro Electric System using Application of Fuzzy with Particle Swarm Optimization Algorithm. In Proceedings of the 2020 International Conference on Artificial Intelligence, Big Data, Computing and Data Communication Systems (icABCD), Durban, South Africa, 6–7 August 2020.

3. Sabahi, K.; Tavan, M. T2FPID Load Frequency Control for a two-area Power System Considering Input Delay. In Proceedings of the 2020 28th Iranian Conference on Electrical Engineering (ICEE), Tabriz, Iran, 4–6 August 2020.
4. Castillo, O.; Melin, P. A review on interval type-2 fuzzy logic applications in intelligent control. *Inf. Sci.* **2014**, *279*, 615–631. [CrossRef]
5. Çam, E.; Kocaarslan, İ. Load frequency control in two area power systems using fuzzy logic controller. *Energy Convers. Manag.* **2005**, *46*, 233–243. [CrossRef]
6. Rasi, D.; Deepa, S. Energy optimization of internet of things in wireless sensor network models using type-2 fuzzy neural systems. *Int. J. Commun. Syst.* **2021**, *34*, e4967. [CrossRef]

Article

Available Transfer Capability Enhancement in Deregulated Power System through TLBO Optimised TCSC

Anurag Gautam [1], Ibraheem [1], Gulshan Sharma [2,*], Pitshou N. Bokoro [2] and Mohammad F. Ahmer [3]

1. Electrical Engineering Department, Jamia Millia Islamia, New Delhi 110025, India; anuragjmi13@gmail.com (A.G.); ibraheem@jmi.ac.in (I.)
2. Department of Electrical Engineering Technology, University of Johannesburg, Johannesburg 2006, South Africa; pitshoub@uj.ac.za
3. Department of Electrical and Electronics Engineering, Mewat Engineering College, Nuh 122107, India; farazahmer007@gmail.com
* Correspondence: gulshans@uj.ac.za

Abstract: Rapid industrial development and innovations in technology bring about the menace of congestion in deregulated power systems (DPS). The transmission lines are continuously working under a stressed condition with reduced power transfer capacity. In this situation, the power losses and voltage deviations at the load buses are increased and hence reduce the system stability. To mitigate congestion, improving available transfer capability (ATC) of the transmission system is one of the most feasible and practical solution. This paper focuses on the implementation of Thyristor Controlled Series capacitor (TCSC) to mitigate congestion by enhancing ATC and via reducing power losses. AC Power Transfer Distribution Factor (ACPTDF) is applied to calculate ATC and to find the location of TCSC. To optimize the TCSC parameter (reactance), a Teaching Learning Based Optimization (TLBO) is proposed in the present work. The proposed optimization is validated on the IEEE 30 Bus system. The results are validated by matching with the results obtained through standard grey wolf optimization (GWO) and particle swarm optimization (PSO) techniques. The results show that despite two antagonistic objectives of ATC enhancement and power loss reduction, TLBO outperformed the other optimization techniques under different contingency conditions. The overall ATC of the IEEE 30 Bus system for the bilateral transactions is enhanced by 11.86%. Further active and reactive power losses are reduced by 16.7% and 29.6% for DPS.

Keywords: available transfer capacity; deregulated power system; TCSC; teaching learning based optimization; congestion; ac power transfer distribution factor

Citation: Gautam, A.; Ibraheem; Sharma, G.; Bokoro, P.N.; Ahmer, M.F. Available Transfer Capability Enhancement in Deregulated Power System through TLBO Optimised TCSC. *Energies* 2022, 15, 4448. https://doi.org/10.3390/en15124448

Academic Editor: Abu-Siada Ahmed

Received: 9 May 2022
Accepted: 1 June 2022
Published: 18 June 2022

Publisher's Note: MDPI stays neutral with regard to jurisdictional claims in published maps and institutional affiliations.

Copyright: © 2022 by the authors. Licensee MDPI, Basel, Switzerland. This article is an open access article distributed under the terms and conditions of the Creative Commons Attribution (CC BY) license (https://creativecommons.org/licenses/by/4.0/).

1. Introduction

Initially, the vertically integrated power system worked as a monopoly under the local public sector energy authorities. A single entity was controlling all the operations of generation, transmission, and distribution. With the advancement of technology in developing countries, the demand for power increased manifolds, making power management very difficult with the monopolized arrangement. This necessitated amendments in the regulated industry to deregulate the power system, hence a deregulated power system (DPS) came into existence. Thus, the DPS was adopted by a number of countries to meet growing power demand, maintain voltage profile at buses, reduction in power losses, enhance the security margin and to make tariff policies as per the requirement of the customers. In a deregulated environment, the power system is unbundled into generation companies (GENCOS), transmission companies (TRANSCOS), and distribution companies (DISCOS). These separate entities are further segregated under two power market models: the independent system operator (ISO) model, where the GENCOS and DISCOS are governed and controlled by ISO by encouraging healthy competition between different market players. The second model is the transmission system operator (TSO) model, which ensures unbiased open access right to entitled market players to use the power transmission network [1]. The deregulated power system has numerous advantages such as the freedom for the consumer to select the most economic generator for their needs. However, in this system, the power generation and demands have very complex patterns. Heavy power transactions in the competitive markets resulted in the overloading of the transmission lines. To supply the demanded power, the transmission lines are working near or at their thermal stability limits. This creates congestion in the system [2]. To avoid this congestion the power transfer capability of the lines must be known before any transaction is made. To manage a safe and economical transaction, the Available Transfer Capability (ATC) of the system must be calculated at regular periods. ATC can be defined as the capability of a line to transfer power securely above the base case power for its utilization in other commercial activities [3]. The effective transmission to the demanded power can be ensured if ATC is calculated at regular intervals and the documented data for the same is made available all the time to the market players. Precisely calculated ATC helps in accurately forecasting the future up-gradation of the system. As up-gradation of the power system involves huge capital, so to make the system economically reasonable ATC enhancement is a feasible option market is left with. When ATC is over-valued, the system become unstable and if the ATC is undervalued there is an economical loss due to underutilization of the power resources [4]. ATC evaluation and enhancement is very crucial when system stability and security is concerned.

2. Literature Survey

Previously, to enhance ATC of a power system, the infrastructure expansion was suggested, which included addition of parallel transmission lines to the estimated overloaded lines. These methods have now become obsolete due to geographical constraints and the invention of new semiconductor technologies such as FACTS devices [5,6]. Different FACTS devices are implemented for the objective of ATC enhancement [7]. The influence of different series and shunt FACTS devices on ATC enhancement under N-1 contingency condition has been validated on the standard power systems. The effective enhancement of ATC not only depends on the type of FACTS devices used, but also on the location of these devices. The FACTS devices are very expensive. To make their implementation economically viable, these must be optimized for location together with their respective parameters. The FACTS devices may either be located on the basis of the system condition or may be located on the basis of certain optimization techniques with an aim to maximize ATC [8]. Static Var Compensator (SVC) and TCSC are implemented to enhance ATC by applying cat swarm optimization (CSO). The ATC value is calculated using the continuous power flow (CPF) method in [9]. ATC enhancement is proposed by implementing SSSC with the proposed multi-agent system (MAS). MAS together with power controllers is used as a smart system for enhancing ATC. Various FACTS devices are located to implement Power Transfer Distribution Factors (PTDFs) for enhancing the ATC of the system [10]. The effect on the ATC value of the system has been proposed by employing Static Synchronous Compensator (STATCOM) in the system [11]; however, this method is less economical due to high cost of STATCOM. Other FACTS devices, Unified Power Flow Controller (UPFC), SSSC, and STATCOM are suggested to effectively enhance ATC in the system and to replace the costly method to lay down the new transmission lines. The parameters of these devices have been optimized by applying Particle Swarm Optimization (PSO) [12]. The improvement of ATC is not only proposed by a single FACTS device, but also by a combination of two or more FACTS devices. Implementing multiple FACTS devices is very costly and is suitable for very large and complex systems only. Series FACTS controller, TCPAR FACTS Controller, and TCSR FACTS Controller are employed in IEEE 5 Bus system to mitigate congestion. The TCSR FACTS Controller was found most effective in mitigating congestion [13]. A generalized unified power flow controller (GUPFC) and interline power flow controller (IPFC) are implemented for N-1 contingency case applying constant P, Q load model, and ZIP load model for the power system [14]. A Hybrid Mutation Particle

Swarm Optimization (HMPSO) is proposed to improve the ATC in the system by suitably locating and optimizing TCSC and SVC parameters [15]. In a multi-objective function including ATC enhancement, voltage deviation minimization, and active power loss reduction, the weighting factors are calculated employing fuzzy logic. To achieve the objective, a self-adaptive PSO is applied to optimize the location and parameter setting of TCSC [16]. A flower pollination algorithm is proposed to optimize TCSC parameters with the aim to enhance the ATC of the system. The location of the device is optimized using ACPTDF as a sensitivity factor in [17].

In the view of the above discussion, this paper contributes a method to determine the optimized location and size of TCSC with a prime objective of enhancement of ATC in DPS. Two completely antagonistic objectives of ATC maximization and active power loss minimization are solved in this research work. TLBO algorithm is applied first time to optimize the objective function under base case and N-1 contingency conditions. The proposed algorithm is validated on IEEE 30 Bus system. A comparison of results is done with those obtained by applying GWO and PSO techniques and the application outcomes are shown to verify the effectiveness of the proposed work over GWO and PSO. This research further contributes in mitigating congestion without changing the operational costs of the system, which in turn results in social welfare of both the generators and the consumers. As ATC is enhanced, the lines are transmitting power very near to the thermal and voltage limits without effecting the security of the system.

3. Method and Materials

3.1. ATC Calculation

ATC of a system depends upon a number of parameters such as total transfer capability (TTC), transmission reliability margin (TRM), capacity benefit margin (CBM), and existing transmission commitments (ETC). ATC of a system can be mathematically represented as [18,19]:

$$\text{ATC} = \text{TTC} - \sum(\text{ETC}, \text{TRM}, \text{CBM}) \quad (1)$$

TTC is the major power that can be transferred between the seller and the buyer without the violation of security constraints during normal and contingency conditions also. ETC is calculated by the traditional power flow calculations. TRM is considered to be 10% of the TTC value and the value of CBM is taken as zero for simplicity in calculations [15]. There are different methods to calculate ATC in a system. Basically, three methods are implemented to calculate ATC.

1. Repeated Power Flow (RPF) method;
2. Calculation based on optimization method;
3. Linear sensitivity factor based method.

In this paper, linear sensitivity factor-based method is applied to calculate ATC [20]. These factors are Power Transfer Distribution Factors. These can be calculated either by applying AC load flow analysis or DC load flow analysis. Depending on the outcome required AC or DC load flow analysis is done. In DC load flow analysis, transmission power losses are neglected. The voltage is considered constant and angle variations are considered negligible. This method takes a very small time for the calculations but the outcome is not very accurate. On the other hand, AC load flow-based ATC calculation is time-consuming, and complex but gives nearly accurate results.

This paper applies AC load flow analysis. The PTDFs thus calculated are called ACPTDF. In the interconnected power system, let us consider a bus p and bus n. These two buses are connected directly and also through indirect paths. Whenever there is a power transaction between sending bus p and receiving bus n, the power flow is affected in other transmission lines as well. Let us consider such an indirect path between buses l and

k. The ratio of the change in power flow between buses l and k due to a power transaction between buses p and n is termed ACPTDF.

$$\text{ACPTDF} = \frac{\Delta P^{lk}}{\Delta P^{pn}} \quad (2)$$

ACPTDF can be calculated in the following steps:
1. Considering m node test system with $1, \ldots, g$ are PV buses and $g+1, \ldots, m$ are
2. PQ buses.

The value of ΔP^{lk} (change in power flow in any random line lk) is calculated by NR load flow analysis.

$$\begin{bmatrix} \Delta \delta \\ \Delta |V| \end{bmatrix} = \begin{bmatrix} J_A & J_B \\ J_C & J_D \end{bmatrix}^{-1} \times \begin{bmatrix} \Delta P \\ \Delta Q \end{bmatrix} \quad (3)$$

The elements of the Jacobian matrix in Equation (3) are determined in the follows:

$$\left. \begin{array}{l} J_A = \frac{\partial P_{lk}}{\partial \delta_{lk}} = V_l V_k Y_{lk} \sin(\theta_{lk} + \delta_k - \delta_l) \\ J_B = \frac{\partial P_{lk}}{\partial \delta_{lk}} = -V_l V_k Y_{lk} \sin(\theta_{lk} + \delta_k - \delta_l) \\ J_C = \frac{\partial P_{lk}}{\partial V_l} = V_k Y_{lk} \cos(\theta_{lk} + \delta_k - \delta_l) - 2V_k Y_{lk} \cos \theta_{lk} \\ J_D = \frac{\partial P_{lk}}{\partial V_k} = V_l Y_{lk} \cos(\theta_{lk} + \delta_k - \delta_l) \end{array} \right\} \quad (4)$$

From Equations (3) and (4):

$$\Delta P_{lk} = \begin{bmatrix} \frac{\partial P_{lk}}{\partial \delta_2} & \cdots & \frac{\partial P_{lk}}{\partial \delta_n} & \frac{\partial P_{lk}}{\partial V_{g+1}} & \cdots & \frac{\partial P_{lk}}{\partial V_n} \end{bmatrix} \begin{bmatrix} \Delta \delta_2 \\ \vdots \\ \Delta \delta_n \\ \Delta |V_{g+1}| \\ \vdots \\ \Delta |V_n| \end{bmatrix} \quad (5)$$

Now the transacted power between buses l and k is, $\Delta P_l = +P_t$ and $\Delta P_k = -P_t$. Hence, Equation (5) can be rewritten as:

$$\Delta P_{lk} = \begin{bmatrix} \frac{\partial P_{lk}}{\partial \delta_2} & \cdots & \frac{\partial P_{lk}}{\partial \delta_n} & \frac{\partial P_{lk}}{\partial V_{g+1}} & \cdots & \frac{\partial P_{lk}}{\partial V_n} \end{bmatrix} J^{-1} \begin{bmatrix} \Delta \delta_2 \\ \vdots \\ \Delta \delta_n \\ \Delta |V_{g+1}| \\ \vdots \\ \Delta |V_n| \end{bmatrix} = \text{ACPTDF} \times P_t \quad (6)$$

If P_{lk} is the real power flow in line $l-k$; TL_{lk}^{max} is the maximum transaction limit of line $l-k$; P_{lk-pn}^m is the maximum constrained power transaction between p-n; N_s is the total number of load buses in system.
Then,

$$P_{lk-pn}^m = \begin{cases} \frac{TL_{lk}^{max} - P_{lk}}{\text{ACPTDF}_{lk,pn}} & \forall \ \text{ACPTDF}_{lk,pn} > 0 \\ \infty & \forall \ \text{ACPTDF}_{lk,pn} = 0 \\ \frac{-TL_{lk}^{max} - P_{lk}}{\text{ACPTDF}_{lk,pn}} & \forall \ \text{ACPTDF}_{lk,pn} < 0 \end{cases} \quad (7)$$

and

$$\text{ATC}_{pn} = \min\left\{ P_{lk-pn}^m \forall \ lk \in N_s \right\} \quad (8)$$

3.2. Reactance Model of TCSC under Consideration

TCSC is one of the very versatile reactance-based-series FACTS devices with variable inductive or capacitive reactance, jX_{TCSC}. The capability of TCSC to promptly manipulate the line reactance makes it a suitable device to be implemented for mitigating congestion. Here, the transmission line is modelled with lumped parameters as a classical π model. A TCSC connected between bus p and n is shown in Figure 1.

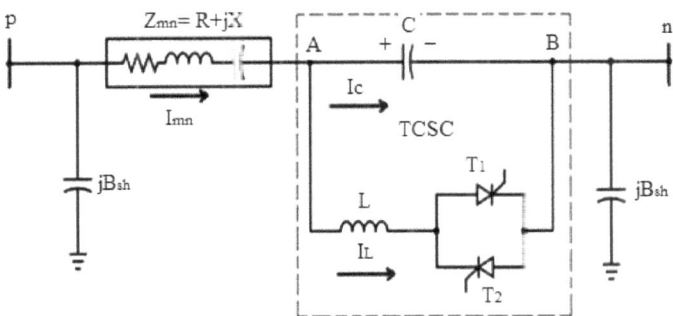

Figure 1. Schematic of TCSC model.

Equation (9) determines the change in magnitude of line admittance, Δy_{pn}:

$$\Delta y_{pn} = y'_{pn} - y_{pn} \tag{9}$$

$$y'_{pn} = (g_{pn} + jb_{pn})' \tag{10}$$

$$y_{pn} = (g_{pn} + jb_{pn}) \tag{11}$$

Δy_{pn} is the change in admittance of line p–n; y'_{pn} is the new admittance of the line p–n after implementation of TCSC; y_{pn} is the original admittance of the line p–n before implementation of TCSC.

In Equations (10) and (11);

$$\left. \begin{array}{l} g_{pn} = \frac{r_{pn}}{r_{pn}^2 + x_{pn}^2} \\ b_{pn} = -\frac{x_{pn}}{r_{pn}^2 + x_{pn}^2} \\ g'_{pn} = \frac{r_{pn}}{r_{pn}^2 + (x_{pn} + x_{TCSC})^2} \\ b'_{pn} = -\frac{x_{pn} + x_{TCSC}}{r_{pn}^2 + (x_{pn} + x_{TCSC})^2} \end{array} \right\} \tag{12}$$

g'_{pn} and b'_{pn} represents modified conductance and susceptance, respectively, of the line p–n. Thus, with the application of TCSC, the system admittance matrix is manipulated, which in turn modifies the quantities in the NR load flow.

3.3. Objective Function

(1) The major objective is to maximize ATC of the test system.

$$ATC_{max} = \min \left\{ P^m_{lk-pn} \forall \; lk \in N_s \right\} \tag{13}$$

(2) The second objective is to minimize active power loss:

$$\text{Min } P_L = \sum_{i=1}^{N_L} g'_i \left\{ ((V_p - V_n)^2 + V_p V_n (\theta_p - \theta_n)^2 \right\} \tag{14}$$

The multi-objective function can now be formed as:

$$f = w_1 * (ATC_{max}) + w_2 * (\text{Min } P_L) \tag{15}$$

w_1 and w_2 are the weight factors. When the value of ATC is enhanced, the value of active power loss increases automatically. Thus, the objective function is solved for two cases:
- For ATC maximization;
- For Power loss minimization.

3.4. Constraints

The function is subjected to the following constraints:

(a) Equality constraints:

$$P_{Gi} - \sum_{j=1}^{N_b} V_i V_j Y_{ij} \cos(\delta_i - \delta_j - \theta_{ij}) = P_{Di} \quad (16)$$

$$Q_{Gi} - \sum_{j=1}^{N_b} V_i V_j Y_{ij} \sin(\delta_i - \delta_j - \theta_{ij}) = Q_{Di} \quad (17)$$

here, P_{Gi} and Q_{Gi} are the real and reactive power generations at ith bus, respectively; P_{Di} and Q_{Di} are the real and reactive power demands at ith bus, respectively; V_i, δ_i, and V_j, δ_j are voltage magnitudes and their corresponding angles at ith and jth buses, respectively. Y_{ij}, θ_{ij} is the admittance of line between ith and jth bus and its corresponding angle; N_b is the number of buses.

(b) Inequality constraints:

$$P_{Gi}^{min} \leq P_{Gi} \leq P_{Gi}^{max} \forall\ i = 1, 2, \ldots, n_g \quad (18)$$

$$Q_{Gi}^{min} \leq Q_{Gi} \leq Q_{Gi}^{max} \forall\ i = 1, 2, \ldots, n_g \quad (19)$$

$$V_b^{min} \leq V_b \leq V_b^{max} \forall\ i = 1, 2, \ldots, n_b \quad (20)$$

(c) Practical constraints of the TCSC [21]:

$$-0.8 X_{pn}(p.u.) \leq X_{TCSC} \leq 0.2\ X_{pn}(p.u.) \quad (21)$$

3.5. TLBO Algorithm

Teaching Learning based Optimization is based on the teacher–student relation in a class for imparting knowledge to the student. Hence, TLBO comprises two main factors, the 'Teachers Phase' and the 'Learners Phase'. Here, the class room size denotes the space for search. The teacher plays part of 'Influencer' to influence students for obtaining 'Effective' learning output from them [22]. The algorithm operates in two phases:

3.5.1. Teacher Phase

In this phase, for enhancing the average output of the class of learners, teacher delivers information to the class (learners). Therefore, this phase deals with imparting learning through the teacher only. Here, it is assumed that the teacher is more knowledgeable (better fitness) as compared to the learner. At the end of the simulation, the teacher will be replaced by the best learner. This can be mathematically explained as:

$$\mathbf{D_m = rand[X_{teacher} - T_f(X_m)]} \quad (22)$$

In Equation (22), D_m is difference between teachers' best result ($X_{teacher}$) and the existing mean result of the learner (X_m). T_f represents the teaching factor. Specific value of T_f is not mentioned in literature but more accurate results are obtained if value is considered between (1, 2). T_f can be calculated as in Equation (23):

$$T_f = \text{round}\ \{1 + \text{rand}[0, 1](2 - 1)\} \quad (23)$$

rand is any random value in (0, 1). Next, the solution is updated through Equation (24) as:

$$X_{updated} = X_{existing} + D_m \qquad (24)$$

The updated value obtained here is substituted to calculate new fitness function value. For current fitness better than the former, the new solution is memorized. These updated values act as the input for the next phase, i.e., learner phase.

3.5.2. Learner Phase

This phase includes the learners who only interact among themselves with their randomly selected partner. In this phase, the learner has two inputs from the teacher and from the randomly selected partner. A learner will improve his/her information if the randomly selected partner is more knowledgeable then him/her. Consider two learners, X_1 and X_2.

The fitness functions for X_1 and X_2 are evaluated and compared. Updated values for minimization function are evaluated by Equation (25).

$$\left. \begin{array}{l} X_{updated} = X_{current} + rand(X_1 - X_2) \\ X_{updated} = X_{current} + rand(X_2 - X_1) \end{array} \right\} \qquad (25)$$

Here, $X_{updated}$ is the updated value of the results of learner for minimizing the fitness function.

The steps involved in application of TLBO are given in Figure 2.

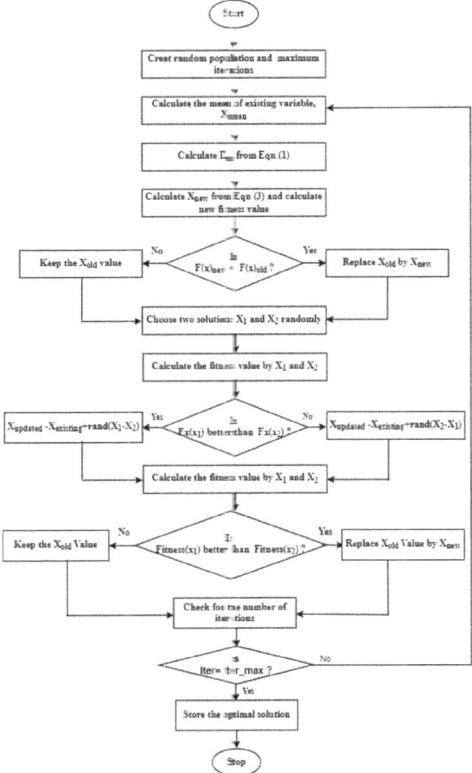

Figure 2. Operations of TLBO algorithm.

4. Steps Involved in the Implementation of Algorithm in Objective Function

The method applied is carried out in two parts:

4.1. For ATC Maximization

1. TLBO driven optimal power flow (OPF) is carried out.
2. Considering prime objective of ATC maximization, TCSC is optimized for suitable location and size by using ACPTDF and TLBO.
3. Line flows are obtained by NR with OPF.
4. A bilateral transaction between seller bus and buyer bus is created and NR is applied to calculate new line flows.
5. Change in active power flow (ΔP^{lk}) is calculated.
6. ACPTDF values are obtained by using Equation (2).
7. ATC is then calculated by applying Equation (8) for nearby and farthest bus connected to the common generator bus (sending bus).

4.2. For Power Loss Minimization

For this case also, the same steps are involved as in case of ATC maximization with a difference in weight factor (w_2) as given in Equation (15) for minimization of active power loss.

The proposed algorithm is validated on standard IEEE 30 Bus system as shown in Figure 3.

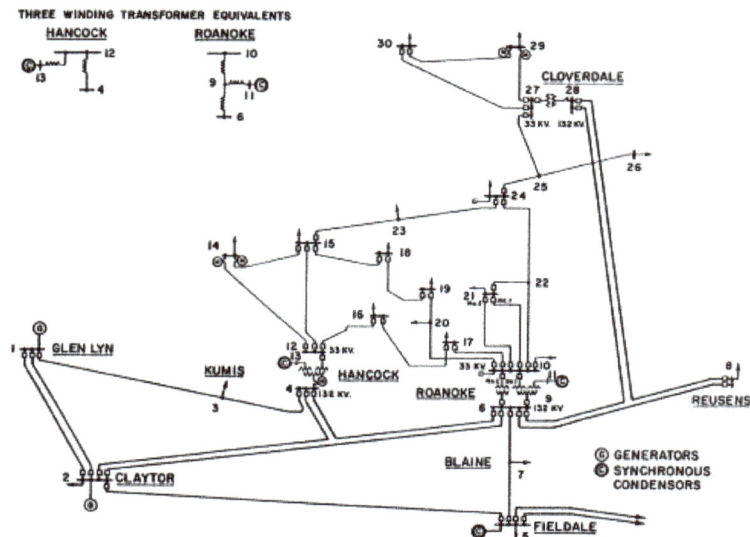

Figure 3. IEEE 30 BUS System [23].

This system comprises of 30 buses, 41 interconnected lines and 6 generators. Out of 30 buses, 24 buses are load buses. Bus number 1 is considered as slack bus. Bilateral transactions between the generator buses, i.e., 2, 5, 8, 11, and 13 and the load buses situated nearest and farthest to these buses is carried out. The generator buses are considered as sender/seller buses while the load buses are considered as receiver/buyer buses.

The ATC values calculated for base case without the application of TCSC by considering effects of generator at particular bus are shown in Figures 4–8. It is observed that the ATC value for nearby bus is higher as compared to the ATC for line connected at the farthest bus.

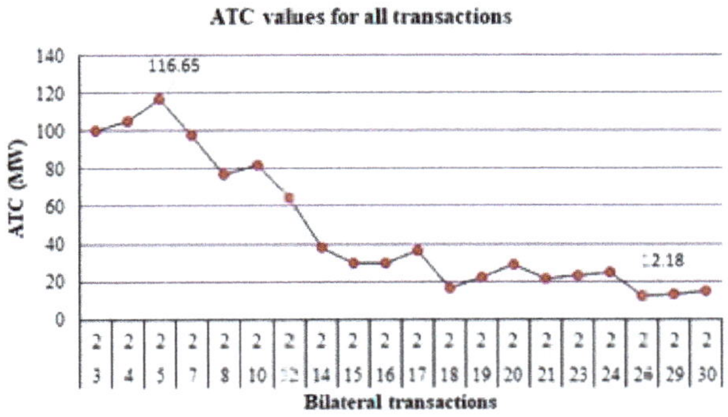

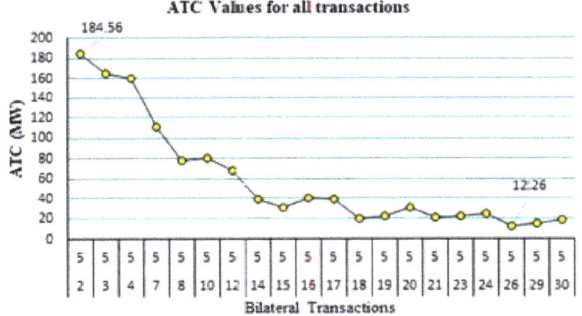

Figure 4. ATC values for all transactions with respect to generator at bus 2.

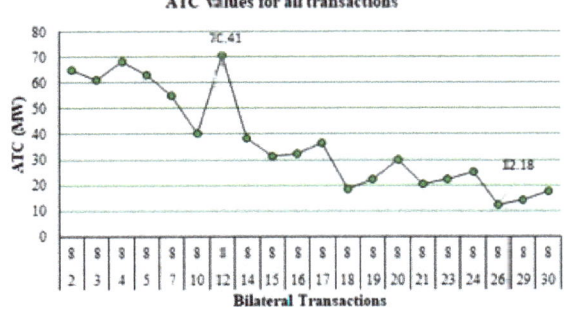

Figure 5. ATC values for all transactions with respect to generator at bus 5.

Figure 6. ATC values for all transactions with respect to generator at bus 8.

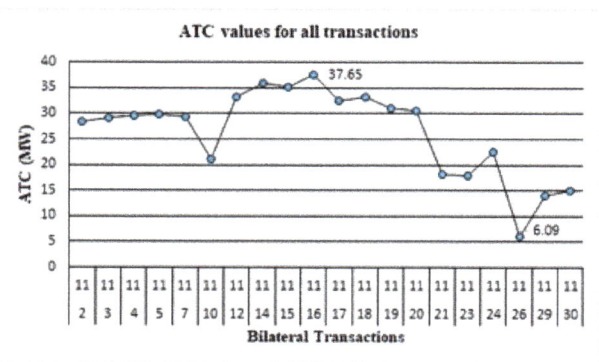

Figure 7. ATC values for all transactions with respect to generator at bus 11.

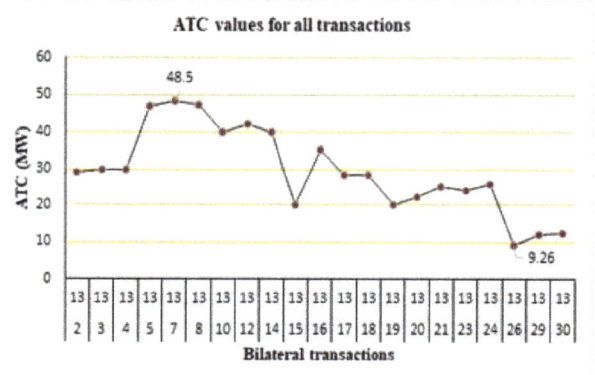

Figure 8. ATC values for all transactions with respect to generator at bus 13.

5. Result Analysis

The proposed TLBO algorithm is validated on IEEE 30 Bus test system for two cases.

5.1. For ATC Maximization

Figure 9 elaborates the enhanced ATC values when TCSC is optimized with the TLBO technique. From the results, it is observed that the ATC value for different bilateral transactions is enhanced. For transaction between 2–5, the ATC value is increased from the base case value of 116.5 MW to 125.13 MW and for the transaction 2–26, the ATC value is increased from 12.18 MW in base case to 14.53 MW with TLBO optimized TCSC. Similarly, significant ATC enhancement is achieved for both nearby bus and far away bus for each bilateral transaction. The ATC value for distant situated lines is increased significantly and can be concluded from Table 1. The proposed algorithm not only has increased the ATC value for nearby situated lines, but also for distant situated lines and hence enhance the overall ATC of the test system. With the enhancement of ATC, the active power losses also increase due to increased power flow through the lines during bilateral transactions.

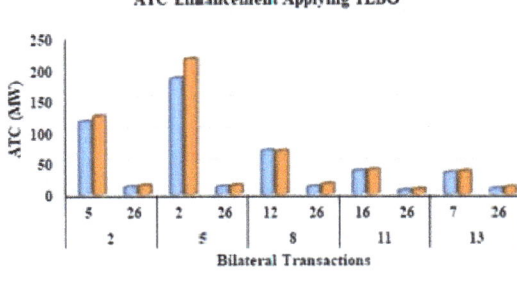

Figure 9. ATC enhancement using TLBO with and without TCSC.

Table 1. Consolidated results for ATC maximization.

Bilateral Transactions		ATC (MW)				Active Power Loss (MW)			
		OPF	OPF with TCSC			OPF	OPF with TCSC		
Seller	Buyer		GWO	PSO	TLBO		GWO	PSO	TLBO
2	5	116.65	120.75	121	125.13	7.01	7.13	7.25	7.3
	26	12.18	13.18	12.23	14.53	7.01	7.13	7.45	7.32
5	2	184.56	198.04	211.3	215.13	7.03	7.45	7.14	7.23
	26	12.26	13.45	12.75	14.45	7.09	7.19	7.22	7.29
8	12	70.41	69.57	67.9	75.07	7.04	7.17	7.17	7.25
	26	12.18	14.23	13.05	16.57	7.08	7.21	7.24	7.32
11	16	37.65	38.8	37.3	39.58	7.11	7.24	7.2	7.3
	26	6.09	6.45	6.35	8.01	7.01	7.78	7.25	7.32
13	7	48.5	36.89	47.5	50.23	7.05	7.69	7.24	7.23
	26	9.26	9.46	10.24	11.46	7.08	7.97	7.26	7.29

Figure 10 compares the percentage increment of ATC for both nearby and distant lines. It can be seen that maximum ATC enhancement obtained by the proposed TLBO algorithm is 36% for the bilateral transaction between buses 8–26, which is a line at the distant location. For the same transaction, the percentage enhancement of ATC by GWO and PSO is 16.83% and 7.14%, respectively. Additionally, for transaction between 2–26, 5–26, 11–26, and 13–26 (transactions for faraway lines) are enhanced by 19%, 16.56% 31.52%, and 23.75%, respectively, by the proposed algorithm. Total ATC in base case for the bilateral transactions considered is 509.74 MW in base case. This increases to 520.84 MW with GWO, 539.64 MW with PSO, and 570.16 MW with the TLBO optimized TCSC. Here, an approximate enhancement of ATC with proposed TLBO algorithm is 11.853%.

The reactive power losses are also calculated here and given in Figure 11. The results show that with the enhancement of ATC, the active power losses are increased, but the reactive power losses are decreased. Figure 11 presents the comparative analysis of reactive power losses reduction by the proposed TLBO, GWO and PSO algorithms. It is observed that maximum reduction in reactive power loss is obtained with the application of TLBO optimized TCSC. The average power loss in base case for all the transactions considered is 30.481 MVAr. With the application of TLBO, GWO, and PSO, the average reactive power loss is reduced to 21.44 MVAr, 28.556 MVAr, and 28.593 MVAr, respectively. This means that the reactive power losses are reduced by 29.65%, 6.28%, and 6.19% by the application of TCSC optimized by TLBO, GWO, and PSO, respectively, and hence the results of TLBO are better than GWO and PSO.

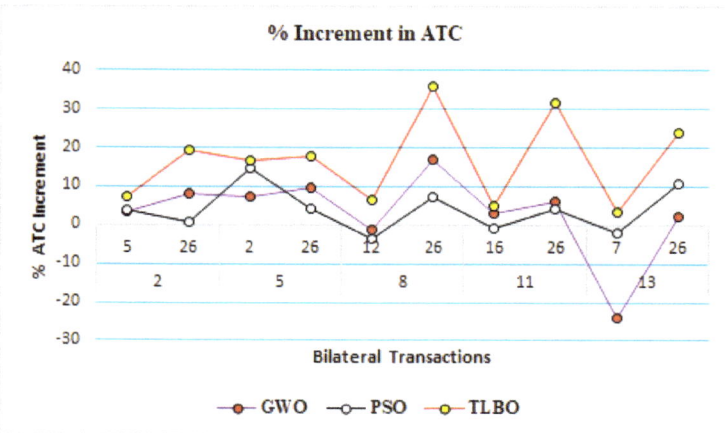

Figure 10. Comparative percentage increment in ATC by GWO, PSO and TLBO.

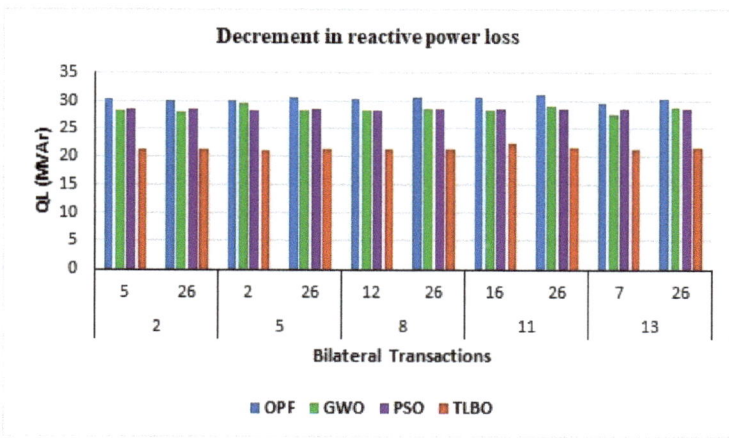

Figure 11. Comparative decrement in reactive power loss by TLBO, GWO, and PSO.

5.2. *For Active Power Loss Minimization*

Figure 12 shows the variation of active power loss in the test system for the bilateral transactions. It is observed that as compared to the base case without TCSC, the active power loss is reduced significantly. When the weight factor (w_2) in Equation (15) is changed, the active power losses are reduced. This also results in the reduction of ATC of the system. The maximum power loss reduction is obtained in the transaction 5–26 where the active power loss is reduced from the base case value of 7.09 MW to 4.97 MW. Similarly, in the transaction 13–7, the active power loss is reduced from base case value of 7.05 MW to 4.80 MW. In all the transactions considered, the average base case active power loss is 7.05 MW, which is reduced to an average of 5.25 MW after TLBO optimized TCSC is implemented in the system. This reduction is approximately 25%, which is quite significant when system efficiency is concerned. Reduced power losses mitigate the congestion in the lines.

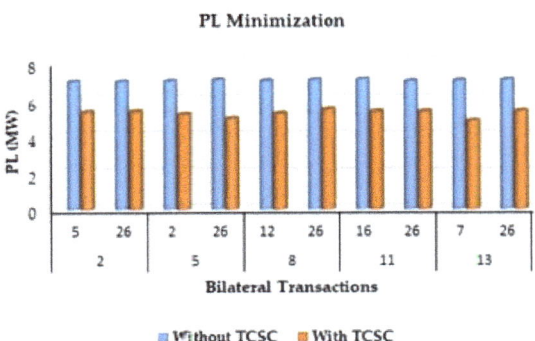

Figure 12. Active power loss minimization by applying TLBO with and without TCSC.

A detailed comparative analysis between the results obtained by GWO, PSO, and TLBO algorithm is presented in Table 2. From Table 2, it is observed that with the objective to reduce active power losses, the ATC of the system is also reduced.

Table 2. Consolidated results for active power loss minimization.

Bilateral Transactions		ATC (MW)				Active Power Loss (MW)			
		OPF	OPF with TCSC			OPF	OPF with TCSC		
Seller	Buyer		GWO	PSO	TLBO		GWO	PSO	TLBO
2	5	116.65	98.99	90.67	100.23	7.01	6.000	6.06	5.34
	26	12.18	10.33	8.34	10.85	7.01	6.000	6.78	5.37
5	2	184.56	156.61	165.57	164.83	7.03	6.017	5.89	5.22
	26	12.26	10.4	9.45	10.64	7.09	6.069	5.63	4.97
8	12	70.41	59.75	45.89	60.87	7.04	6.026	5.32	5.25
	26	12.18	10.33	7.54	10.58	7.08	6.060	5.98	5.48
11	16	37.65	31.95	32.78	32.56	7.11	6.086	5.56	5.36
	26	6.09	5.16	6.13	5.04	7.01	6.000	6.04	5.36
13	7	48.5	41.15	32.43	45.92	7.05	6.034	5.67	5.32
	26	9.26	7.85	9.54	8.13	7.08	6.060	6.87	5.37

Taking example of transaction between 2–5, the base case ATC is 116.65 MW. With the power loss minimization objective, the ATC value is reduced to 98.99 MW in GWO, 90.67 MW in PSO, and it reduces to 100.23 MW in case of TLBO application. Hence, it can be depicted that TLBO optimized TCSC in the system results in less reduction of ATC as compared to GWO and PSO. The total ATC of the system for the bilateral transactions considered is 509.74 MW. This reduces to 432 MW in case of GWO, 408.3 MW in case of PSO, and 453.65 MW in case of TLBO.

Figure 13 indicates the percentage reduction in active power loss achieve with the implementation of TCSC optimized by GWO, PSO and TLBO for all the bilateral transactions considered. It can be depicted from Figure 13, that overall active power reduction for the bilateral transactions under consideration is 3.15% in case of GWO, 3.822% in case of PSO, and it is 16.76% when TLBO is applied to locate and optimize the size of TCSC.

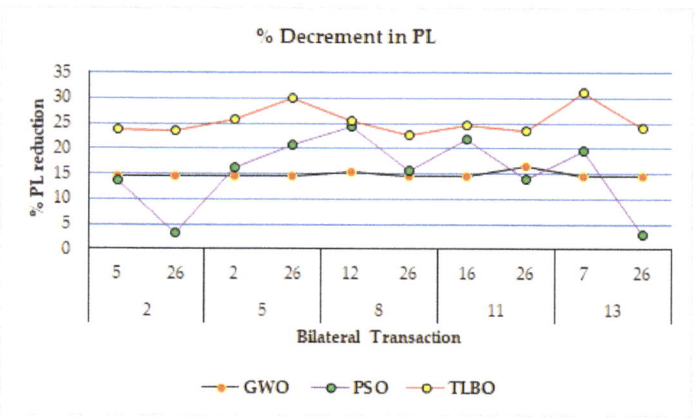

Figure 13. Comparative percentage decrement in PL by GWO, PSO and TLBO.

Table 3 shows the results of line number and size of TCSC well optimize through TLBO technique for ATC enhancement and power losses minimization.

Table 3. Consolidated results for TCSC location and size.

Bilateral Transactions		ATC Enhancement		Power Loss Minimization	
Seller	Buyer	TCSC Loc.	TCSC Size	TCSC Loc.	TCSC Size
2	5	15	−0.7574	6	−0.60243
	26	20	0.1896	9	0.1024
5	2	2	0.1270	36	0.2000
	26	9	−0.776	19	0.1069
8	12	16	−0.0708	21	−0.0986
	26	36	−0.0789	36	−0.6704
11	16	6	0.1025	6	0.1069
	26	20	0.0754	36	0.1128
13	7	35	−0.7348	26	−0.7658
	26	5	0.200	5	0.1869

The results as presented in Table 1 clearly indicates that as compared to PSO and GWO the ATC value obtained by proposed TLBO algorithm is higher. ATC is not only enhanced for the near bus, but also is increased significantly for the far bus. Similarly, Table 2 validates the effectiveness of TLBO in reducing system active power loss significantly as compared to the two other algorithms presented in the paper. This proves effectiveness of TLBO to mitigate congestion by reducing power loss and enhancing ATC over PSO and GWO for IEEE 30 bus system.

6. Conclusions

This paper proposes a novel method to enhance ATC of the DPS using the application of TLBO and TCSC. The problem is formulated and processed with the help of MATLAB software. The method is implemented using ACPTDF as the sensitivity factor to calculate ATC. The algorithm is applied effectively to significantly enhance the ATC of DPS and to reduce active power losses by a significant value of 11.86% and 16.7%. With the enhancement of ATC, the congestion in the line is reduced, which in turn increased the efficiency and overall stability of the power system. As the need to lay down alternate parallel lines in the system is eliminated, the system becomes more economically viable. The results obtained strongly suggest and prove that TLBO has outperformed the GWO and PSO for various considered cases. Moreover, the implementation of TCSC well optimize through TLBO resulted in ATC reduction when active power loss minimization is the main objective of the work. The ATC values for far away lines are also enhanced in comparison to the base case and through those obtained from GWO and PSO techniques. A detailed investigations and comparative analysis strongly suggest the application of TLBO for deregulated and modern power system.

Author Contributions: All authors planned the study, contributed to the idea and field of information; introduction, A.G.; software, A.G. and G.S.; analysis, A.G., I and G.S.; conclusion, A.G. and G.S.; writing—original draft preparation, P.N.B. and M.F.A. writing—review and editing, I. and G.S.; supervision. A.G., I., G.S., P.N.B. and M.F.A.; All authors have read and agreed to the published version of the manuscript.

Funding: This research received no external funding.

Institutional Review Board Statement: Not applicable.

Informed Consent Statement: Not applicable.

Data Availability Statement: Not applicable.

Conflicts of Interest: The authors declare no conflict of interest.

Nomenclature

DPS	Deregulated power system
GENCOS	Generation companies
TRANSCOS	Transmission companies
DISCO	Distribution company
ISO	Independent system operator
FACTS	Flexible Alternating Current transmission system
TCSC	Thyristor controlled series capacitor
TLBO	Teaching learning based algorithm
PSO	Particle swarm optimization
GWO	Grey wolf optimization
ATC	Available transfer capability
TTC	Total transfer capability
ETC	Existing transmission commitments
TRM	Transmission reliability margin
CBM	Capacity benefit margin
PTDF	Power transmission distribution factor
CPF	Continuous power flow
NR	Newton Raphson
OPF	Optimal power flow
P_{Gi} and Q_{Gi}	real and reactive power generations at ith bus respectively
P_{Di} and Q_{Di}	real and reactive power demands at ith bus, respectively
V_i, δ_i	voltage magnitude and corresponding angle of ith bus, respectively
Y_{ij}, θ_{ij}	admittance of line between ith and jth bus and its corresponding angle
N_b	number of buses
X_{pu}	per unit reactance
P_L	Active power loss

References

1. Silva, R.; Alves, E.; Ferreira, R.; Villar, J.; Gouveia, C. Characterization of TSO and DSO Grid System Services and TSO-DSO Basic Coordination Mechanisms in the Current Decarbonization Context. *Energies* **2021**, *14*, 4451. [CrossRef]
2. Khan, M.T.; Siddiqui, A.S. Congestion management in deregulated power system using FACTS device. *Int. J. Syst. Assur. Eng. Manag.* **2017**, *8*, 1–7. [CrossRef]
3. Shweta, N.V.; Prakash, V.; Kuruseelan, S.; Vaithilingam. C. ATC Evaluation in a Deregulated Power System. *Energy Procedia* **2017**, *117*, 216–223. [CrossRef]

4. Mohammed, O.O.; Mustaf, M.W.; Mohammed, D.S.S.; Otuoze, A.O. Available transfer capability calculation methods: A comprehensive review. *Electr. Energy Syst.* **2019**, *29*, e2846. [CrossRef]
5. Alekhya, B.; Rao, J.S. Enhancement of ATC in a deregulated power system by optimal location of multi-facts devices. In Proceedings of the International Conference on Smart Electric Grid (ISEG), Oshawa, ON, Canada, 12–15 August 2014; pp. 1–9. [CrossRef]
6. Nireekshana, T.; Rao, G.K.; Raju, S.S. Available transfer capability enhancement with FACTS using Cat Swarm Optimization. *Ain Shams Eng. J.* **2016**, *7*, 159–167. [CrossRef]
7. Acharya, N.; Mithulananthan, N. Locating series FACTS devices for congestion management in deregulated electricity markets. *Electr. Power Syst. Res.* **2007**, *77*, 352–360. [CrossRef]
8. Sawhney, H.; Jeyasurya, B. Application of unified power flow controller for available transfer capability enhancement. *Electr. Power Syst. Res.* **2004**, *69*, 155–160. [CrossRef]
9. Pandey, R.K.; Kumar, K.V. Multi agent system driven SSSC for ATC enhancement. In Proceedings of the National Power Systems Conference (NPSC), Bhubaneswar, India, 19–21 December 2016; pp. 1–6. [CrossRef]
10. Kumar, A.; Kumar, J. ATC determination with FACTS devices using PTDFs approach for multi-transactions in competitive electricity markets. *Int. J. Electr. Power Energy Syst.* **2013**, *44*, 308–317. [CrossRef]
11. Kumar, R.; Kumar, A. Impact of STATCOM Control Parameters on Available Transfer Capability Enhancement in Energy Markets. *Procedia Comput. Sci.* **2015**, *70*, 515–525. [CrossRef]
12. Bavithra, K.; Raja, S.C.; Venkatesh, P. Optimal Setting of FACTS Devices using Particle Swarm Optimization for ATC Enhancement in Deregulated Power System. *IFAC-Pap. Online* **2016**, *49*, 450–455. [CrossRef]
13. Gupta, M.; Gupta, A.K.; Sharma, N.K. A Comparative Study of Series Compensation, TCPAR and TCSR FACTS Controllers for Mitigation of Congestion. *J. Inst. Eng. (India) Ser. B* **2020**, *101*, 717–728. [CrossRef]
14. Kumar, A.; Kumar, J. ATC enhancement in electricity markets with GUPFC and IPFC—A comparison. *Int. J. Electr. Power Energy Syst.* **2016**, *81*, 469–482. [CrossRef]
15. Farahmand, H.; Rashidinejad, M.; Mousavi, A.; Gharaveisi, A.A.; Irving, M.R.; Taylor, G.A. Hybrid Mutation Particle Swarm Optimisation method for Available Transfer Capability enhancement. *Int. J. Electr. Power Energy Syst.* **2012**, *42*, 240–249. [CrossRef]
16. Kumar, Y.N.; Bala, A. Self-adaptiveness in particle swarm optimisation to enhance available transfer capability using thyristor controlled series compensation (TCSC). *Sadhana* **2018**, *43*, 152.
17. Venkatraman, K.T.; Paramasivam, B.; Chidambaram, I.A. Optimal allocation of TCSC devices for the enhancement of ATC in deregulated power system using flower pollination algorithm. *J. Eng. Sci. Technol.* **2018**, *13*, 2857–2871.
18. North American Electric Reliability Council (NERC). *Available Transfer Capability Definition and Determination (Report June 1996)*; North American Electric Reliability Council: Atlanta, GA, USA, 1996.
19. Ejebe, G.C.; Tong, J.; Waight, J.G.; Frame, J.G.; Wang, X.; Tinney, W.F. Available transfer capability calculations. *IEEE Trans. Power Syst.* **1998**, *13*, 1521–1527. [CrossRef]
20. Gautam, A.; Sharma, P.R.; Kumar, Y. Sensitivity based ATC Maximization by Optimal Placement of TCSC applying Grey Wolf Optimization. In Proceedings of the 3rd International Conference on Recent Developments in Control, Automation & Power Engineering (RDCAPE), Noida, India, 10–11 October 2019; pp. 313–318. [CrossRef]
21. Vetrivel, K.; Ganapathy, S.S.; Rao, K.U. Congestion Management and ATC Enhancement in Deregulated Power System Using Optimal Allocation of FACTS Devices. *Middle-East J. Sci. Res.* **2017**, *25*, 22–33. [CrossRef]
22. Rao, R.V. Review of applications of TLBO algorithm and a tutorial for beginners to solve the unconstrained and constrained optimization problems. *Decis. Sci. Lett.* **2016**, *5*, 1–30.
23. Al-Roomi, A.R. *Power Flow Test Systems Repository*; Dalhousie University, Electrical and Computer Engineering: Halifax, NS, Canada, 2015.

Article

A Fuzzy-PSO-PID with UPFC-RFB Solution for an LFC of an Interlinked Hydro Power System

Milan Joshi [1], Gulshan Sharma [2,*], Pitshou N. Bokoro [2] and Narayanan Krishnan [3]

1. Department of Electrical and Computer Engineering, University of New Mexico, Albuquerque, NM 87106, USA; milanjoshi7@unm.edu
2. Department of Electrical Engineering Technology, University of Johannesburg, Johannesburg 2006, South Africa; pitshoub@uj.ac.za
3. Department of Electrical and Electronics Engineering, SASTRA Deemed University, Thanjavur 613401, India; narayanan.mnit@gmail.com
* Correspondence: gulshans@uj.ac.za

Abstract: An LFC plays a vital part in passing on quality electric energy to energy consumers. Furthermore, with cutting-edge designs to move to modern and pollution-free energy generation, it may be conceivable to have a major hydropower in the future. Hydro plants are not suitable for continuous load alteration due to the large response time of hydroturbines. Hence, this paper shows a novel control design for an LFC of a hydro-hydro interlinked system based on joint actions of fuzzy logic with PID effectively optimized through particle swarm optimization (PSO) resulting in a Fuzzy-PSO-PID. The outcome of Fuzzy-PSO-PID is evaluated for step load variation in one of the regions of hydropower, and the outcomes of Fuzzy-PSO-PID are compared with a recently published LFC with respect to integral time absolute error (ITAE) value, values of PID, and graphical outcomes to show the impact of the proposed LFC action. The numerical results show that the ITAE value (0.002725) obtained through the proposed design is minimum in comparison to error values achieved through other LFC actions, and the pickup values obtained on these error values are considered to achieve the desired LFC. However, there is still scope for LFC enhancement as responses of hydropower are sluggish with higher oscillations; hence the UPFC and RFB are integrated into the interlinked hydro-hydro system, and the application outcomes are evaluated again considering the non-linearity, standard load alteration, random load pattern, and in view of parametric alterations. It is seen that the ITAE value reduces to 0.002471 from 0.002725 when UPFC is connected to the tie-line, and it further reduces to 0.001103 when a UPFC-RFB combination is used with Fuzzy-PSO-PID for a hydro leading system. The positive impact of the UPFC-RFB for hydropower is also seen from the application results.

Keywords: LFC; ACE; fuzzy logic; PSO; FACTS; RFB; UPFC

1. Introduction

The electrical time and demand network is currently developing, giving cost-effective power generation with superior electrical energy delivery to the demand system despite the diverse working constraints of the network. To make this network more reliable, secure, and economical, the electrical system operates in an interconnected model. This illustrates that tie-lines connect power generation zones. By extension, these control ranges have power interchange as characterized by the transmission network operator. As a result, every control zone can meet its own load demand to maintain the fixed power interchange via the AC tie-line. The mismatch between electrical energy generation and demand causes the framework frequency and the AC tie-line power to deviate from the predefined system value, affecting power delivery to various customers. The LFC strategy controls this change in frequency and tie-power. The adjustment of the indirect structure of these amounts is well-defined as the Area Control Error (ACE) in LFC [1,2].

The LFC controllers' role is to connect electrical power generated with electrical energy consumption to maintain or specific ACE deviations to zero. A primary action during this part would be to set up an LFC described as a conventional control hypothesis [3–5] In either circumstance, the gain of a standard LFC controller's action is limited for a specific situation and incapable of supplying the desired control movement under arranged working circumstances. Subsequently, the investigators and researchers are working hard and exploring the various controller design and plans such as optimal control, control based on few states, variable structure control, decentralized control, etc. to ensure a productive LFC for the electrical generation system [6–10]. In [6], the authors have shown the design of a variable structure LFC for an interlinked hydrothermal with a variable structure LFC for a multi-area system in [7]. A Dual mode controller with the impact of non-linearity was well covered in [8]. In [9,10], the impact of merging wind turbines and their assessments for LFC using all states and few states for the LFC were shown The artificial-intelligence-based LFC [11] designs, especially artificial neural networks and fuzzy logic, are increasing day by day to solve the LFC problem in a two-area or multi-area power system considering various types of power generation in view of the regulated or deregulated power system. From the above two techniques, the fuzzy logic technique is much more popular these days due to its better capability to deal with differing types of instabilities and system non-linearities and hence may prove to be beneficial concerning variable loading conditions of power systems [12–16]. The LFC technique based on hybrid evolutionary fuzzy PI is discussed in [12]. The effective Takagi–Sugeno base controller for LFC is demonstrated in [13] such that, the created fuzzy framework runs perfectly for non-linearity as well as parametric changes. The fuzzy methodology for interconnected an LFC is presented in [14]. The type-2 fuzzy-based research considering GRC non-linearity is demonstrated in [15]. In [16], a fuzzy gain scheduling control for an LFC using a genetic algorithm (GA) has already been presented. Some limitations in GA, such as convergence issues and trapping in local minima while finding optimal results of the problem, have been observed by researchers. The change in the optimization method, namely PSO, has significantly reduced the issues associated with the application of GA. PSO has fewer chances of becoming trapped in local minima for the same degree of execution, and it also takes very little calculation time [17]. Furthermore, the majority of control methodologies are based on thermal–thermal control models or hydrothermal LFC models. Hydroturbines have a much longer response time than thermal turbines; the LFC composition lacks an appropriate and guide control arrangement for hydropower systems. As a result, the LFC yield of the hydro-overseeing framework is slow and has persistent motions [18–20]. Furthermore, modern and developing society is approaching cleaner and renewable energy sources; hence hydropower may generate bulk electrical power or limited power, i.e., micro-hydro. As a result, the interconnected power framework may become more complex, with hydro-leading control regions included.

The cutting-edge and rapid advancement of power electronics industries has prompted the advancement of the Flexible Alternating Current Transmission System (FACTS) to provide a solution to various technical limitations of the power system. FACTS has the ability to regulate (i.e., active and reactive) power and hence can improve the output of the power system such as frequency and tie-line responses to a great extent. A redox flow battery (RFB) is additionally within the framework of FACTS and may well be a fast-acting useful capacity mechanism that can give the capacity in development to the dynamic essentialness of generator rotors. This can effectively dampen the electromechanical oscillations of the system by sharing the unexpected changes within the power demand [21,22]. At any point, there is a sudden rise within the control area (i.e, area-1). The power in an RFB is immediately discharged via the control conversion mechanism which consists of an inverter/rectifier. Basically, it adapts to the system through the quick delivery of loads. But it is not conceivable to put an RFB in each area or region of interlinked regions due to financial reasons, and subsequently unified power flow control (UPFC) may prove to be a viable solution. It is very cheap and can be presented in a course of action with a tie-line in position to progress the execution of the electrical system [23,24]. In this way, the simulated performance of an RFB and UPFC is used to improve the output of an LFC for hydro-leading system in a more cost-effective and quicker manner. Given the above discussion, the current research work is set to;

- Create a linear design of the hydro-leading system using an interconnected approach for LFC studies. The hydro framework is divided into hydro zones and connected via an AC tie-line.
- Develop a fuzzy logic control with two inputs for a hydro-leading system and then use the fuzzy outputs as PID inputs.
- Determine the PID gains using PSO by selecting the appropriate error definition, i.e., Integral Time Multiplied Absolute Error (ITAE). The performance of the PSO is evaluated by performing it for 100 iterations and using the outcome of the 100th iteration to obtain the final control action, which is Fuzzy-PSO-PID.
- Validate the Fuzzy-PSO-PID result for regular load variation from one of the control zones, and the result is compared with a recently published LFC to pick up the value of PID, ITAE, and through a graphical LFC.
- The results of Fuzzy-PSO-PID are good with regard to earlier published LFC outcomes. However, it still needs enhancement; hence the UPFC and the RFB are added to the hydro model, and the output is observed again considering load alteration, random load pattern, and parametric alterations from the original values.
- At last, all results are concluded to show the benefits of Fuzzy-PSO-PID, UPFC and RFB integration with regard to the present research work.

2. Model Details of the Hydro System

This is an interlinked system that uses hydroturbines in each area or region linked through an AC tie-line. Figure 1 shows the model of a hydro-hydro system. In Appendix A, all necessary system values are listed. For workspace programmers using SIMULINK modeling, MATLAB software edition (R2022a) was utilized to study the output of an LFC for the considered system. Two hydro plants with mechanical governors are utilized to increase or decrease the power generation as per the requirement of each area. To make frequency domain calculations easier, each element is represented using transfer function blocks. The next sections discuss the transfer functions of each region for a two-area hydro-hydro system. Figure 2 shows the transfer function model used for simulation and investigations for an LFC.

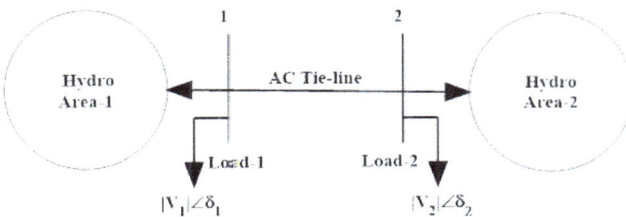

Figure 1. An AC tie-line connects area-1 with area-2 with each area generated power via a hydropower plant.

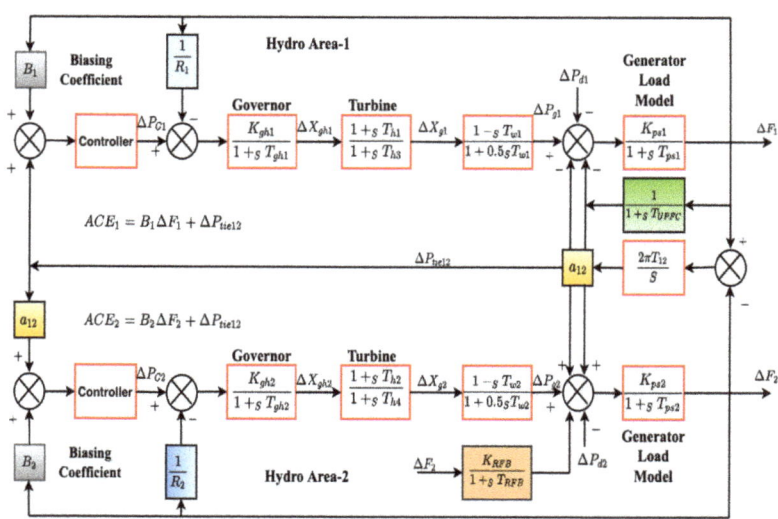

Figure 2. LFC model with hydropower in area-1 interlinked via an AC tie-line to another hydropower plant.

3. Modelling of Fuzzy-PSO-PID

The design and implementation of advanced control activity is a fuzzy sequence including PID. In addition, the pickup of PID is determined with results of an optimization technique identified as Particle Swarm Optimization (PSO) resulting in a Fuzzy-PSO-PID. The performance of the Fuzzy-PSO-PID depends on the value of K_P, K_I, and K_D. Hence, in arranging an absolute LFC activity, K_P, K_I, and K_D pickup must be taken suitably to attain extra beneficial active performance for the framework of the closed-loop. In this investigation, a new tuning methodology called PSO is used to actuate the foremost promising result of screen pickups to remove better dynamic accomplishment of the Fuzzy-PSO-PID screen LFC. The numerical description of PID is:

$$K(s) = K_P + \frac{K_I}{s} + K_D \quad (1)$$

where K_P = proportional gain, K_I = integral gain, and K_D = derivative gain.

K_P, K_I, and K_D are the pickups of the control activity, and the LFC output exceedingly depends upon this amount. These amounts are chosen for the PSO optimization handle detailed in this article. On the other hand, the FLC is composed of four leading components: the fuzzification, the fuzzy acceptance model, the run showing up the range, and the defuzzification. The Fuzzy-PSO-PID has two input signals, an area control error (ACE) and a derivative of (ACE), and one output signal. The developed structure of the Fuzzy-PSO-PID is given in Figure 3.

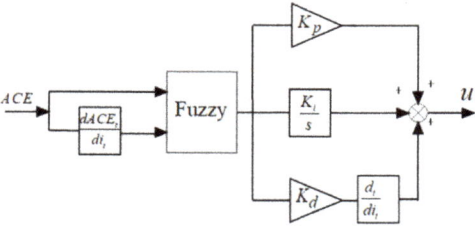

Figure 3. A schematic diagram of the fuzzy tuning.

3.1. Fuzzification

Fuzzification is the component by which a crisp value of the results changes over transform fuzzy value by utilizing points of interest within the information base. Different sorts of bends are present in various broadly utilized regions within the fuzzification strategy within the history as Gaussian, triangle, and trapezoidal membership functions (MFs). In any case, in the present investigation, triangle MFs are preferred due to effortlessness and balance. The Fuzzy-PSO-PID maintains two information signals namely, (1) area control error (ACE) and (2) derivative of (ACE), and one creation yield. Each information and output becomes five contributions (every one MF is triangular). The step middle way of MFs is likely applied, i.e., [−1 to 1].

3.2. Fuzzy Inference System

The input MFs, the fuzzy in the case of next etymological laws, and the output MF are formed of the fuzzy inference design (fis). The fis regulation is completed in four stages. A fundamental level is now to fuzzify the appropriate data crisp factors that further are in here as ACE as well as a subsidiary of ACE. It defines the degree to which that necessary information can reach the specific basic fuzzy locations by MFs. The specific action is how the evaluation or inference is displayed. The fuzzified information data are achieved on the heralds of the fuzzy laws. The fuzzy law includes the formation of a rule base to achieve the desired output. The rules are formed on an IF-THEN condition to reach a decision on the basis of the level of inputs with expert knowledge. The decision is taken on the basis of the min–max concept in the inference engine, and finally, the output need to be converted back to real values before applying to the plant from crisp values. A fourth stage is the defuzzification of the accumulated individual fuzzy location, which is performed with the help of the center of gravity method.

3.3. Allocation of Region of Inputs

The feature of the control laws might be a little more challenging than MFs, based upon the responsibility and approaching the required action. The laws are included for them to be used in the composition of K_P, K_I, and K_D. Shifting inputs and outputs individually has 5 MFs and 25 rules to obtain a fuzzy output. If-then laws are used in the following way: If ACE is NB-1 and a dACE is NB-1, the output is NB-1.

Table 1 contains the entire rule framework.

Table 1. Rule structure for the Fuzzy-PSO PID.

ACE \ d_{ACE}	NB_{-1}	NS_{-1}	Z_0	PS_1	PB_1
NB_{-1}	NB_{-1}	NB_{-1}	NS_{-1}	NS_{-1}	Z_0
NS_{-1}	NB_{-1}	NS_{-1}	NS_{-1}	Z_0	PS_1
Z_0	NB_{-1}	NS_{-1}	Z_0	PS_1	PB_1
PS_1	NS_{-1}	Z_0	PS_1	PS_1	PB_1
PB_1	Z_0	PS_1	PS_1	PB_1	PB_1

3.4. Defuzzfying the Output Value

The crisp amount is defuzzified by the well-popular technique called the centroid.

3.5. Objective Function

The proper selection of the objective function within the technique of a modern heuristic optimization technique-based controller plays a significant role in achieving the required target with the minimum of effort. The error definitions available and well tested in the history of LFC are Integral of Absolute Error (IAE), Integral of Squared Error (ISE), Integral of Time Absolute Error (ITAE) and Integral of Time multiplied Squared Error (ITSE). The ITAE degree decreases the settling duration, which IAE and ISE-based tuning do not achieve. The ITAE type also reduces the maximum overshoot. ITSE-based control provides a sweeping control product for an instantaneous variation into state problem which is not useful in case the control determines the condition of seeing. It necessitates that ITAE may perform a major supportive accurate work in an LFC point by point. Concurrently, ITAE is used as an error definition in this work to optimize the measured value and calculate the pickups of Fuzzy-PSO-PID [19]. Expression for the ITAE objective work is depicted in Equation (2).

$$J = ITAE = \int_0^{t_{sim}} (|\Delta F_1|) + (|\Delta F_2|) + (|\Delta P_{tie12}|) \cdot t dt \qquad (2)$$

where ΔF_1 and ΔF_2 = the framework frequencies of region 1 and region 2, ΔP_{tie12} = the incremental alteration in tie-line power regions 1 and 2, and t_{sim} = the simulation period.

The issue objectives continue these PID component boundaries. In this system, the organized problem is capable to be established such as catching subsequent optimization problems. Depending upon the performance record, the J optimization problem can be signified as: Reduction J restrained to:

$$K_P^{min} \leq K_P \leq K_P^{max} \qquad (3)$$

$$K_I^{min} \leq K_I \leq K_I^{max} \qquad (4)$$

$$K_D^{min} \leq K_D \leq K_D^{max} \qquad (5)$$

where K_P^{min} and K_P^{max} = proportional gain with minimum and maximum limit, K_I^{min} and K_I^{max} = integral gain with minimum and maximum limit, and K_D^{min} and K_D^{max} = derivative gain with minimum and maximum limit.

3.6. PSO Algorithm

This advanced approach highlights various goals of interest; it is fundamental, quick, and can be coded in a few lines. Other than that, this research has several advantages over human evolution and genetic algorithms. Each particle recalls its best course of action (adjacent best) as well as the bunch's best organization (around the world best). Some other benefit of PSO is that the beginning population of a PSO is kept up; there is no need for utilizing supervisors toward the state, and it is a time and memory-storage-consuming designation. In extension, PSO is based upon "helpful cooperation" connecting particles, differentiating from the genetic algorithms, which are based on "the survival of the fittest".

PSO begins with a population of self-assertive organizations referred to as "particles" in a D-dimensional interior. The *i*th particle is described by $Xi = (xi_1, xi_2, \ldots, xi_D)$. Various particles maintain a record of their hyperspace organization, which also is related to the most appropriate organization they have obtained so distantly. The standard of eligibility for particles (*pbest*) is also saved as $Pi = (pi_1, pi_2, \ldots, pi_D)$. PSO keeps a record of the common best standard (*gbest*), and its region, received in this way distant by any particle inside the populace. PSO covers for every step by varying the speed with which every particle approaches its *pbest* and *gbest*. The speed of particles is talked to as $Vi = (vi_1, vi_2, \ldots, vi_D)$. Ramping up is weighted by such a variable term, and confined subjective numbers are generated for increasing speed forward into *pbest* and *gbest*. The balanced speed and place of every particle can really be calculated using the actual speed as well as separations from $pbest_{j,g}$ to $gbest_g$ as shown in the following conditions [17]:

$$v_{j,g}^{(t+1)} = w \times v_{j,g}^{(t)} + c_1 \times r_1() \times (pbest_{j,g} - x_{j,g}^{(t)}) + c_2 \times r_2() \times (gbest_g - x_{j,g}^{(t)}) \qquad (6)$$

$$x_{j,g}^{(t+1)} = X_{j,g}^{(t)} + v_{j,g}^{(t+1)} \qquad (7)$$

As for $j = 1, 2, \ldots, n$ and $g = 1, 2, \ldots, m$. Where n = value of particles inside the swarm, m = number of elements for the vectors v_j and x_j, t = value of times (generations), $v_{j,g}^{(t)}$ = the *g*th element of the velocity of particle *j* at iteration t, $v_g^{min} \leq v_{j,g}^{(t)} \leq v_g^{max}$, w = inertia weight calculate, $c_1; c_2$ = cognitive and communicative speeding up factors independently, $r_1; r_2$ = irregular values reliably passed on inside the run $(0, 1)$, $x_{j,g}^{(t)}$ = the gth component of the position of particle j at cycle t, $pbest_j$ = pbest of particle j, and $gbest$ = gbest of the particle.

The execution steps of PSO are given in Figure 4.

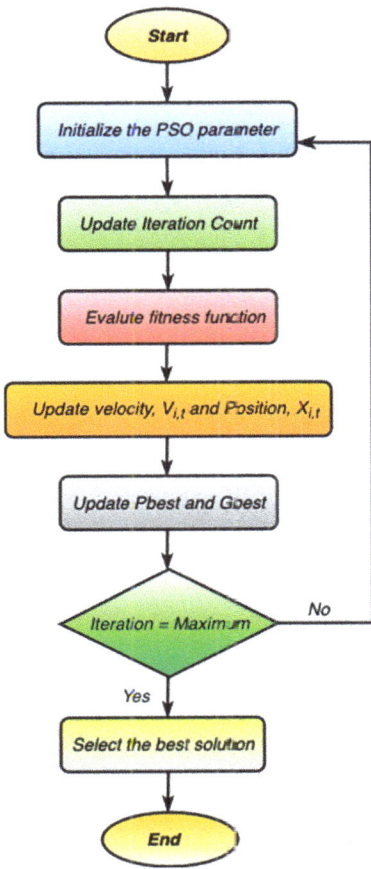

Figure 4. The execution steps of PSO.

3.7. UPFC Modeling

The settling time of the LFC system responses for hydropower is longer than 20 s. This is because hydroturbines have a faster response time, and as a result, UPFC is set up in series with the tie-line. Figure 2 depicts a point-by-point design of UPFC interconnected between two interconnected hydro regions, with a schematic model of UPFC in Figure 5 and its corresponding linear transfer function block in Figure 6.

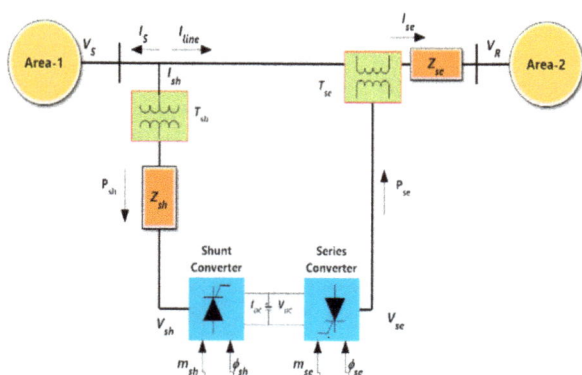

Figure 5. Model of UPFC integration in interlinked hydro systems.

In [21], the numerical calculations of UPFC are developed, and the combined power (i.e., complex power) at the receiving end of the line is calculated as:

$$P_{real} - jQ_{reactive} = \overline{V_r^*} \overline{I}_{line} = \overline{V_r^*}\{(\overline{V}_s + \overline{V}_{se} - \overline{V}_r)/j(X)\} \tag{8}$$

and,

$$\overline{V}_{se} = |V_{se}|\angle(\delta_s - \phi_{se}) \tag{9}$$

In the equation, the Vse implies the voltage magnitude in series, and ϕ_{se} is the series phase angle. By assembling Equation (8), the real value can be written as:

$$P_{real} = \frac{|V_s||V_r|}{X}\sin\delta + \frac{|V_s||V_{se}|}{X}\sin(\delta - \phi_{se}) = P_0(\delta) + P_{se}(\delta, \phi_{se}) \tag{10}$$

In the above equation, in case $V_{se} = 0$, it means that real power is uncompensated, and the UPFC series voltage can regulate between 0 and V_{se} to its maximum value. Finally, the phase angle can be varied from 0° to 360°. In an LFC, the UPFC operation can be written in the form of single order gain and time constant with frequency deviation of area-1 as input and altered power from the UPFC as [21];

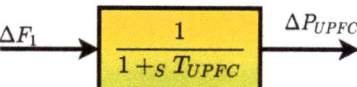

Figure 6. UPFC transfer function.

$$\Delta P_{UPFC}(s) = \left\{\frac{1}{1 + sT_{UPFC}}\right\}\Delta F_1(s) \tag{11}$$

T_{UPFC} = UPFC time constant.

3.8. RFB Modeling

The RFB is a rechargeable battery whose life is not affected if it is charged more than once or released frequently, and it has a quick response time for unexpected load variations. When it comes to load leveling, the RFB is more valuable for the performance of the LFC, and it also helps to maintain power quality. The RFB operation time is fast enough to be considered zero secs through LFC research. The model of the RFB for an LFC is as taken after [22] and given by Figure 7:

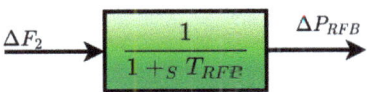

Figure 7. Transfer function of RFB.

$$\Delta P_{RFB}(s) = \left\{\frac{K_{RBF}}{1 + sT_{RFB}}\right\} \Delta F_2(s) \quad (12)$$

4. Simulation of Fuzzy-PSO-PID and Its Analysis

The present work is set to study and investigate the two-area interlinked power system with hydroturbines in each area or region and interlinked via an AC tie-line. The idea is to analyze the performance for such a type of power system to illustrate and propose a convincing arrangement of Fuzzy-PSO-PID to determine the execution of control for an LFC under different system working conditions. Researchers are avoiding the hydro-leading system model as the time taken by hydroturbines to reply for load alteration is significantly higher than other types of turbines and hence affect the frequency and power deviations with higher overshoot, more settling time, and a larger steady state error.

The Fuzzy-PSO-PID controller is presented in each area or region, and the obtained output (u) is fed to each area or region so that power can be up or down per its requirement and hence achieve the required LFC. Furthermore, GRC non-linearity is also considered in each area to see the impact of this non-linearity on LFC output as it is seen in the past literature that LFC output is limited by considering the GRC non-linearity. At first, the fuzzy logic action is designed for the hydro system, and the ACE and the rate of change of ACE are taken as input to the fuzzy logic system. The two inputs of the fuzzy system are scaled with the help of the membership function to convert original values into crisp values. The rule base of the fuzzy logic system for two input and one control output, i.e., 'u' having a total of 25 rules, is given in Table 1. The output is defuzzified to get back the original value from the crisp value and fed as input to PID. The best pickup value of PID is obtained by running the PSO technique. The ITAE is selected to achieve the best pickup value for PID, and these values are selected by achieving the ITAE minimum value. The pickup values of PID are selected through the PSO concerning minimum and maximum constraints as set in the optimization technique. The PSO technique is executed for 100 iterations and the best solution achieved after 100 iterations concerning the minimum value of ITAE and corresponding pickup values of PID are mentioned in Table 2 and used to check LFC standards. The Fuzzy-PSO-PID result is evaluated for standard load deviation (0.01 p.u.) in area-1, and the application outcomes are compared with recent LFC studies [19].

Table 2. Numerical outcomes of the Fuzzy-PSO PID.

Methods	K_P	K_I	K_D	ITAE
Classical PID [19]	−0.12	−0.091608	0.0393	41.1935015
Pessen PID [19]	−0.14	−0.050381	0.05502	46.5603916
Some overshoot PID [19]	−0.066	0.050381	0.05764	38.0953828
No overshoot PID [19]	−0.04	−0.030538	0.034933	31.388228
Fuzzy-PSO-PID	−1.0684	−0.0591	−0.1305	0.002725

Table 3. Numerical outcomes of Fuzzy-PSO PID+UPFC+RFB.

Models	K_P	K_I	K_D	ITAE
Fuzzy-PSO PID	−1.0684	−0.0591	−0.1305	0.002725
Fuzzy-PSO PID+UPFC	−1.0684	−0.0591	−0.1305	0.002471
Fuzzy-PSO PID+UPFC+RFB	−1.0684	−0.0591	−0.1305	0.001103

Table 4. Comparison performance overshoot, undershoot, and settling time shown in Figure 8a–c.

	Overshoot (Hz)			Undershoot (Hz)			Settling Time (S)		
	ΔF_1	ΔF_2	P_{tie12}	ΔF_1	ΔF_2	P_{tie12}	$\Delta F1$	$\Delta F2$	P_{tie12}
Fuzzy PSO	0.02136	0.02293	0.00343	−0.04975	−0.0498	−0.00992	33.10	34.48	49.99
Fuzzy-PSO-PID+UPFC	0.02345	0.02395	0.001803	−0.04748	−0.0457	−0.00754	32.42	33.90	49.55
Fuzzy-PSO-PID+UC+RFB	0.009904	0.009585	0.001562	−0.02248	−0.01495	−0.00450	30.91	25.99	49.24

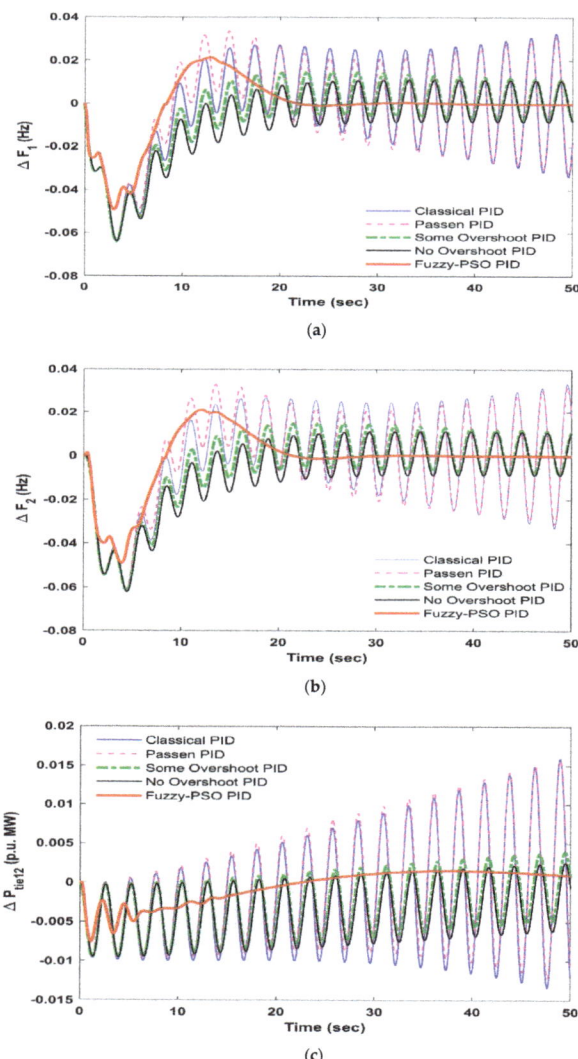

Figure 8. (a–c) LFC results in 1% load alteration in region-1.

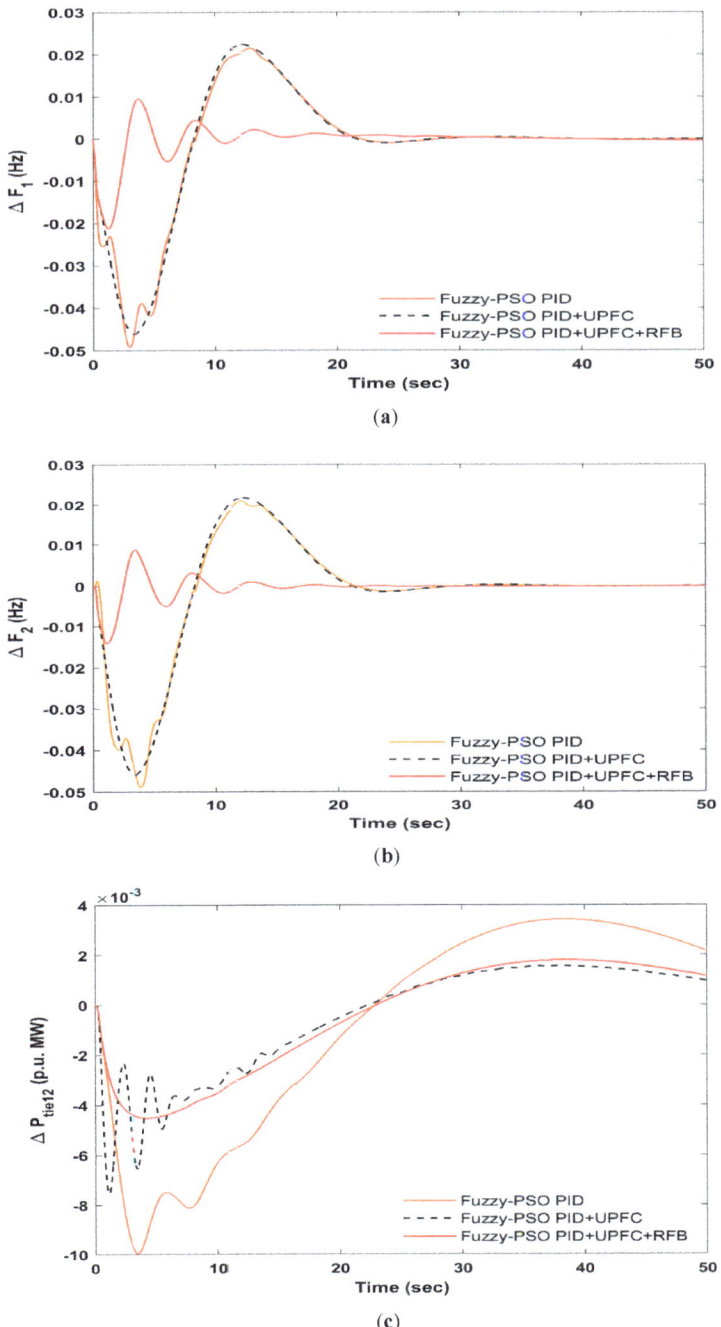

Figure 9. (a–c) LFC results for 1% load alteration in region-1.

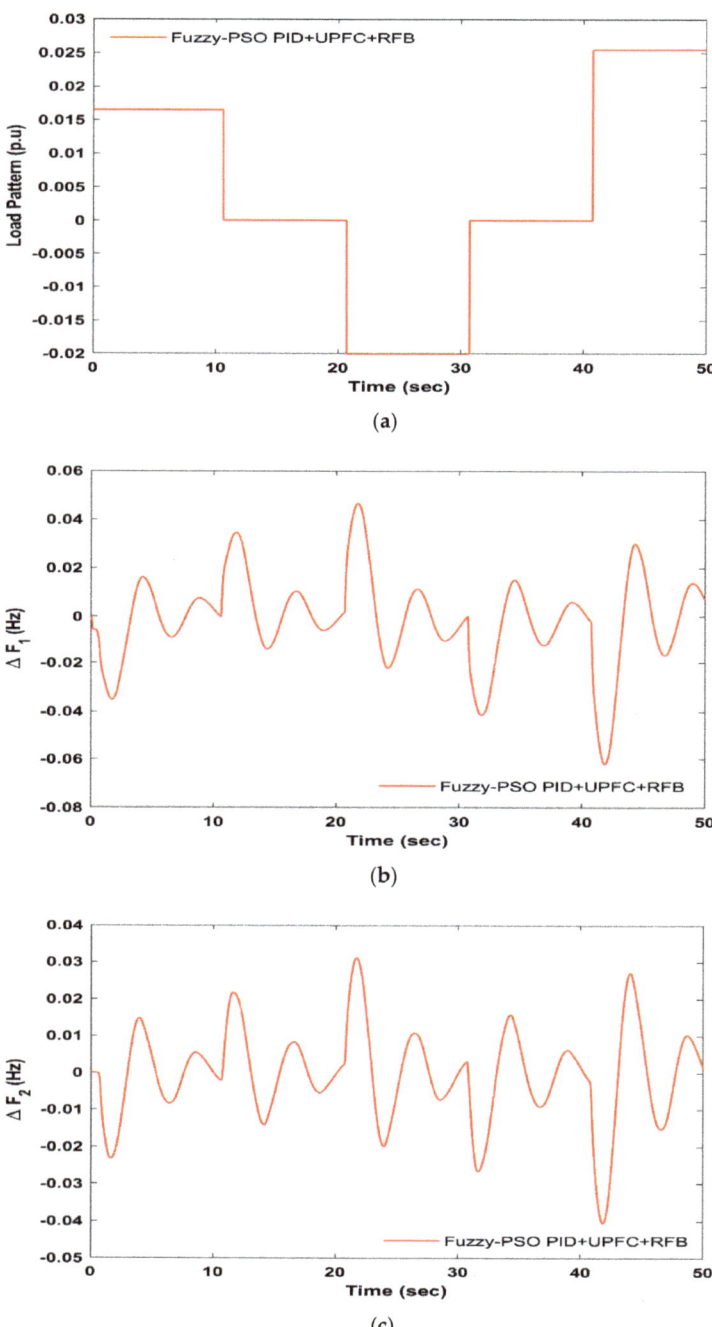

Figure 10. Cont.

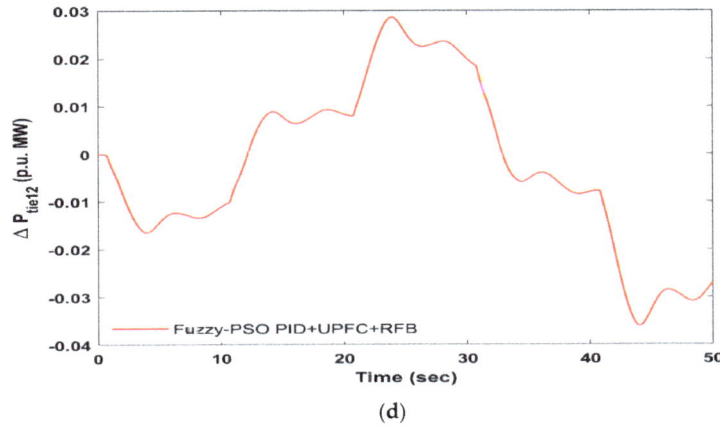

(**d**)

Figure 10. (**a**–**d**) LFC results for continuous load alteration for 50 s.

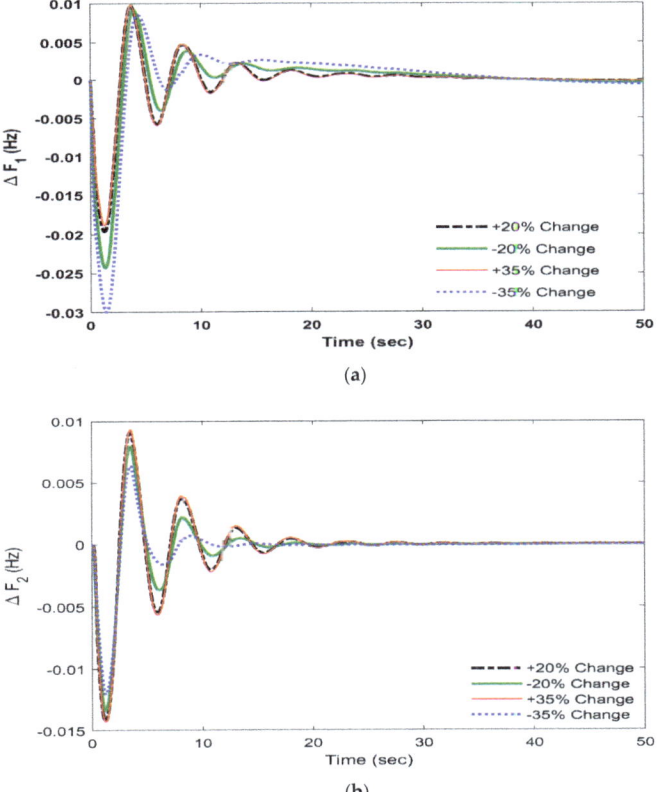

Figure 11. *Cont.*

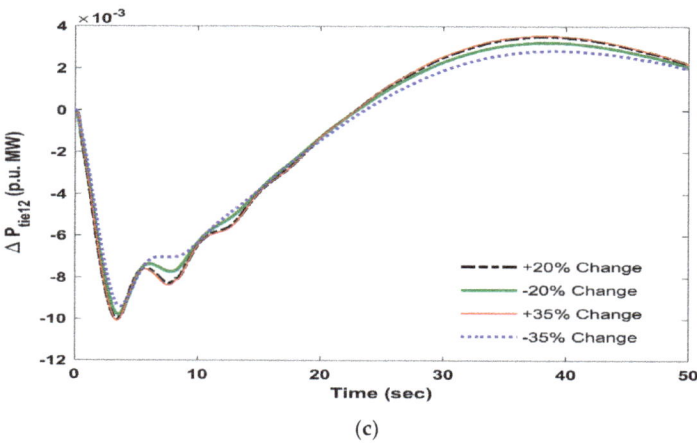

(c)

Figure 11. (a–c) LFC results for sensitivity analysis.

From Table 2 results, it is observed that ITAE is reduced to the best value, i.e., 0.002725, which shows the financial aspects of the LFC strategy implementation. It is also observed that the value of ITAE is quite less in comparison to Classical PID (41.1935015), Pessen PID (46.5603916), Some overshoot PID (38.0953828), and No overshoot PID (31.388228). The graphical LFC outcomes are given in Figure 8a–c, and it is shown that frequency and tie-power deviations have a minimum value of the first peak and settle back to the original value very quickly after load alteration. The same outcome or near to Fuzzy-PSO-PID outcomes for an LFC are not achieved through other LFC techniques which are Classical PID, Pessen PID, Some overshoot PID, and No overshoot PID. It is also discovered that LFC outcomes are completely free of oscillations, which is not conceivable with the other LFC behavior for the hydro-leading system.

Still, there is a scope for further enhancement, particularly UPFC links with a tie-line in series connection and an RFB connected in region-2. Now the studies are extended to see the impact of these FACTS in an LFC for the hydro-hydro system. The pickup value of the Fuzzy-PSO-PID and the value of the ITAE are matched with the Fuzzy-PSO-PID with the UPFC and the Fuzzy-PSO-PID with the UPFC-RFB. These results are tabulated in Table 3. It is seen that UPFC integration has resulted in a reduction of ITAE (i.e., 0.002471) and further reduction seen with joint efforts from UPFC-RFB (i.e., 0.001103). Figure 9a–c shows the LFC results when UPFC only as well as UPFC with an RFB is connected in the hydropower sytem at suitable location and the performance of the Fuzzy-PSO-PID is evaluated again for load change in region-1, and it is observed that combination of the Fuzzy-PSO-PID with the UPFC and an RFB outperforms other LFC results in view of reduced overshoot, better settling time, and oscillation free system results. LFC results are also matched by considering the overshoot, undershoot, and settling time for frequency deviations of each area and for tie-power deviations. Numerical results are listed in Table 4, and it is seen that results obtained via the Fuzzy-PSO-PID with UPFC-RFB is best when compared with other investigated LFC results.

The random load pattern is applied to hydro power system for 50 s, and the LFC results are revealed in Figure 10a–d. It is realized that the Fuzzy-PSO-PID with UPFC and an RFB culminates in taking after the random load pattern continuously and reducing the frequency and tie-power deviations successfully to zero value continuously over the period of 50 s.

In addition, the affectability examination of the Fuzzy-PSO PID with UPFC and an RFB control surveyed by changing the original parameters (i.e., T_{12} coefficient of tie-line synchronizing), and T_g (governor's response time) over the extensive run from standard parameters, and the results are shown in Figure 11a–c. From the results of Figure 11a–c, it is clearly seen that the changing system parameters from standard parameters. whether positive or negative change from the original value, hardly affect the system output. and all LFC results are reaching back to nominal value after load alteration. Hence, it can be said that the Fuzzy-PSO-PID with UPFC and an RFB combination is robust and promising in reaching the LFC standards for a hydro-leading power system.

5. Conclusions

This research paper proposes a novel strategy of linking the fuzzy logic technique with PID, effectively evaluated through PSO resulting in an advanced and robust design known as the Fuzzy-PSO-PID for an LFC of the hydro-leading system. The proposed Fuzzy-PSO-PID arrangement is attempted for standard load variation in one of the regions of the hydro-leading system, and the positive outcomes are shown over other LFC actions. The following conclusions are drawn from the research work carried out:

- Fuzzy-PSO-PID is agreeably sufficient to meet the LFC guidelines of hydropower systems with regards to ITAE value, pickups of PID, and LFC responses in comparison to other LFC actions. Still, it needs change and enhancement to have superior LFC designs for such systems.
- The synchronization of UPFC with an AC tie-line as well as an RFB in region-2 of a hydro-leading system with the Fuzzy-PSO-PID has very well covered the frequency and tie-line power variations of the hydro system to an extraordinary degree also in the presence of non-linearity.
- The reduction in ITAE achieved via the Fuzzy-PSO-PID with UPFC and an RFB demonstrates the importance and regulates the reasonability of the current research work.
- The sensitivity analysis and random load pattern for the Fuzzy-PSO-PID with UPFC-RFB shows that the Fuzzy-PSO-PID design is reasonably good, clear, and capable of supporting the LFC output of a hydro-leading system.
- In the present research work, triangular MFs are used for FL. However, diverse MFs can affect the LFC output, and it needs to be explored further. The Type-2 FL can be a better solution for an LFC of a hydro-leading system.
- The research work can be extended to multi-areas with a hydro-leading system in a regulated and deregulated environment. Furthermore, the design of a proposed LFC can further improve by using other and advanced optimization techniques.
- The energy storage devices will play a major role in improving the LFC action further.
- The output of the Fuzzy-PSO-PID can be evaluated again using OPAL-RT and other real-time software in comparison to standard MATLAB software.

Author Contributions: All authors planned the study, contributed to the idea and field of information; introduction, M.J.; software. M.J. and G.S.; analysis, M.J., G.S. and P.N.B.; conclusion, M.J. and G.S.; writing—original draft preparation, P.N.B. and N.K.; writing—review and editing, N.K. and G.S.; supervision, G.S., P.N.B. and N.K. All authors have read and agreed to the published version of the manuscript.

Funding: This research received no external funding.

Institutional Review Board Statement: Not applicable.

Informed Consent Statement: Not applicable.

Data Availability Statement: Not applicable.

Conflicts of Interest: The authors declare no conflict of interest.

Abbreviations

ΔF_i	Alteration in frequency of area i (i = 1, 2)
ΔP_{tie12}	Alteration in Tie-line power (p.u. MW)
$2\Pi T_{12}$	Tie-line power synchronizing coefficient (p.u.MW/Hz)
ΔP_{gi}	Power generation alteration (p.u. MW)
ΔP_{di}	Load alteration (p.u. MW/Hz)
ΔP_{ci}	Alteration in speed changer position
R_i	Speed regulation factor (Hz/p.u. MW)
K_{ghi}	Governor gain
T_{ghi}	Governor time constant (s)
T_{hi}	Time constant associated with hydro governor
T_{wi}	Hydro turbine time constant (s)
K_{psi}	Gain of power system
T_{psi}	Time constant of power system (s)
a_{12}	Area size ratio co-efficient
B_i	Frequency bias constant (p.u. MW/Hz)
ACE_i	Area Control Error
ΔP_{UPFC}	Power alteration of UPFC
K_{UPFC}	UPFC gain
T_{UPFC}	UPFC time constant (s)
ΔP_{RFB}	RFB power alteration
K_{RFB}	RFB gain
T_{RFB}	RFB replying time (s)
n	Dimension of search space
C_1, C_2, R_1, R_2	PSO random parameters
p_{best}	Positions best
g_{best}	Global best

Appendix A

Area-1	Area-2	Data\Value
B_1	B_2	0.425
R_1	R_2	3.0
K_{gh1}	K_{gh2}	1
T_{gh1}	T_{gh2}	0.6
T_{h1}	T_{h2}	5
T_{h3}	T_{h4}	32
T_{w1}	T_{w2}	1
K_{ps1}	K_{ps2}	20
T_{ps1}	T_{ps2}	3.76
$2\Pi T_{12}$		0.545
a_{12}	a_{12}	1

RFB Device		
K_{RFB}	0.67	
T_{RFB}	0	

UPFC Device	
T_{UPFC}	0.01

PSO Parameter Name	Value
Max Generation	100
Population in Swarm	50
c_1, c_2	1.5, 0.12
w_{max}, w_{min}	0.9, 0.4

References

1. Sharma, G.; Narayanan, K.; Davidson, I.; Akindeji, K. Integration and Enhancement of Load Frequency Control Design for Diverse Sources Power System via DFIG Based Wind Power Generation and Interonnected via Parallel HVDC/EHVAC Tie-Lines. *Int. J. Eng. Res. Afr.* **2020**, *46*, 106–124. [CrossRef]
2. Joshi, M.; Sharma, G.; Davidson, I.E. Load Frequency Control of Hydro Electric System using Application of Fuzzy with Particle Swarm Optimization Algorithm. In Proceedings of the 2020 International Conference on Artificial Intelligence, Big Data, Computing and Data Communication Systems (icABCD), Durban, South Africa, 6–7 August 2020; pp. 1–6. [CrossRef]
3. Panwar, A.; Sharma, G.; Sahoo, S.; Bansal, R. Active Power Regulation of Hydro Dominating Energy System using IDD optimized FPA. *Energy Procedia* **2019**, *158*, 6328–6333. [CrossRef]
4. Sharma, G. Optimal AGC Design for Diverse Sources of Power Generations in each Area Using Output Vector Feedback Control Technique. *Int. J. Eng. Res. Afr.* **2019**, *45*, 99–114. [CrossRef]
5. Bevrani, H. *Robust Power System Frequency Control*; Springer: Cham, Switzerland, 2014. [CrossRef]
6. Hsu, Y.; Chan, W. Optimal variable structure controller for the load-frequency control of interconnected hydrothermal power systems. *Int. J. Electr. Power Energy Syst.* **1984**, *6*, 221–229. [CrossRef]
7. Ali-Hamouze, Z.; Magide, Y.A. Variable structure load frequency controllers for multi area power system. *Electr. Power Energy Syst.* **1995**, *15*, 22–29.
8. Velusami, S.; Chidambaram, I. Decentralized biased dual mode controllers for load frequency control of interconnected power systems considering GDB and GRC non-linearities. *Energy Convers. Manag.* **2007**, *48*, 1691–1702. [CrossRef]
9. Ibraheem, K.N.; Sharma, G. Study on dynamic participation of wind turbines in AGC of power system. *Electr. Power Component Syst.* **2014**, *43*, 44–55.
10. Sharma, G.; Nasiruddin, I.; Niazi, K.R. Optimal Automatic Generation Control of Asynchronous Power Systems Using Output Feedback Control Strategy with Dynamic Participation of Wind Turbines. *Electr. Power Compon. Syst.* **2015**, *43*, 384–398. [CrossRef]
11. Zeynelgil, H.; Demiroren, A.; Sengor, N.S. The application of ANN technique to automatic generation control for multi-area power system. *Int. J. Electr. Power Energy Syst.* **2002**, *24*, 345–354. [CrossRef]
12. Juang, C.F. Load-frequency control by hybrid evolutionary fuzzy PI controller. *IET Proc.—Gener. Transm. Distrib.* **2006**, *153*, 196–204. [CrossRef]
13. Lee, H.J.; Park, J.B.; Joo, Y.H. Robust load-frequency control for uncertain nonlinear power systems: A fuzzy logic approach. *Inf. Sci.* **2006**, *176*, 3520–3537. [CrossRef]
14. Ilhan, K.; Çam, E. Fuzzy logic controller in interconnected electrical power systems for load-frequency control. *Int. J. Electr. Power Energy Syst.* **2005**, *27*, 542–549. [CrossRef]
15. Sudha, K.; Vijaya Santhi, R. Robust decentralized load frequency control of interconnected power system with Generation Rate Constraint using Type-2 fuzzy approach. *Int. J. Electr. Power Energy Syst.* **2011**, *33*, 699–707. [CrossRef]
16. Arya, Y.; Kumar, N. Fuzzy Gain Scheduling Controllers for Automatic Generation Control of Two-area Interconnected Electrical Power Systems. *Electr. Power Compon. Syst.* **2016**, *44*, 737–751. [CrossRef]
17. Sahu, R.K.; Panda, S.; Chandra Sekhar, G. A novel hybrid PSO-PS optimized fuzzy PI controller for AGC in multi-area interconnected power systems. *Int. J. Electr. Power Energy Syst.* **2015**, *64*, 880–893. [CrossRef]
18. Sharma, G. Performance enhancement of a hydro-hydro power system using RFB and TCPS. *Int. J. Sustain. Energy* **2019**, *38*, 615–629. [CrossRef]
19. Panwar, A.; Sharma, G.; Nasiruddin, I.; Bansal, R. Frequency stabilization of hydro–hydro power system using hybrid bacteria foraging PSO with UPFC and HAE. *Electr. Power Syst. Res.* **2018**, *161*, 74–85. [CrossRef]
20. Joshi, M.K.; Sharma, G.; Davidson, I.E. Investigation of Diverse Sampling Time for LFC of Hydro Power System using Discrete LQR with UPFC and RFB. In Proceedings of the 2020 International SAUPEC/RobMech/PRASA Conference, Cape Town, South Africa, 29–31 January 2020; pp. 1–6. [CrossRef]
21. Sahu, R.K.; Gorripotu, T.S.; Panda, S. A hybrid DE–PS algorithm for load frequency control under deregulated power system with UPFC and RFB. *Ain Shams Eng. J.* **2015**, *6*, 893–911. [CrossRef]
22. Arya, Y.; Kumar, N.; Gupta, S. Optimal automatic generation control of two-area power systems with energy storage units under deregulated environment. *J. Renew. Sustain. Energy* **2017**, *9*, 064105-20. [CrossRef]
23. Chidambaram, I.; Paramasivam, B. Optimized load-frequency simulation in restructured power system with Redox Flow Batteries and Interline Power Flow Controller. *Int. J. Electr. Power Energy Syst.* **2013**, *50*, 9–24. [CrossRef]
24. Rangi, S.; Jain, S.; Arya, Y. Utilization of energy storage devices with optimal controller for multi-area hydro-hydro power system under deregulated environment. *Sustain. Energy Technol. Assess.* **2022**, *52*, 102191. [CrossRef]

Article

Coordinated Control of Wind Energy Conversion System during Unsymmetrical Fault at Grid

Hemant Ahuja [1], Arika Singh [2], Sachin Sharma [3], Gulshan Sharma [4,*] and Pitshou N. Bokoro [4]

1. Ajay Kumar Garg Engineering College, Ghaziabad 201009, Uttar Pradesh, India; ahuja.iitd@gmail.com
2. KIET Group of Institutions, Ghaziabad 201206, Uttar Pradesh, India; arika.singh@kiet.edu
3. Department of Electrical Engineering, Graphic Era Deemed to Be University, Dehradun 248002, Uttrakhand, India; sachineesharma@gmail.com
4. Department of Electrical Engineering Technology, University of Johannesburg, Johannesburg 2006, South Africa; pitshoub@uj.ac.za
* Correspondence: gulshans@uj.ac.za

Abstract: High penetration of wind power into the grid necessitates the coordinated action of wind energy conversion systems and the grid. A suitable generation control is required to fulfill the grid integration requirements, especially during faults. A system using a pair of voltage source converters with a squirrel cage induction generator coupled to a wind turbine is proposed to provide fault ride-through during grid faults. A threefold action is used for providing the effective fault ride-through via coordinated action of the machine side and the grid side converter. The entire wind energy conversion system is controlled such that the wind turbine remains connected even during the faults. To implement the threefold action: (i) A decoupled current controller is placed in the grid side converter, which separately controls the positive and negative sequence currents arising during faults. The grid side converter controller is capable of eliminating the double frequency oscillations at the dc-link voltage and, hence, real power, which arises during the unsymmetrical faults; (ii) Reactive power injection is additionally provided by the grid side converter for better grid support; and (iii) The vector control technique is used in machine side converter along with the droop control to adjust the generator speed and the torque resulting in actuation of the pitch control mechanism to limit power generation without shutdown of the turbine.

Keywords: induction generator; wind energy; inverters; stationary reference frame; synchronous reference frame; pitch angle; converters; grid; STATCOM

1. Introduction

Keeping the wind energy conversion systems (WECS) connected to the grid during short-term faults has been a major requirement from the grid operators, especially in the wake of higher penetration of wind into the grid. The WECS are now expected to behave as an equal partner to conventional power generating plants and provide the enhanced control of wind turbines (WT) that keeps the grid afloat even during disturbances. Fault ride-through (FRT) has now become an important requirement for the power providers to manage diverse grid conditions, which may include a terminal voltage limit and active/reactive power recovery. The requirements listed under wind farm transmission grid codes for different countries have been described in [1]. These grid codes include the importance of fault ride-through during severe grid faults.

Among the variable and fixed speed WECS, variable speed WECS are preferred, because of their various advantages, viz., efficient operation at various wind speeds, decreased mechanical stress, and the possibility of being used as direct (gearless) drive systems [2]. In variable speed, the WECS configured with a squirrel cage induction generator (SCIG) uses a set of two full-rated converters (FRC) that provide the interface between generator and grid. SCIG's ease of availability at low cost, ability to operate in harsh environments, and decreasing cost of power electronics make it a lucrative option to be used for WECS [3]. In [4,5], it is established that STATCOM, dynamic voltage restorers (DVRs), and the static VAR compensators are capable of improving the FRT potential of WECS. In [6], a non-linear controller is incorporated with the grid side converter (GSC) controller that limits the current rise within the safe value and also the power transfer to the grid. In [7], the use of a controlled, but static, braking resistor is presented. The braking resistor is placed in the dc link, which dissipates the extra energy produced during lower voltages, in a PMSG-based WECS. Paper [8] has presented an effective strategy using a DVR, in series between the generator and the grid. During grid faults, the dc-link voltage is controlled by the energy storage system while the grid side converter is used as STATCOM. Paper [9] gives a review of conventional and state-of-the-art methods for analyzing the dynamic stability of WECS. Different transient models are simulated for various WT generator configurations, under different grid conditions. In [10], a voltage source converter with sinusoidal PWM of the STATCOM is used to provide the ride through of faults occurring at the point of common coupling, between the generator and the grid. The use of external devices such as STATCOM, static VAR compensators, DVR, etc., adds to the complexity, and raises the overall expense of the system [11].

A study, in [12], explains a variable speed wind system that uses a different type of power electronic converter (PEC) at the machine side and the grid side, for providing grid synchronization, maintaining power quality, and operation at a unity power factor. Khan et al. [13] have described the LVRT of WECS by suppressing the overvoltage appearing at the dc-link and the active power limitation during an unbalanced fault. The authors, however, did not discuss the oscillations occurring during the unsymmetrical faults. A generalized discussion about the control of WECS is provided for operation under unbalanced network conditions. Sánchez et al. [14] discussed the maximum power point tracking of the WECS. The technique described is a tip speed ratio, which is used for the smaller rating turbine. The correct measurement of wind speed is required to implement an efficient control in the practical scenario. Nasiri et al. [15] have used the sliding mode controllers in the machine side converter (MSC). The authors have used GSC for controlling the dc-link voltage and providing the optimal operation, respectively. An energy shaping controller is described in [16], to alleviate the sub-synchronous control interaction (SSCI) incidents that happen during the asymmetrical exchange of power between the grid and the wind farm. To ensure the asymptotical stability, the insertion of damping in the controller dynamics is proposed. A non-linear controller, which is based on the feedback linearization technique and sliding mode control, is proposed in [17], to mitigate the SSCI in wind farms based on a doubly fed induction generator. A self-regulating control of active plus reactive power, in a distributed generating system, has been presented using the current control of GSC [18]. During the severe voltage drop conditions, the coordinated control becomes active and fixes the rotor speed at its upper limit, so that the input power of MSC is reduced. The surplus power is taken by the super capacitor energy storage system. Ahuja et. al. [19] have presented a coordinated control strategy for both real and reactive powers during grid faults. Considerable research has been carried out in innovating different strategies to enhance the FRT capabilities of WECS. The majority of FRT strategies deal with the fault conditions on the grid by adding extra hardware, which adds to the cost and complexity of the system. Though such strategies are designed to deal with the faults on the grid, they are not able to address the issues related to unbalanced grid conditions effectively. During grid faults, there occurs a difference between power generated and power consumed. It is because the grid is not able to send away the power generated. The voltage at dc-link, therefore, rises due to this imbalance. The situation becomes more complicated when the fault is unsymmetrical, resulting in the appearance of dual-frequency oscillations at the dc-link. The WECS needs to have comprehensive control of the active power as well as the reactive power handling.

This paper proposes a coordinated control strategy for a variable speed SCIG-based WECS with back-to-back connected VSCs between SCIG and grid. The novel features of the proposed strategy include: (i) droop control through MSC, to handle the imbalance in power; (ii) active power control through reconfigurable reference current selection; (iii) injection of reactive power to support the grid during faults; and (iv) removal of dual-frequency oscillations arising in the dc-link voltage as well as the active power via GSC. A dual current controller based on positive and negative components is designed for GSC. This controller is also capable of dealing with unsymmetrical faults. The results achieved through MATLAB/SIMULINK simulations for the proposed control strategy are presented, and compared with the conventional method, to analyze the effectiveness of the control. Aspects of machine modeling and vector control are extracted from textbooks [20–24]. The proposed control strategy is employed on a 1.5 MW WECS using an SCIG, which exhibits effective control during balanced and unbalanced grid conditions, for different wind speeds.

2. Traditional Control Configuration of SCIG-Based WECS

The control implementation of SCIG-based WECS is focused on the MSC and GSC control, as well as the WT control for MPPT, as depicted in Figure 1.

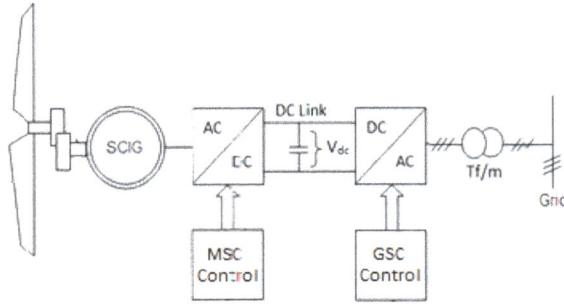

Figure 1. Schematic Diagram of variable speed SCIG-based WECS.

2.1. Wind Turbine Control and Maximum Power Extraction in WECS

A variable speed WT operates at an optimized speed, derived with respect to the wind speed, so that optimal power is captured. Various optimal control algorithms are available. Under nominal speed control range, the WT generally provides MPPT using the power signal feedback control algorithm as it is considered to be one of the effective methods of WT optimal operation [25]. When the wind speed increases beyond a particular value (safe limit), the power is limited to its nominal value. The WT is controlled such that the pitch control is activated during high wind speed conditions and grid disturbances, thereby preventing the over-speeding of WT. The entire control of WT is achieved through MSC control during normal operation and the pitch control during abnormal conditions. The MSC only provides the reference values for WT control.

2.2. Machine Side Converter Control

The technique used for MSC control is based on the vector control, oriented with rotor flux. The rotor flux-oriented control is applied on current-controlled VSI as it provides a faster current control. The controller block diagram is presented in Figure 2.

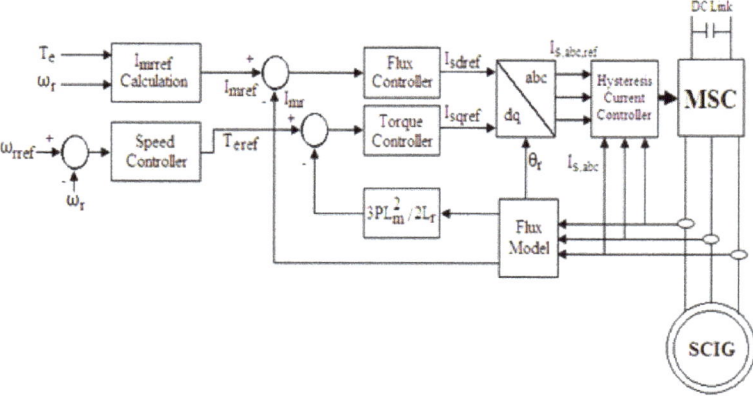

Figure 2. Schematic of the Machine Side Converter control.

The electromechanical torque produced by the generator is stated in terms of rotor flux linkage [26].

$$T_e = \frac{3}{2} P \frac{L_m}{L_r} (\psi_{rd} i_{sq} - \psi_{rq} i_{sd}) \tag{1}$$

In Equation (1), the direct and quadrature axis rotor flux linkages are represented by ψ_{rd} and ψ_{rq} respectively. L_m and L_r represent the magnetizing inductance and the rotor self-inductance, respectively, and P represents the number of poles. i_{sd} and i_{sq} represent the stator currents on the direct and quadrature axis (on a special reference frame), respectively.

The space phasor of rotor flux linkage is aligned with the direct axis making $\psi_{rq} = 0$. By substituting the same in (1), the torque expression becomes modified as

$$T_e = \frac{3}{2} P \frac{L_m}{L_r} \psi_{rd} i_{sq} = \frac{3}{2} P \frac{L_m^2}{L_r} i_{sd} i_{sq} \tag{2}$$

$$|\overline{\psi_{r\psi_r}}| = \psi_{rd} = L_m |\overline{i_{mr}}| = L_m i_{sd} \tag{3}$$

It is seen from Equations (2) and (3) that the torque and flux control can be decoupled by controlling the current components i_{sd} and i_{sq}. This is the basis of the vector control, where i_{sd} controls the flux inside the machine and i_{sq} controls the torque produced. Reference d axis current (i_{sdref}) is decided by the flux controller, while q axis current (i_{sqref}) is generated by the torque controller. $i_{s,abc,ref}$, the three-phase reference current(s), are generated by dq to abc transformation [22]. θ_r is the rotor flux angle required for transforming dq currents to 3-phase currents, estimated by a flux model.

2.3. Grid Side Converter Control

Under usual working conditions, the GSC maintains a constant voltage at the dc-link and also maintains the needed power factor at the grid, by suitably controlling the active and reactive power. A widely used control structure for GSC control is derived from [27–31] and the control configuration is as depicted in Figure 3. In Figure 3, V_{bus} and V_{bus}^* represent the dc-link voltage and its reference value, respectively, whereas θ_{grid} represents the phase angle extracted from the grid voltages using a phase-locked loop. The rest of the symbols used for current, voltage, and power are standard.

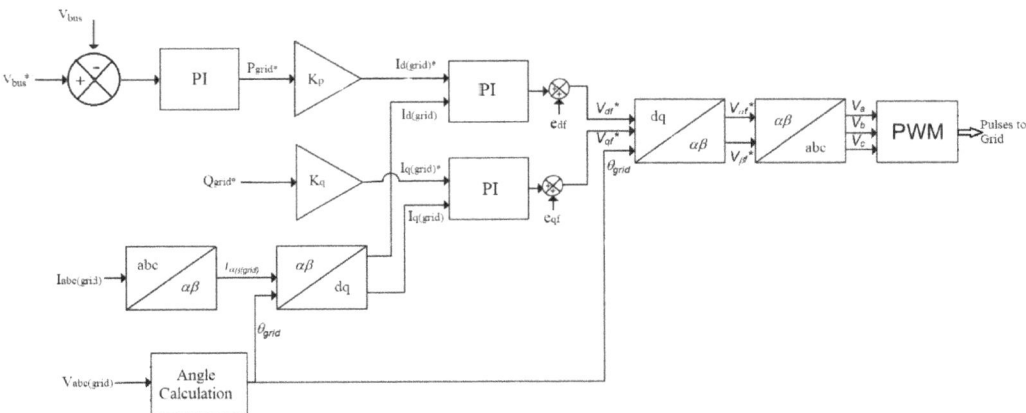

Figure 3. Schematic of the Grid Side Converter control.

A vector-controlled double closed loop structure, associated with the PI controllers, is used in GSC as it provides a satisfactory operation while regulating the DC variables. The vector control technique is based on the grid voltage, and is applied using park transformation, as presented in Figure 3. The DC link voltage is controlled by the outer loop, whereas the inner loop controls the current. The reactive power reference is generally set to provide the unity power factor operation.

3. The Coordinated Control Strategy and the Proposed Controller

A WECS includes three main stages of power conversion, viz., Aerodynamic Control for high winds, Generator Control for optimal operation, and Grid converter control for power conditioning and grid synchronization. These control stages also provide control during abnormal conditions. For a WECS to be capable of providing an effective FRT, these three stages of power conversion should work simultaneously and in coordination. As described in Section 2, the traditional systems provide the optimal operation through MSC and the active and reactive power control through GSC. The GSC also provides the grid synchronization and ride-through of faults at the grid side, particularly the symmetrical faults. The traditional GSC controller is only capable of providing the positive sequence control during faults and fails to suppress the oscillations that arise during unsymmetrical faults. The traditional GSC controller does not have separate control over positive and negative sequences of currents that arise during the unsymmetrical faults.

To improve the performance during FRT of unsymmetrical faults, a positive–negative sequence-based controller is used in GSC. This controller separates out the two sequences and generates the active and reactive power references depending on the severity of the fault. The references are selected such that active power reference is reduced as per the capacity of the grid to absorb power, and reactive power reference is enhanced to support the grid voltage from going further down. In addition, the oscillations arising in the active power and the dc-link are suppressed by the individual control of positive and negative sequences.

The coordinated control strategy is further implemented by sensing the dc-link by MSC, which is not featured in traditional systems. This dc-link voltage, at MSC, initiates the action of the droop controller as described in the following sub-section. When the control targets assigned to MSC and GSC are met and the fault is still persistent, the pitch mechanism is actuated by sensing the speed of the generator.

3.1. The Droop Control Action of the Machine Side Converter

In addition to the traditional MSC control described in Section 2, a de-loading droop as illustrated and highlighted in Figure 4 is used. This de-loading droop provides the coordinated control and is used to adjust the torque to shed power during the fault conditions at the grid side. During grid faults, it is observed that an imbalance occurs in the power generated (P_{gen}) and supplied to the grid (P_{grid}). This happens because the grid is not capable of taking power, while the WT keeps on generating the power. Due to this imbalance of power, the voltage at the dc-link rises. The rate of rise in the dc-link voltage depends on the difference between the power generated by WT, and the power supplied to the grid via GSC. To prevent the excessive dc-link voltage, droop control is used to reduce the WT power so that P_{gen} becomes equal to P_{grid} [32]. The droop controller facilitates the rapid decrease in power through de-loading of the generator by reducing the torque linearly. The torque is reduced until the P_{gen} becomes equal to P_{grid}. The rate of decrease in torque will depend upon the severity of the fault and the difference in power between the grid and the generator. The generator speed sensing is constantly performed to send a signal to Pitch actuators of the WT during an increase in speed beyond a defined value of 1.2 in the proposed work.

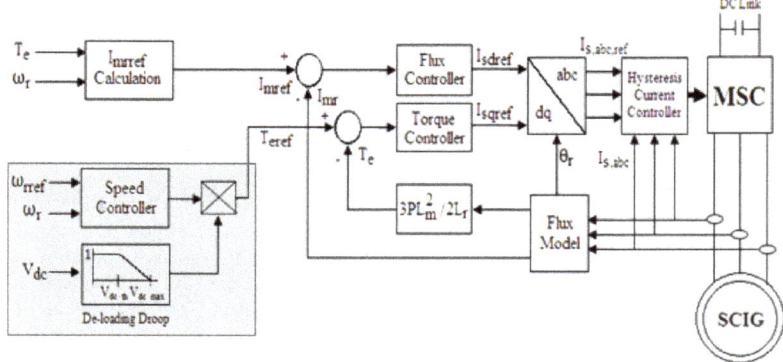

Figure 4. Schematic of Proposed MSC control embedded with a De-Loading Droop.

3.2. The Proposed GSC Controller

When the fault occurs at the grid, the GSC controller controls the active power based on a reconfigurable reference current selector and instantly injects the reactive power for better grid support. In the event of grid unbalancing, the controller also eliminates the dual-frequency oscillations that appear in the active power and the dc-link voltage.

The active power (P) and reactive power (Q), delivered to the grid, are expressed in terms of d-q components of voltages (V_d and V_q) and currents (I_d and I_q) [33] as given below:

$$P = 1.5\,(V_d * I_d + V_q * I_q) = 1.5\,V_d * I_d \tag{4}$$

$$Q = 1.5\,(-V_d * I_q + V_q * I_d) = -1.5\,V_d * I_q \tag{5}$$

To eliminate V_q in the above equations, the d-axis of the reference frame is aligned with the stator voltage phasor. The constant supply voltage (or constant V_d) makes P and Q proportional to I_d and I_q, respectively, in (4) and (5).

The controller has to deal with the positive sequence currents only during the symmetrical faults. However, during unsymmetrical faults, the negative sequence components of the current appear. The interaction of positive and negative components leads to the development of dual-frequency oscillations in P and Q, which navigate to the entire system. The oscillation in active power produces ripples in the dc-link voltage, leading to the malfunctioning of the PLL in providing the right estimation of phase angle (θ). This

'θ' is required for abc-dq transformation. This erroneous estimation of 'θ' results in incorrect transformation, due to which the synchronization of VSI output with the grid is badly affected. These oscillations further affect the control of GSC by the generation of non-sinusoidal current references. These non-sinusoidal references deteriorate the power quality and pilot the tripping of over-current protection. To accurately estimate θ, a low pass notch filter-based PLL is used.

The control scheme proposed in this work mainly derives from the source current references. These reference currents are required by the PWM controller. The control is developed by using a decoupled-dual synchronous reference frame current controller, facilitating the unbalanced current injection. The dual current controller provides decoupled control of positive sequence currents and negative sequence currents. In this proposed work, the required positive and negative sequence components are detected by a positive and negative-sequence control (PNSC) strategy based on a second-order generalized integrator (SOGI) [34]. This system offers the best solution for grid synchronization even during grid faults.

With the unbalanced voltages at the input side, the P and Q can be written as [35]

$$P = P_o + P_{c2}\cos(2\omega t) + P_{s2}\sin(2\omega t) \tag{6}$$

$$Q = Q_o + Q_{c2}\cos(2\omega t) + Q_{s2}\sin(2\omega t) \tag{7}$$

where P_o and Q_o are the average values of instantaneous active and reactive power associated with the MSC. P_{c2}, P_{s2}, Q_{c2}, and Q_{s2} are the active and reactive power oscillation terms caused by the voltage unbalance. The amplitude of these powers is calculated as

$$P_o = 1.5\left(V_d^+ I_d^+ + V_q^+ I_q^+ + V_d^- I_d^- + V_q^- I_q^-\right) \tag{8}$$

$$P_{c2} = 1.5\left(V_d^+ I_d^- + V_q^+ I_q^- + V_d^- I_d^+ + V_q^- I_q^+\right) \tag{9}$$

$$P_{s2} = 1.5\left(V_q^- I_d^+ - V_d^- I_q^+ - V_q^+ I_d^- + V_d^+ I_q^-\right) \tag{10}$$

$$Q_o = 1.5\left(V_q^+ I_d^+ - V_d^+ I_q^+ + V_q^- I_d^- - V_d^- I_q^-\right) \tag{11}$$

$$Q_{c2} = 1.5\left(V_q^+ I_d^- - V_d^+ I_q^- + V_q^- I_d^+ - V_d^- I_q^+\right) \tag{12}$$

$$Q_{s2} = 1.5\left(V_d^+ I_d^- + V_q^+ I_q^- - V_d^- I_d^+ - V_q^- I_q^+\right) \tag{13}$$

Here, the direct and quadrature axis voltages, and the currents are denoted positive by using superscript "+", and negative using superscript "−". As four degrees of freedom exist in the currents (+ and − of I_d and I_q) to be injected by the GSC, only four of the total six power magnitudes defined by the above equations can be controlled for the given grid voltages (+ and − of V_d and V_q). The four power coefficients, neglecting the higher-order reactive power oscillation terms, are therefore considered. The voltage at the dc link is determined by the real power balance, power received, and delivered to the grid. If the active power varies with time, then P_{c2} and P_{s2} in Equations (9) and (10) will not be equal to zero; therefore, the dc-link voltage will fluctuate and a dual-frequency ripple will appear. The coefficients P_{c2} and P_{s2} must therefore be nullified to keep the dc level constant. After nullifying the higher-order active power coefficients and dropping the higher-order reactive power coefficients, the current references may be deduced as

$$\begin{bmatrix} I_d^{+*} \\ I_q^{+*} \\ I_d^{-*} \\ I_q^{-*} \end{bmatrix} = [M4*4]^{-1} * \frac{2}{3} \begin{bmatrix} P_o \\ Q_o \\ 0 \\ 0 \end{bmatrix} = \frac{2P_o}{3D} \begin{bmatrix} V_d^+ \\ V_q^+ \\ -V_d^- \\ -V_q^- \end{bmatrix} + \frac{2Q_o}{3F} \begin{bmatrix} V_q^+ \\ -V_d^+ \\ V_q^- \\ -V_d^- \end{bmatrix} \tag{14}$$

where $D = [(V_d^+)^2 + (V_q^+)^2] - [(V_d^-)^2 + (V_q^-)^2]$ and $F = (V_d^+)^2 + (V_q^+)^2 + (V_d^-)^2 + (V_q^-)^2$.

Under the usual operating conditions at the grid, and when its voltage has dropped to no more than 20% of its nominal value, Q_o is maintained as zero and the active power component P_o is calculated as

$$P_o = K_p(V_{dc}^* - V_{dc}) + K_I \int (V_{dc}^* - V_{dc}) \tag{15}$$

where V_{dc} represents the dc-link voltage while V_{dc}^* represents the reference dc-link voltage, K_p and K_i are the proportional and integral controller gains, respectively. The power is transferred to the grid at the unity power factor (UPF). For limiting the converter current to a safer value, the maximum value of 'P_o' is made equal to $3V^+I_N$. Here 'I_N' represents the rms current of the grid side converter. For the grid voltages ranging 15–80% of its nominal value, the maximum limit of P_o is set to

$$P_o = \sqrt{(3V^+I_N)^2 - (Q_0)^2} \tag{16}$$

where Q_o is calculated as

$$Q_o = 3V^+I_N\left[1 - \frac{V^+}{V_N}\right] \tag{17}$$

The control scheme of GSC, considering the above mathematical modeling, is illustrated in Figure 5. When the voltage at the grid falls below 15%, the P_o is brought to zero, and Q_o is set as $3V^+I_N$. The current limitation, as implemented above, causes an imbalance of power on the dc-link during fault conditions and, therefore, the dc-link voltage tends to rise. In such a case, the de-loading mechanism of MSC is activated, which controls the generated power as explained in Section 3.1. Figure 5 depicts the structure of the proposed controller placed as the GSC. Separate and decoupled controllers for the positive and negative sequences are used. The positive and negative sequences are controlled in positive and negative controllers, respectively.

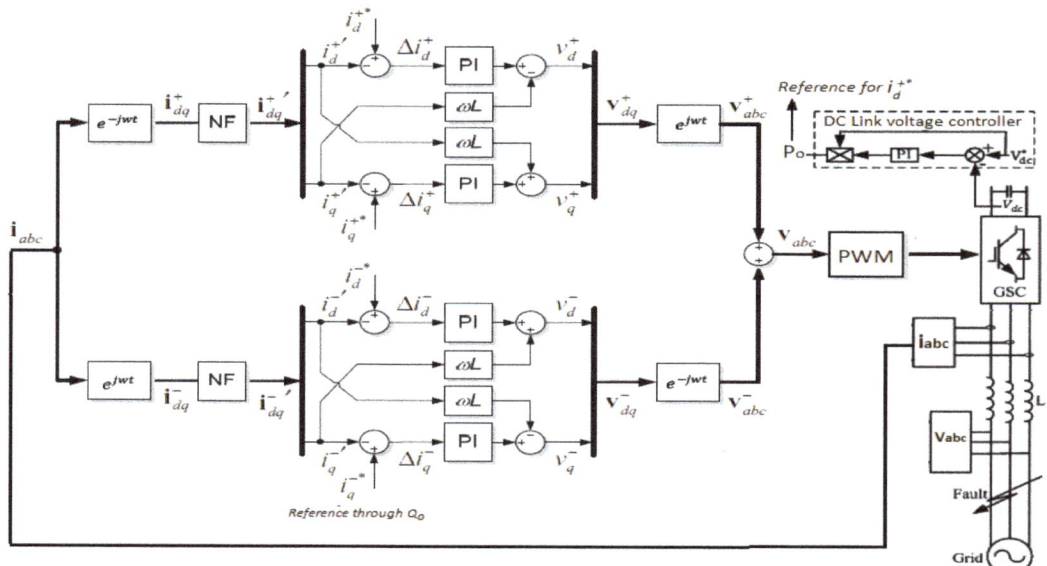

Figure 5. Schematic of Grid Side Converter using Dual Current Controller.

4. Fault Ride-Through Results

The WECS using an SCIG is simulated in MATLAB/Sim Power Systems. A 1.5 MW, 3-phase generator of 690 V is connected to the 11 kV, 50 Hz grid through a transformer. The SCIG parameters considered are as follows:

R_s (Resistance of the stator winding) = 0.007 pu,
R_r (Resistance of the rotor winding) = 0.0072 pu,
L_s (Self Inductance of the stator winding) = 0.18 pu,
L_r (Self Inductance of the stator winding) = 0.16 pu, and
L_m (Mutual Inductance) = 3.2 pu.

The WECS is analyzed for a 3-phase line-to-line (LLL) fault and double line-to-ground (LL-G) fault at the grid (refer to Figure 1) for a duration of half a second. The circuit breaker operation time, located close to the fault, may vary from a half cycle to 25 cycles in the case of a 50 Hz system. Hence, the fault duration of maximum 0.5 s (25 cycles) is considered to observe the absolute response and analyze the effectiveness of the proposed controller.

In this section, the simulation results of WECS during the fault, and the pre and post-conditions of faults are presented for a conventional controller, i.e., without FRT and with the proposed controller. The analysis is carried out for comparing the performance of the proposed control strategy with the conventional strategy at 8.5 m/s and 10.5 m/s wind speeds.

The two wind speeds are specifically chosen to present the issues arising at rated and lower wind speeds. Figures 6 and 7 shows the fault behavior of the system during a symmetrical (LLL) fault, with the proposed and the conventional controller for the above-mentioned two wind speeds. Before the occurrence of a fault, the voltage and current at the point of generator-grid interconnection (GGI), dc-link voltage, generator speed, and the active power delivered to the grid are observed approximately to 1 pu.

A symmetrical (LLL) fault was initiated at 1.5 s for a duration of 0.5 s, i.e., till 2.0 s. During the fault, the voltage at GGI is reduced below 0.1 pu (refer to Figure 6a) indicating the severity of the stress on the grid. Figure 6b illustrates that the generator current during a symmetrical (LLL) fault, without FRT, increases to around 3 pu, which is highly detrimental for the power converters. The currents are increased due to the inability of the grid to absorb the generated power. This increase in current is associated with the accumulation of power at the DC link. During the dip in voltage at the grid, the active power transferred to the grid is reduced in proportion to the dip in voltage, as observed in Figure 7b.

The generated power, however, remains the same; therefore, an imbalance in active power is seen as the rise in dc-link voltage. The SCIG speed remains more or less the same, as seen in Figure 6d. The voltage at the dc-link rises approximately to 1.85 and 1.55 pu for the wind speeds, 10.5 m/s and 8.5 m/s, respectively, as depicted in Figure 7a.

This abnormal rise in dc-link voltage indicates a power insertion in the dc-link capacitor, and this power is not able to evacuate to the grid. In some conventional systems, this rise is provided using a crow-bar to dissipate the excess power, which is highly criticized in the literature.

Referring to Figure 7a, the dc-link voltage starts rising during the three-phase (LLL) fault. The proposed controller incorporates a de-loading controller that becomes actuated during the fault and reduces the generator torque, causing the generator speed to rise, as seen in Figure 6d. The rise in speed activates the pitch control mechanism to limit the generator speed to its bounded value of 1.2 pu (Figure 6d). The pitch control of WT increases the pitch angle to shed the power. The reduction in torque, through the droop controller, regulates the generator power, which causes a reduction in current. For a wind speed of 10.5 m/s, as shown in Figure 6c, converter current is controlled within the safe limit to nearly 1.5 pu. The dc-link voltage is maintained at a constant of 1.2 pu by activating the de-loading droop within 0.06 s. The combined effect of the droop controller in MSC and the pitch control of WT ensures a power balance between the WECS and the grid.

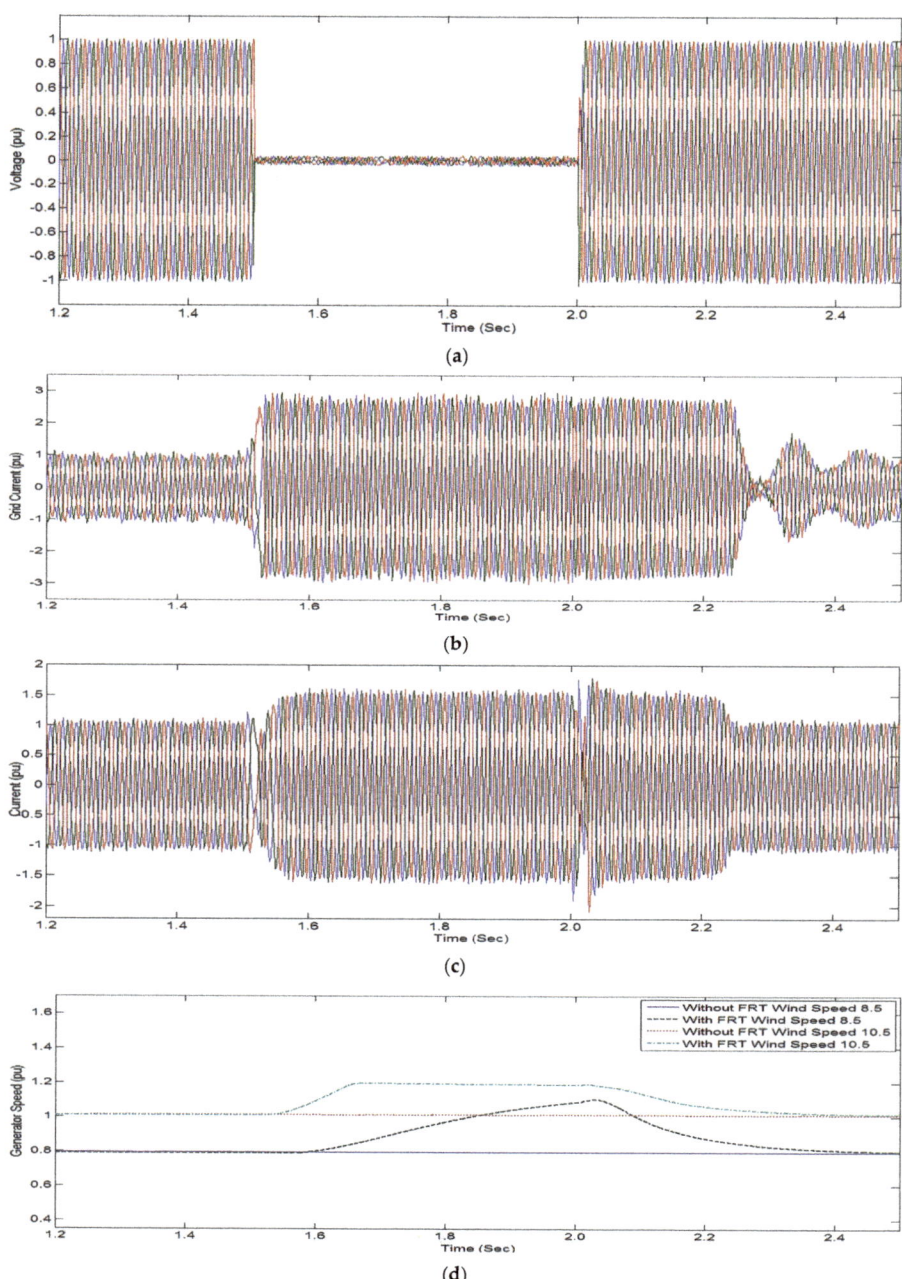

Figure 6. Behavior of WECS during symmetrical fault, with and without FRT control for 8.5 m/s and 10.5 m/s wind speeds: (**a**) Voltage at point of GGI; (**b**) Current through the point of GGI—without FRT control at 10.5 m/s; (**c**) Current through the point of GGI—with proposed strategy at 10.5 m/s; (**d**) Generator Speed.

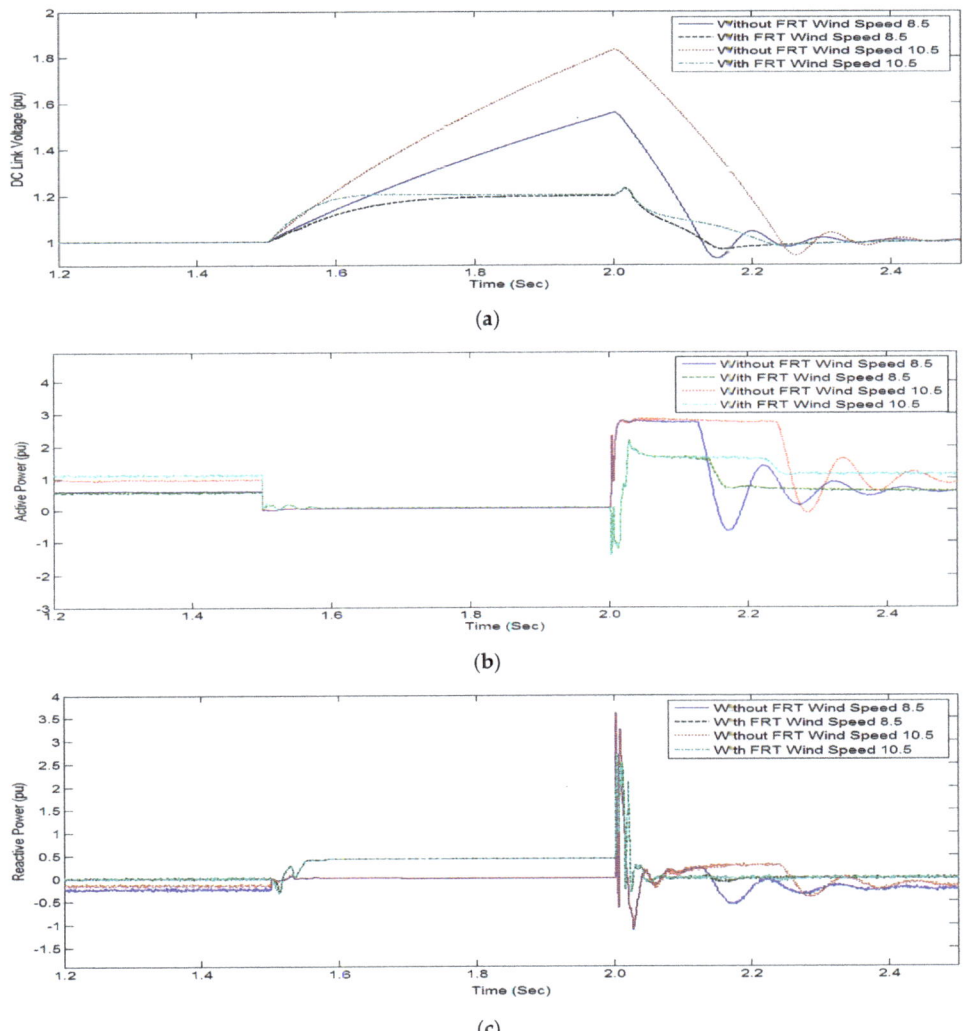

Figure 7. Behavior of WECS during symmetrical fault, with and without FRT control for 8.5 m/s and 10.5 m/s wind speeds: (**a**) dc-link Voltage, (**b**) Active Power supplied, (**c**) Reactive Power supplied by Generator.

The effect of the fault on active and reactive powers is illustrated in Figure 7b,c. Figure 7b exemplifies that the active power during the LLL fault drops to nearly zero, as the grid is not capable of accepting power. It could be seen from Figure 7c that reactive power remains zero during the fault and the system takes more than 0.5 s to regain after fault clearance without FRT. With FRT, active power is further reduced, and required reactive power is supplied to the grid during the fault. The injected reactive power supports the grid voltage and helps the grid to regain a faster stability (approx. 0.2 s). The reactive power injection during the fault shows that the GSC works in STATCOM mode during faults.

This section further thrashes out the behavior of SCIG-based WECS under an LL-G fault, with conventional controls as well as a proposed control. During unsymmetrical faults, the presence of negative sequence components and a drop in grid voltage causes the current magnitude at the grid to rise significantly. The de-loading droop and pitch mechanism acts to control the current magnitude as in previous cases, while delivering nearly the same amount of average power in this case.

The results in Figures 8 and 9 demonstrate the generator speed, voltage at the dc-link, active and reactive power during the fault, pre-fault, and after applying the unsymmetrical (LL-G) fault. The results are illustrated for both the wind speeds (8.5 m/s and 10.5 m/s), with the conventional as well as the proposed controls. The double frequency oscillations are observed in the dc-link voltage, active power, and reactive power during unsymmetrical faults, as revealed in Figures 8b and 9a,b. The major problems observed during the symmetrical fault include the increase in dc-link voltage and the converter current. A long time to regain system stability is also observed. Double frequency oscillations in the dc-link voltage and power (both P and Q) are observed as the major issues in the case of an unsymmetrical (LL-G) fault.

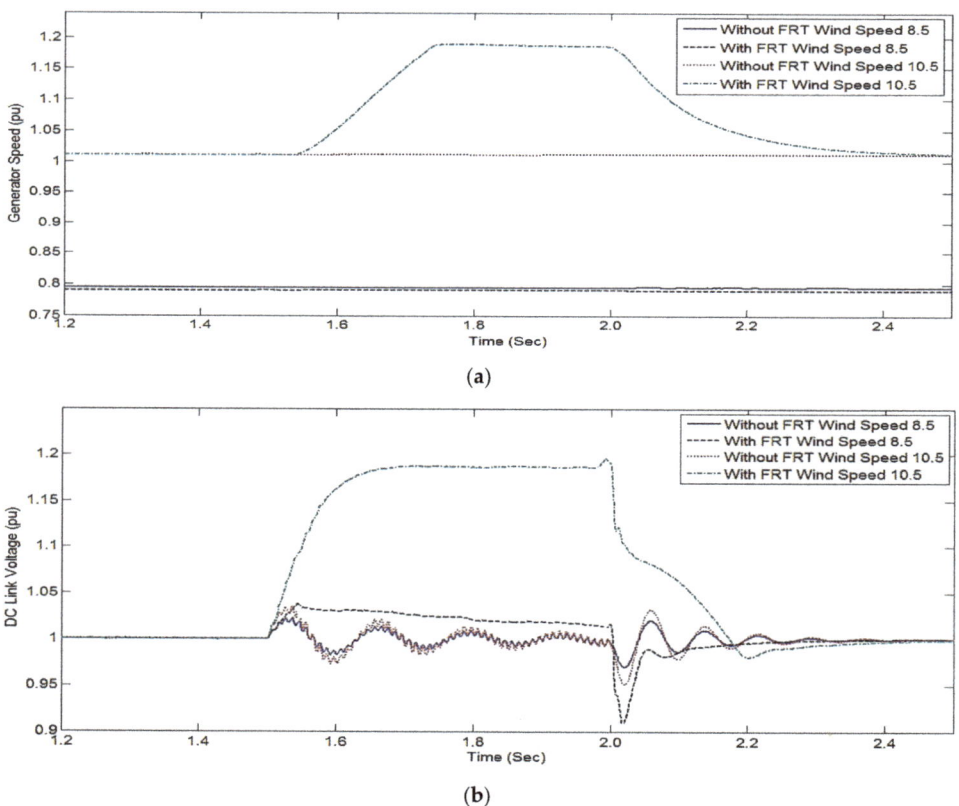

Figure 8. Behavior of WECS during unsymmetrical fault, with and without FRT control for 8.5 m/s and 10.5 m/s wind speeds: (**a**) Generator Speed (**b**) dc-link Voltage.

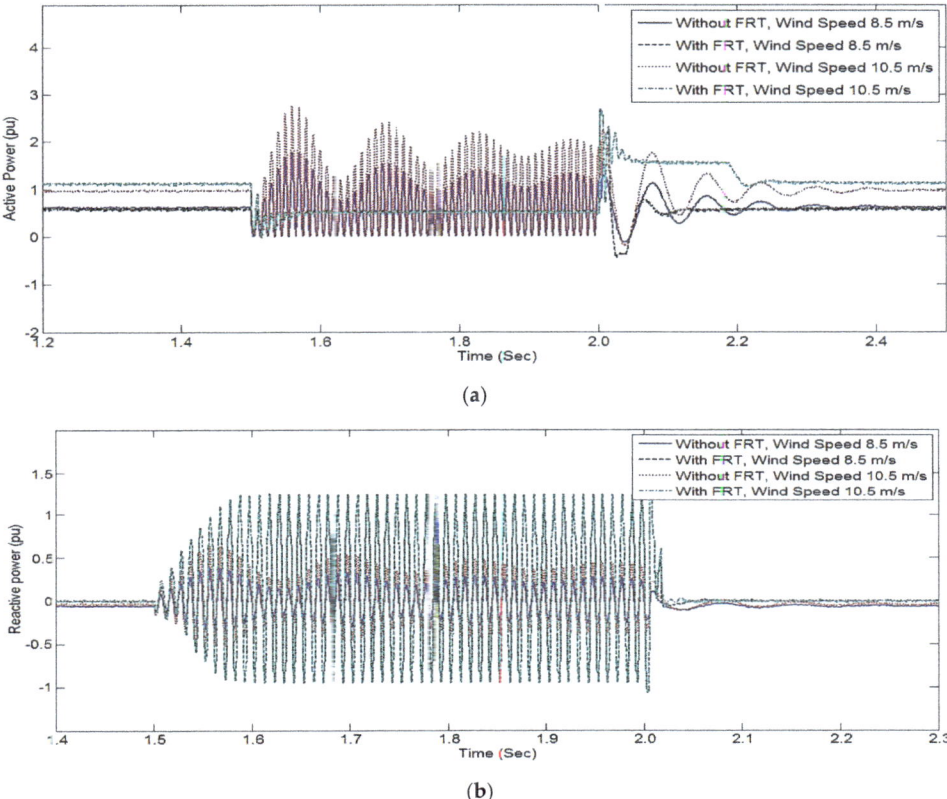

Figure 9. Behavior of WECS during unsymmetrical fault, with and without FRT control for 8.5 m/s and 10.5 m/s wind speed: (a) Active power supplied, (b) Reactive power supplied by the generator.

De-loading control provided through MSC effectively limits the dc-link voltage to 1.2 pu. VA of the converter is limited to 1.5 pu, thereby limiting the active and reactive powers. When the generator speed rises, the pitch control mechanism is activated, thereby limiting the generator speed within a safer value, which is set as 1.2 pu.

As observed in Figure 9a,b, during unsymmetrical faults, the second harmonic oscillations in active power and the dc-link voltage are totally suppressed by the decoupled control of positive and negative sequence components of currents. With the proposed control scheme, the double frequency oscillations in active power have been completely eliminated (Figure 9a). Though the oscillations are observed in reactive power during the fault, the system is seen regaining faster with FRT being implemented (Figure 9b). The ripples in dc-link voltage have also been eliminated along with active power, as shown in Figure 8b.

Overall, a system with a double current control scheme is suggested, which uses two synchronous reference frames rotating at 50 Hz, but in opposite directions. As the negative and positive sequences appear as dc in their own frames, each can be measured separately by using a 100-Hz notch filter. The currents of negative sequence and positive sequences are controlled independently. The independent control of the sequences regulates the active power completely and helps to achieve a constant dc-link voltage.

5. Conclusions

The performance analysis of a WECS using an SCIG is presented in this paper, considering the normal as well as fault conditions at the grid side. A coordinated fault ride-through strategy is proposed for WECS, providing an effective solution for alleviating diverse problems arising during grid disturbances. Conventional control, as well as the proposed control strategies for the SCIG-based WECS, was conceived, designed, modeled and simulated. It is realized that with the use of a conventional controller, the grid fault raises the dc-link voltage. This raised voltage can be highly detrimental to the power devices in the converter. Additionally, the dual-frequency oscillations are produced in the dc-link voltage and the power due to the occurrence of negative sequence components in the system during unbalanced grid conditions. This deteriorates the power quality of the system, particularly at the grid interface. The novel control strategy, based on unbalanced current injection, effectively controls the negative and positive sequence components separately. The dual-frequency oscillations in power and dc-link voltage arising during the unsymmetrical faults are also completely corrected along with providing the flow of active and reactive power. The reactive power support during the fault enables the WECS to regain a faster stability. Moreover, the de-loading of the generator during fault conditions prevents excessive voltage rise at the dc-link. The pitch control mechanism manages the generator speed when it goes beyond the maximum allowed. The proposed control strategy provides a better performance in mitigating the unsymmetrical fault, as compared to the conventional system.

Author Contributions: All authors planned the study, and contributed to the idea and field of information. All authors have read and agreed to the published version of the manuscript.

Funding: This research received no external funding.

Institutional Review Board Statement: Not Applicable.

Informed Consent Statement: Not Applicable.

Data Availability Statement: Not applicable.

Conflicts of Interest: The authors declare no conflict of interest.

References

1. Hansen, A.D.; Das, K.; Sørensen, P.; Singh, P.; Gavrilovic, A. European and Indian Grid Codes for Utility Scale Hybrid Power Plants. *Energies* **2021**, *14*, 4335. [CrossRef]
2. Ramirez, D.; Martinez, S.; Carrero, C.; Platero, C.A. Improvements in the grid connection of renewable generators with full power converters. *Renew. Energy* **2012**, *43*, 90–100. [CrossRef]
3. Behabtu, H.A.; Coosemans, T.; Berecibar, M.; Fante, K.A.; Kebede, A.A.; Van Mierlo, J.; Messagie, M. Performance Evaluation of Grid-Connected Wind Turbine Generators. *Energies* **2021**, *14*, 6807. [CrossRef]
4. Ramirez, D.; Martinez, S.; Blazquez, F.; Carrero, C. Use of STATCOM in wind farms with fixed-speed generators for grid code compliance. *Renew. Energy* **2012**, *37*, 202–212. [CrossRef]
5. Reddy, K.V.R.; Babu, N.R.; Sanjeevikumar, P. A Review on Grid Codes and Reactive Power Management in Power Grids with WECS. In *Advances in Smart Grid and Renewable Energy*; Springer: Berlin/Heidelberg, Germany, 2017; pp. 525–539.
6. Hawkins, N.; McIntyre, M.L. A Robust Nonlinear Controller for PMSG Wind Turbines. *Energies* **2021**, *14*, 954. [CrossRef]
7. Conroy, J.; Watson, R. Low-voltage ride-through of a full converter wind turbine with permanent magnet generator. *IET Renew. Power Gener.* **2007**, *1*, 182–189. [CrossRef]
8. Nguyen, T.H.; Lee, D.-C. Advanced Fault Ride-Through Technique for PMSG Wind Turbine Systems Using Line-Side Converter as STATCOM. *IEEE Trans. Ind. Electron.* **2012**, *60*, 2842–2850. [CrossRef]
9. Ukashatu, A.; Mekhilef, S.; Mokhlis, H.; Seyedmahmoudian, M.; Horan, B.; Stojcevski, A.; Bassi, H.; Rawa, M.J.H. Transient Faults in Wind Energy Conversion Systems: Analysis, Modelling Methodologies and Remedies. *Energies* **2018**, *11*, 2249.
10. Mosaad, M.I. Model reference adaptive control of STATCOM for grid integration of wind energy systems. *IET Electr. Power Appl.* **2018**, *12*, 605–613. [CrossRef]
11. Nasiri, M.; Milimonfared, J.; Fathi, S. A review of low-voltage ride-through enhancement methods for permanent magnet synchronous generator based wind turbines. *Renew. Sustain. Energy Rev.* **2015**, *47*, 399–415. [CrossRef]
12. Chatterjee, S.; Chatterjee, S. Review on the techno-commercial aspects of wind energy conversion system. *IET Renew. Power Gener.* **2018**, *12*, 1581–1608. [CrossRef]

3. Khan, A.; Ahmad, H.; Ahsan, S.M.; Gulzar, M.M.; Murawwat, S. Coordinated LVRT Support for a PMSG-Based Wind Energy Conversion System Integrated into a Weak AC-Grid. *Energies* **2021**, *14*, 6588. [CrossRef]
4. García-Sánchez, T.; Mishra, A.; Hurtado-Pérez, E.; Puché-Panadero, R.; Fernández-Guillamón, A. A Controller for Optimum Electrical Power Extraction from a Small Grid-Interconnected Wind Turbine. *Energies* **2020**, *13*, 5809. [CrossRef]
5. Nasiri, M.; Milimonfared, J.; Fathi, S.H. Robust Control of PMSG-based Wind Turbine under Grid Fault Conditions. *Indian J. Sci. Technol.* **2015**, *8*, 1–13. [CrossRef]
6. Li, P.; Wang, J.; Xiong, L.; Huang, S.; Ma, M.; Wang, Z. Energy-Shaping Controller for DFIG-Based Wind Farm to Mitigate Subsynchronous Control Interaction. *IEEE Trans. Power Syst.* **2020**, *36*, 2975–2991. [CrossRef]
7. Li, P.; Xiong, L.; Wu, F.; Ma, M.; Wang, J. Sliding mode controller based on feedback linearization for damping of sub-synchronous control interaction in DFIG-based wind power plants. *Int. J. Electr. Power Energy Syst.* **2018**, *107*, 239–250. [CrossRef]
8. Yan, X.; Yang, L.; Li, T. The LVRT Control Scheme for PMSG-Based Wind Turbine Generator Based on the Coordinated Control of Rotor Overspeed and Supercapacitor Energy Storage. *Energies* **2021**, *14*, 518. [CrossRef]
9. Ahuja, H.; Sharma, S.; Singh, G.; Arvind, S.; Singh, A. Coordinated Fault Ride Through Strategy for SCIG based WECS. In Proceedings of the 2nd IEEE Conference on Computational Intelligence and Communication Technology, Ghaziabad, India, 12–13 February 2016; pp. 1–6.
10. Vas, P. *Vector Control of AC Machines*; Clarendon Press: Oxford, UK, 1990.
11. Boldea, I. *Variable Speed Generators*; CRC Press Taylor & Francis Group: Boca Raton, FL, USA, 2006.
12. Boldea, I.; Nasar, S.A. *Vector Control of AC Drives*; CRC Press: Boca Raton, FL, USA, 1992.
13. Vas, P. *Sensorless Vector and Direct Torque Control*; Oxford University Press: Oxford, UK, 1998.
14. Munteanu, I.; Bratcu, A.L.; Cutululis, N.-A.; Ceanga, E. *Optimal Control of Wind Energy Systems*, 1st ed.; Springer: London, UK, 2008.
15. Manaullah, M.; Sharma, A.K.; Ahuja, H.; Bhuvaneswari, G.; Balasubramanian, R. Control and Dynamic Analysis of Grid Connected Variable Speed SCIG Based Wind Energy Conversion System. In Proceedings of the 2012 Fourth International Conference on Computational Intelligence and Communication Networks, Mathura, India, 3–5 November 2012; pp. 588–593.
16. Bekiroglu, E.; Yazar, M.D. MPPT Control of Grid Connected DFIG at Variable Wind Speed. *Energies* **2022**, *15*, 3146. [CrossRef]
17. Ramasamy, T.; Basheer, A.A.; Tak, M.-H.; Joo, Y.-H.; Lee, S.-R. An Effective DC-Link Voltage Control Strategy for Grid-Connected PMVG-Based Wind Energy Conversion System. *Energies* **2022**, *15*, 2931. [CrossRef]
18. Pura, P.; Iwański, G. Rotor Current Feedback Based Direct Power Control of a Doubly Fed Induction Generator Operating with Unbalanced Grid. *Energies* **2021**, *14*, 3289. [CrossRef]
19. Ma, Y.; Yang, L.; Zhou, X.; Yang, X.; Zhou, Y.; Zhang, B. Linear Active Disturbance Rejection Control for DC Bus Voltage Under Low-Voltage Ride-Through at the Grid-Side of Energy Storage System. *Energies* **2020**, *13*, 1207. [CrossRef]
20. Ma, Y.; Yang, X.; Zhou, X.; Yang, L.; Zhou, Y. Dual Closed-Loop Linear Active Disturbance Rejection Control of Grid-Side Converter of Permanent Magnet Direct-Drive Wind Turbine. *Energies* **2020**, *13*, 1090. [CrossRef]
21. Singh, A.; Ahuja, H.; Bhadoria, V.; Singh, S. Control Implementation of Squirrel Cage Induction Generator based Wind Energy Conversion System. *J. Sci. Ind. Res.* **2020**, *79*, 306–311.
22. Ramtharan, G.; Arulampalam, A.; Ekanayake, J.; Hughes, F.; Jenkins, N. Fault ride through of fully rated converter wind turbines with AC and DC transmission systems. *IET Renew. Power Gener.* **2009**, *3*, 426–438. [CrossRef]
23. Amin, M.M.N.; Mohammad, O.A. Vector oriented control of voltage source PWM inverter as a dynamic VAR compensator for wind energy conversion system connected to utility grid. In Proceedings of the Twenty Fifth Annual IEEE Conference and Exposition on Applied Power Electronics, Palm Springs, CA, USA, 21–25 February 2010; pp. 1640–1650.
24. Rodriguez, P.; Teodorescu, R.; Candela, I.; Timbus, A.V.; Liserre, M.; Blaabjerg, F. New Positive-sequence Voltage Detector for Grid Synchronization of Power Converters under Faulty Grid Conditions. In Proceedings of the 37th IEEE Power Electronics Specialists Conference, Jeju, Korea, 18–22 June 2006.
25. Song, H.-S.; Nam, K. Dual current control scheme for PWM converter under unbalanced input voltage conditions. *IEEE Trans. Ind. Electron.* **1999**, *46*, 953–959. [CrossRef]

Article

Super-Twisting Algorithm-Based Virtual Synchronous Generator in Inverter Interfaced Distributed Generation (IIDG)

Sudhir Kumar Singh [1], Rajveer Singh [1], Haroon Ashfaq [1], Sanjeev Kumar Sharma [2], Gulshan Sharma [3,*] and Pitshou N. Bokoro [3]

1 Department of Electrical Engineering, Jamia Millia Islamia, New Delhi 110025, India
2 Department of Electrical Engineering, JSS Academy of Technical Education, Noida 201301, India
3 Department of Electrical Engineering Technology, University of Johannesburg, Johannesburg 2006, South Africa
* Correspondence: gulshans@uj.ac.za

Abstract: The significant proliferation of renewable resources, primarily inverter interfaced distributed generation (IIDG) in the utility grid, leads to a dearth of overall inertia. Subsequently, the system illustrates more frequency nadir and a steeper frequency response. This may degrade the dynamic frequency stability of the overall system. Further, virtual inertia has been synthetically developed in IIDG, which is known as a virtual synchronous generator (VSG). In this work, a novel STO-STC-based controller has been developed, which offers flexible inertia following system disturbance. The controller is based on the super-twisting algorithm (STA), which is a further advancement in the conventional sliding mode control (SMC), and has been incorporated in the control loop of the VSG. In this scheme, two steps have been implemented, where the first one is to categorize all states of the system using a super-twisting observer (STO) and further, it is required to converge essential states very quickly, exploiting a super-twisting controller (STC). Thus, the STO-STC controller reveals a finite-time convergence to the numerous frequency disturbances, based on various case studies. The performance of the controller has been examined in the MATLAB environment with time–domain results that corroborate the satisfactory performance of the STO-STC scheme and that illustrate eminence over the state of the art.

Keywords: virtual inertia emulation; virtual synchronous generator (VSG); inverter interfaced distributed generation (IIDG); sliding mode control (SMC); super-twisting algorithm (STALG); super-twisting control (STC)

1. Introduction

The increasing penetration of renewable energy sources, mainly IIDG, has endangered frequency stability concerns in the low inertia system, due to the unavailability of the rotating mass [1]. Thus, inertia emulation is urgently required in IIDG to assist with frequency regulation [2]. Commonly, synchronous machine (SM) droop characteristics have been developed virtually in IIDG, which is popularly known as a virtual synchronous generator (VSG). Therefore, VSG-IIDG is essentially required in frequency regulation. A VSG control scheme has been presented in [3,4], in which the swing equation of an actual SM has been utilized. Moreover, in recent years, several VSG control schemes were presented, in which the active power extracted from IIDG has been exploited to regulate the frequency deviation, the rate of change of frequency (RoCoF), and the settling time. Nevertheless, IIDG has some active power constraints that mainly depend upon maximum power point tracking (MPPT). Further, various control schemes have been developed that are based on droop control [5,6], in which the droop characteristics are employed to regulate a fraction of the active power in accordance with the frequency deviation. In addition, the various parameters (inertia constant, damping constant, and droop slope) required to regulate the active power supplied through IIDG have been optimized through a number of presented schemes [7,8]. Nonetheless, for simplicity, there has been barely two parameters (inertia and damping) that have been incorporated in the basic design of VSG. In addition, there is an inherent trade-off between a selection of various parameters and the accomplishment of the required aspect viz. RoCoF, the settling time, the frequency regulation, and the minimization of the frequency nadir. Further, there will be no alternative approach to regulating all of the discussed aspects simultaneously, depending upon the selection of the parameters. Lastly, it is quite problematic to tune the various parameters of VSG to achieve all of the requirements at the same time, since a variation to any of them can negatively affect another performance.

1.1. Motivation of Work

In recent years, VSG control schemes have been extensively studied and investigated, especially for IIDG, considering both the grid-connected and islanded modes. VSG-IIDG has been designed to support the frequency dynamics, and to minimize the output power and frequency oscillations. Several control techniques have been developed to suppress the frequency and power oscillations [9]. The proper selection of the inertia and damping constants can effectively reduce both of these oscillations. Further, to have a proper acquaintance with VSG design, small-signal modeling has been discussed in several papers [8,10] that provide more precise parameters for the emulated inertia. Further, a VSG based on an adaptive linear quadratic regulator has been proposed in [11]. A VSG scheme discussed in the literature has considered only a few IIDGs in the proposed test system, which lacks authenticity; thus, reliability is still an open problem. Secondly, the effectiveness of previously presented schemes in both standalone and grid-connected modes still requires intense study. The grid-connected IIDG has been recognized to be more sensitive to voltage sag/swell and exposure to unbalanced conditions. The VSG design should be well-operated in both the modes, and lastly, the controller performance is required to be more robust, as per the variation of various parameters.

1.2. Literature Survey on Recent State of Art

A self-tuning algorithm (STALG) in VSG has been proposed in [2]. Further work in self-tuning was given in [12], which was developed based on a RoCoF, and it provides the optimal virtual inertia (VI) via the proper selection of the inertia and damping constants. Since IIDG is a non-linear system, owing to the power–voltage characteristics, many control schemes were therefore developed based on the linearized model with proportional-integral viz. PI controller for inertia emulation. Subsequently, linearized feedback non-linear control schemes have been discussed in the literature, which provide enhanced performance over linear control design. Nonetheless, the non-linear design performs well on optimal parameter selection and an accurate operating point, which is slightly difficult to design. Furthermore, an AI-based VSG has been presented in [13], and further, a neural-fuzzy VSG (N-F-VSG) was designed in [14], which is a more reliable structure for enhancing the frequency dynamics. Recently, a model predictive control (MPC) has been portrayed in the literature, which has been proven to be an excellent VSG design [15,16]. The fuzzy-based VSG (Fuzzy-VSG), which computes the correction factor required to alter the governor output during a sudden disturbance, was discussed in [17]. The improved MPC scheme has been suggested in [18], which was a superior design for emulating the virtual inertia.

1.3. Contribution

As per Table 1, several eminent controllers have been organized, where M. A. Torres [2] suggested a self-tuning virtual synchronous machine that tunes the inertia constant as per the severity of disturbance. Further, J. Alipoor demonstrated the concept of alternating the moment of inertia in [3]; however, it only tunes the constants (inertia and damping) with a slothful response. M. H. Ravanji et al. suggested a swing equation-based virtual inertia, in which the DFIG wind turbines participate in frequency regulation, although it addresses only frequency oscillations and RoCoF. In [13], a fuzzy-based controller has been suggested, but it exhibits a sluggish response after the disturbance. A. Karimi et al. in [17] presents a fuzzy-based controller and computes the correction term needed to adjust the governor output power during a disturbance; however, as in [13], the same controller suffers from a sluggish response. As per the power voltage characteristics, renewable generation is a non-linear system, and consequently, non-linear controllers have been introduced in the literature. In addition to this, artificial intelligence (AI)-based VSG is projected in [13]; however, non-linear controllers need accurate parameter identification for a superior performance; therefore, a model prediction controller [18] has been further explored to determine the optimal power requirement by RES during the disturbance. It determines the real power for IIDG in real-time, using a predictive framework. However, the performance evaluation of the suggested MPC suffers in the identification of the optimal real power set point. Primarily, the proposed work focuses on a comparison with the eminent state of the art, as discussed above, and reveals the performance of the proposed controller. Conventional constant parameter (CP-VSG) and zero VI have also been taken into consideration in numerous case studies. A time–domain analysis obtained on an IEEE 14 bus test system developed in MATLAB, and the simulation results, corroborate the superiority of a proposed STO-based STC over the current state of the art. Time–domain specifications such as settling time, DC voltage variation, and generator-1 settling time have been exploited for numerous popular schemes viz. fuzzy-based VSG, AI-based VSG, and MPC schemes subjected to three phase faults on bus-12. These comparisons have been compiled in later sections. Furthermore, the normalized power supplied during the disturbance has also been plotted for various schemes, as discussed above, and compared with this STO-STC scheme. The super-twisting observer can work in the presence of any kind of bounded disturbances and converge in finite time. Further, if one state of the system is known, then also it can track all the information of the system in finite time. Thus, it reduces the required number of sensors in the system.

Table 1. Comparison of current work eminence and state of the art.

Existing State of the Art	Controller Description	Attributes
M. A. Torres et al., 2014 [2]	Self-tuning virtual synchronous machine	Energy storage system for inertia emulation
J. Alipoor et al., 2015 [3]	Alternating inertia-based virtual SG	Tunes inertia and damping constants
M. H. Ravanji et al., 2017 [4]	Swing equation-based virtual inertia	DFIG-based wind turbines
C. A. B Karim et al., 2018 [13]	VSG based on fuzzy controller	Distributed generation in microgrid
S. Wang et al., 2019 [15]	Advanced control solutions	Enhanced resilience in distribution system
A. Karimi et al., 2020 [17]	Fuzzy-based VSM	Regulated governor output and correction term
A. A-Idowu et al., 2021 [16]	MPC (model predictive control)	Optimal power set points
Proposed work	STC based on STO	Enhance frequency dynamics in multi-machine system

The key outcomes of this work have been given below:

1. Super-twisting is a novel scheme to identify all the states in finite time $t < T_0$ with the minimum number of required sensors.
2. The super-twisting controller converges very quickly and provides the accurate real power required for inertia emulation under finite disturbance $d1$.
3. This novel controller efficiently improves the frequency dynamics.
4. The suggested controller illustrates a superior performance over the recent popular controllers viz. model prediction, fuzzy, or self-tuning controllers.

2. Mathematical Modeling of Proposed VSG Dynamics

In this section, the VSG-IIDG dynamic equation has been developed in the d-q reference frame [19], which is based on two control loops corresponding to the active power and reactive power exchange.

2.1. Non-Linear Dynamics of VSG-IIDG

The 3-Φ active power is basically computed as [20,21], which is given below:

$$P_e = v_d i_d + v_q i_q \tag{1}$$

whereas v_d, v_q, i_d, i_q are voltages and currents in inverter terminals on the d-q reference frame. The transformation from a-b-c to d-q requires the phase angle θ, which is computed from the second-order generalized integrator, SOGI PLL [22]. In a conventional synchronous machine (SM), a well-known inertia equation viz. swing equation of rotor dynamics has been given as follows.

$$J \omega_m \dot{\omega}_m = P_m - P_e - D(\omega_0 - \omega_m) \tag{2}$$

where, ω_0 is the nominal angular frequency of VSG-IIDG, ω_m is the measured angular frequency obtained through SOGI-PLL, J is the moment of inertia, and D is the damping coefficient [23]. Further, a high value of D brings the measured frequency more quickly to the nominal value. P_m is the governor output power. Further, as with conventional SM, the governor model is included in this modeling, as given in Equation (3):

$$P_m = P_0 + K_g(\omega_0 - \omega') \tag{3}$$

where, P_0 is the set active power and K_g is the governor droop coefficient. ω' is the angular frequency generated by the swing equation. Furthermore, one more loop corresponds to the reactive power Q_e for voltage droop control. The reactive power on the d-q coordinates are given as [20,21]:

$$Q_e = v_d i_q - v_q i_d \tag{4}$$

The output voltage of the VSG-IIDG is governed by the following dynamics:

$$V = (K_P + \frac{K_I}{s})(Q^* - Q_e) \tag{5}$$

K_p, K_I are PI (proportional and integral gains). Q^* is the reactive power reference. In addition, the voltage droop generates reference Q^*, which is given by:

$$Q^* = Q_0 + K_q(V_0 - V_r) \tag{6}$$

where K_q is the droop coefficient and Q_0 is the set reactive power. Further, V_r is the rms value of the inverter side voltage (PCC), and V_0 is the reference voltage. Now, upon differentiating Equation (5), we get:

$$\dot{V} = K_P \dot{Q}^* + K_I Q^* - K_P \dot{Q}_e - K_I Q_e$$

Thus,

$$\dot{V} = K_P \dot{Q}^* + K_I Q^* - K_P \dot{Q}_e - K_I Q_e \tag{7}$$

$\dot{Q}^*$ can be neglected with respect to the frequency dynamics; therefore:

$$\dot{V} = -K_P \dot{Q}_e + K_I(Q^* - Q_e) \tag{8}$$

Now, Equations (2) to (8) illustrate the non-linear behavior of VSG-IIDG. Therefore, a non-linear controller such as the sliding mode control (SMC) is required to handle these non-linearities, which enhances the performance of VSG-IIDG, as shown in Figure 1. In this paper, the super-twisting viz. ST algorithm has been developed and incorporated to improve the inertial response in both the grid-connected and autonomous modes. The designed ST algorithm produces a correction factor, as shown in Figure 1, which is the final modification in the governor output power or input power to the swing; Equation (2). The input variables to the ST algorithm are $\omega_m, \dot{\omega}_m$.

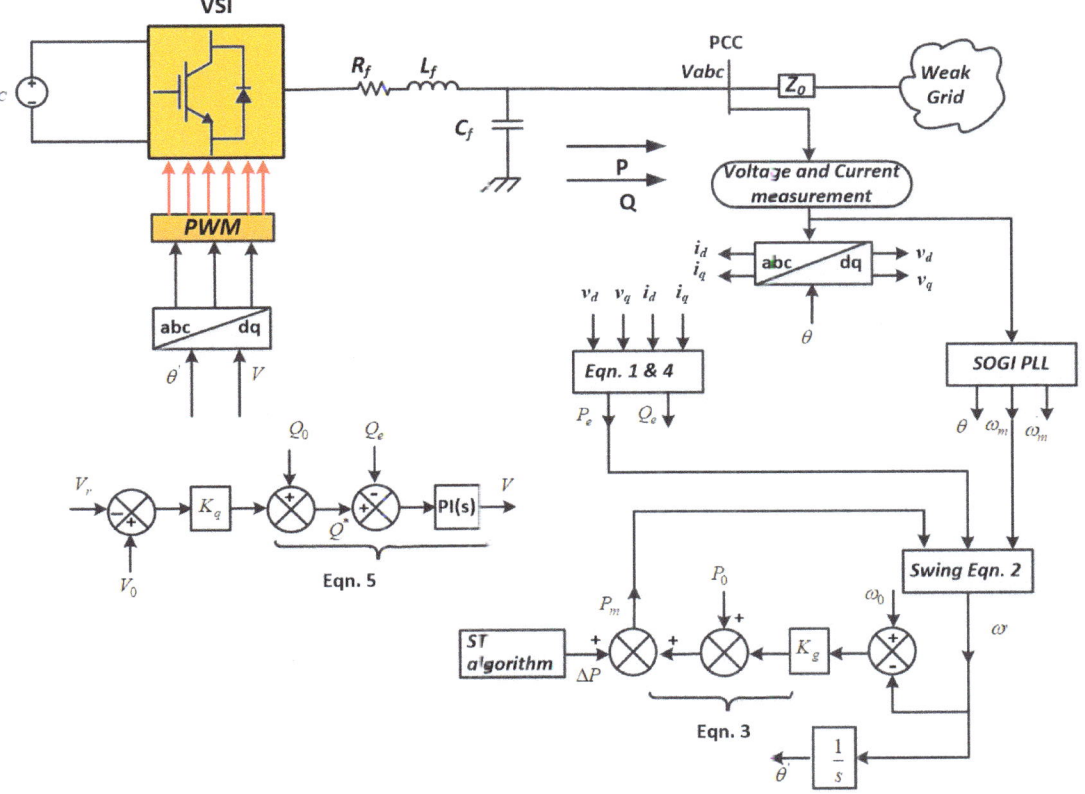

Figure 1. Super-twisting algorithm-based VSG-IIDG.

2.2. VSG-IIDG Scheme

To design the super-twisting control (STC), we need the information for both $\omega_m, \dot{\omega}_m$ in finite time under the finite disturbance. The super-twisting observer (STO) has been required for this purpose. The initial knowledge regarding this control has been taken from [19] and [24].

2.3. STC Based on STO

Consider the mathematical modeling of the dynamic system:

$$\left.\begin{array}{l} \dot{x}_1 = x_2 \\ \dot{x}_2 = u + d_1 \end{array}\right\} \qquad (9)$$

The output of the system is given as $\sigma = x_1$, and $d1$ is a finite external disturbance. The design of STC is based on the other states of the system, rather than the output information. STO has been proven to be the best observer [25], which can compensate for time-varying

(TV) disturbances within finite time $t < T_0$. STO is based on a high-order SMC (HOSMC). The STO dynamics [26,27] for the estimation of the system states (a copy of the dynamic system) have been evaluated as:

$$\left.\begin{aligned}\dot{\hat{x}}_1 &= \xi_1 + \hat{x}_2 \\ \dot{\hat{x}}_2 &= \xi_2 + u\end{aligned}\right\} \qquad (10)$$

In STO design, we neglected the unknown $d1$; further, ξ_1, ξ_2 are correction terms. The error variables are defined as $e_1 = x_1 - \hat{x}_1$ and $e_2 = x_2 - \hat{x}_2$.
Our objective is to estimate x_1, x_2 in finite time, $t < T_0$. Furthermore,

$$x = \begin{bmatrix} x_1 \\ x_2 \end{bmatrix} = \begin{bmatrix} \omega_m \\ \dot{\omega}_m \end{bmatrix} \qquad (11)$$

Now, the derivative of the error, $\dot{e}_1 = \dot{x}_1 - \dot{\hat{x}}_1 = x_2 - \hat{x}_2 - \xi_1 = e_2 - \xi_1$
Similarly, $\dot{e}_2 = \dot{x}_2 - \dot{\hat{x}}_2 = u + d_1 - u - \xi_2 = d_1 - \xi_2$
Selecting ξ_1, ξ_2 in such a way that $e_1, e_2 \to 0 \,\forall\, t \leq T_0$; thus, as per [25], the correction terms will be as given as $\xi_1 = k_1|e_1|^{\frac{1}{2}}sign(e_1)$ and $\xi_2 = k_2 sign(e_1)$. In these equations, e_2 is missing due to the lack of information regarding this error variable, and further, the correction factors are based on e_1 only. Now, STA based on e_1, e_2 is given as follows:

$$\left.\begin{aligned}\dot{e}_1 &= -k_1|e_1|^{\frac{1}{2}}sign(e_1) + e_2 \\ \dot{e}_2 &= -k_2 sign(e_1) + d_1\end{aligned}\right\} \qquad (12)$$

It has been assumed that the finite disturbance $|d_1| < \delta_0$. Now, based on the literature available [19,25], if we choose $k_1 = 1.5\sqrt{\delta_0}$ and $k_2 = 1.1\delta_0$, which have assured convergence, $e_1, e_2 \to 0 \,\forall\, t \leq T_0$, and finally:

$$\left.\begin{aligned}x_1 &= \hat{x}_1 \\ x_2 &= \hat{x}_2\end{aligned}\right\} \forall\, t \leq T_0 \qquad (13)$$

Therefore, in a second-order system, using a higher-order sliding mode observer viz STO, one can estimate numerous states of the system in finite. This observer exposes an acceptable performance in terms of finite time robust estimation. Now, the next section discusses the controller design.

2.4. ST Control Algorithm and Design

For the controller design, a relative degree of two is required. However, STC would be applicable for a relative degree of one only. Accordingly, the sliding manifold has been modified to obtain a required relative degree of one.

$$\left.\begin{aligned}S &= cx_1 + x_2 \\ \dot{S} &= c\dot{x}_1 + \dot{x}_2 = cx_2 + u + d_1\end{aligned}\right\} \qquad (14)$$

Let the controller be given as $u = -cx_2 + \psi(t)$ as shown in Figure 2, where, according to the ST algorithm, $\psi(t)$ is defined as:

$$\psi(t) = -\lambda_1|S|^{\frac{1}{2}}sign(S) - \lambda_2 \int_0^t sign(S)d\tau \qquad (15)$$

where $\lambda_1 = 1.5\sqrt{\delta_0}$ and $\lambda_2 = 1.1\sqrt{\delta_0}$. Further, Δ is defined on the finite disturbance as $\delta_0 \geq |\dot{d}_1|$. After the control input, we get:

$$\left. \begin{array}{l} \dot{x}_1 = S - cx_1 + e_2 \\ S = ce_2 - -\lambda_1 |S|^{\frac{1}{2}} sign(S) - \lambda_2 \int_0^t sign(S) d\tau + k_2 sign(e_1) \end{array} \right\} \quad (16)$$

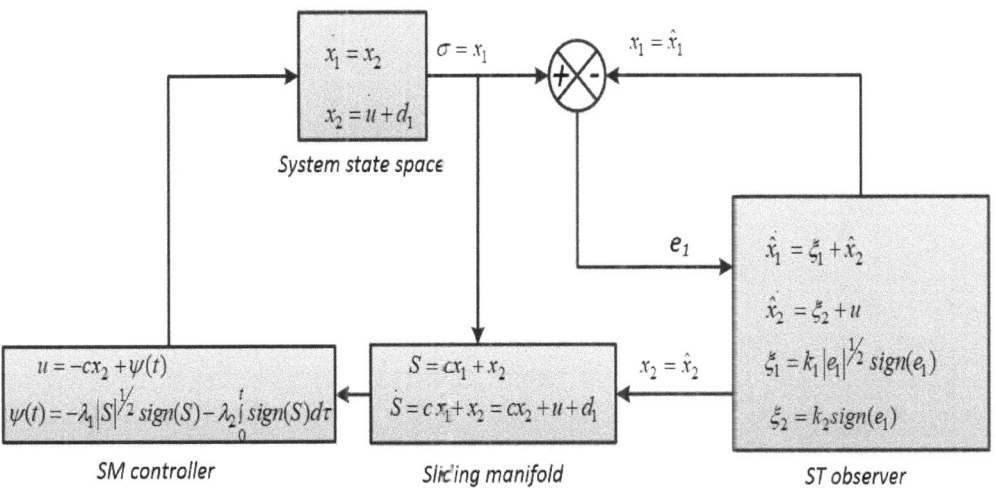

Figure 2. Sliding mode observer-based sliding mode controller.

Furthermore, the Lyapunov stability of the suggested STO-based STC has been dependent upon the error dynamics e_1, e_2. Since the correction factors are based on e_1 only, the Lyapunov function has been chosen for σ. For STC, the scaled function has been taken as $\dot{\sigma} = -k_1 |\sigma|^{\frac{1}{2}} sign(\sigma)$. Now, assuming the Lyapunov function as follows:

$$\|V\| = \frac{1}{2}\sigma^2, \sigma \in \Re \quad (17)$$

V is continuous and positively defined ($V > 0$). However, it is non-differentiable at $V(0) = 0$. The error dynamics should be globally asymptotically stable, based on the selected ST gains.

Now,

$$\|\dot{V}\| = \sigma\dot{\sigma} = -k_1|\sigma|^{\frac{1}{2}}\sigma sign(\sigma) = -k_1|\sigma|^{\frac{3}{2}} < 0 \quad (18)$$

where $\sigma sign(\sigma) = |\sigma|$. Thus, it has been concluded that the error dynamics converge to zero in finite time $t \leq T_2$. To obtain the effectiveness of the STO-based STC controller, here, we tried to observe and explore the attributes of the suggested controller. Consider a very simple physical system that is a moving car system driven by some controlled input (u). The system state space can be written based on Newton's law, and the displacement vector is given as $x1$ and the velocity vector $\dot{x}_1 = x_2$. Further, at the equilibrium point, $\dot{x}_1 = x_2 = 0$. This simple system has been examined with the proposed controller, where the disturbance is $d_1 = 0.5 \sin \omega t$ with $\omega = 1$ rad/s. The sliding surface has been chosen as $S = x_1 + \frac{1}{357.56}x_2$. For simulation purposes, the controller gains have been given as follows.

- STO gains: $k_1 = 2.1, k_2 = 2.2$;
- STC gains: $\lambda_1 = 2.2, \lambda_2 = 1.55$;
- Constant $c = 1$, sampling time (MATLAB) = 1 ms.

The twisting observer and controller algorithm have been written in a MATLAB function, and the prime objective of the controller is to bring the vehicle to the desired position, starting from the zero position. The simulation results have been portrayed in Figure 3. As per the results obtained in Figure 3, the suggested controller works well with a good precision of 10^{-3}.

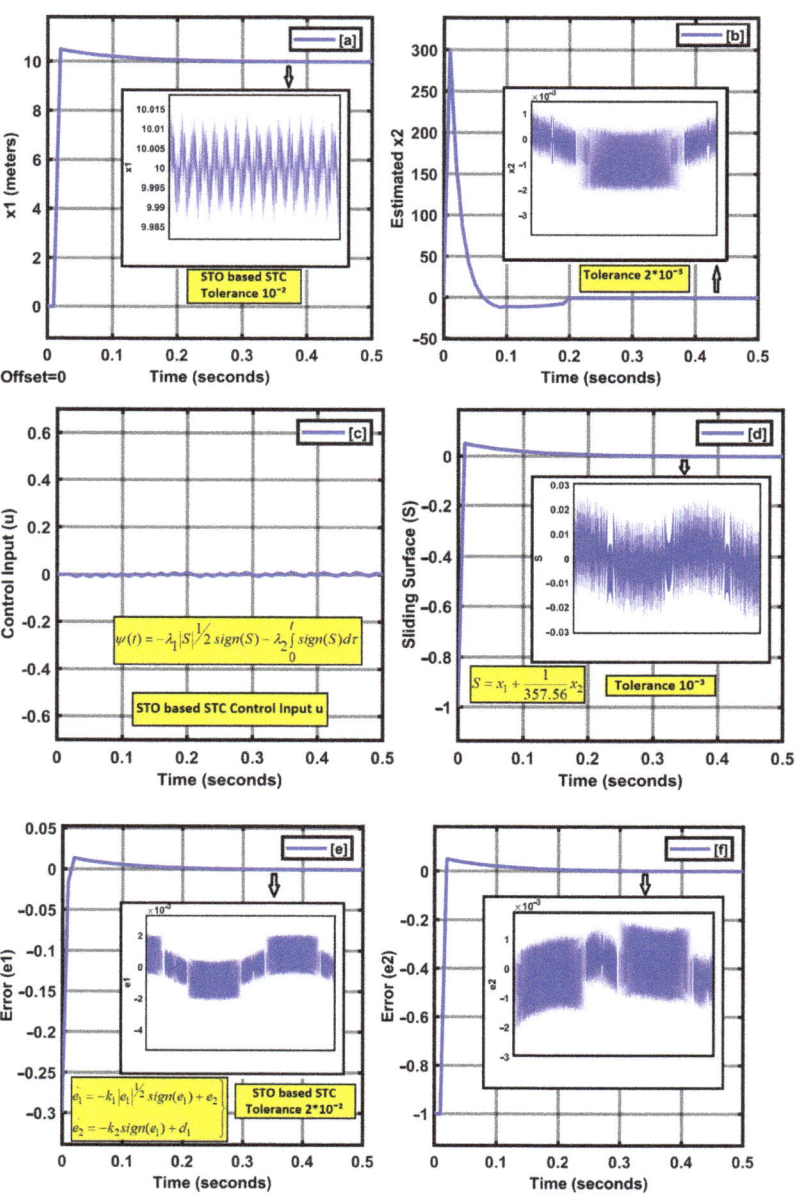

Figure 3. (**a**) Position vector x_1 evolution, (**b**) estimated state vector x_2 evolution, (**c**) control input (u) evolution, (**d**) sliding surface (S) evolution, (**e**) observer error (e_1) of state x_1, (**f**) observer error (e_2) of state x_2.

Furthermore, the same controller has been utilized for virtual inertia emulation on the IEEE 14 bus test system, with the output $\sigma = \Delta P$, where ΔP is the additional power required for the virtual inertia emulation, as per the frequency deviation $\Delta \omega_m$ and the rate of change of frequency $\dot{\omega}_m$.

In this paper, a sudden load change has been taken as 2 pu. Therefore, $\delta_0 = 2$ has been selected. The STO and STC gains have been estimated earlier in this section.

3. Test System Configuration and Setup

In this paper, the IEEE 14 bus system has been taken [23] as a test system. A 415 V photovoltaic system (IIDG) 0.5 MW has been incorporated at bus-12 through a 15 MVA transformer. Several synchronous generators are connected at various buses in isolation, or with some local demands.

Furthermore, various static and dynamic loads have been connected at different buses, as shown in Figure 4. At bus-9, the static Var compensator of 10 MVAR has been coupled. A 50 MW wind turbine generator (WTG) is connected at bus-14 to further supply a high demand at bus-13. In the test system, to obtain sufficient information on the voltage, power flow and angle, power flow analysis has been conducted, and the whole test system has been designed and developed on a MATLAB/Simulink MathWorks® platform. The open-circuit voltage of PV generation is 800 V, with a 600 A short circuit current. The fuses are usually incorporated to protect the PV generation, as shown in Figure 4, and are incorporated at various places. Now, for the inertia emulation, MPPT (maximum power point tracking) does not play a significant role, and it is omitted in the development of the MATLAB detailed model of a test system. The case study has been executed based on various factors, viz. sudden load change and intentional islanding. Further, during simulation, the temperature or irradiance variation have been omitted for simplicity, and a detailed model has been simulated based on ode 23tb, with a maximum step size of 1 ms. The voltage source inverter is an IGBT-based 2 level bridge inverter, which is efficiently regulated through pulse modulation. In this work, STO-based STC accurately measures ΔP based on the nominal frequency deviation and the rate of change of frequency $(\omega_m, \dot{\omega}_m)$, owing to the sudden bounded disturbance d_1. The super-twisting observer and the controller algorithm have been already given in Equations (9)–(18), and consequently, the code is written in a MATLAB function. The efficacy of the proposed controller has been verified by numerous simulation results. The system parameters used in building the test system in MATLAB have been organized in Table 2.

Table 2. Test system parameters.

System Parameters	Values
DC-link voltage (V_{DC})	800 V
Nominal frequency (f_n)	50 Hz
Output inverter voltage	450 V
Inertia constant (H)	4, 7, 10 s
Maximum irradiance	1000 W/m^2
Temperature (T)	38 °C
Parallel strings	76
Series modules	23
Open circuit voltage (V_{oc})	36.3 V/Module
Short circuit current (I_{sc})	7.84 A/Module
Nominal voltage	835 V(DC)
Nominal current	598 A(DC)
Max power/module	213.15 W
Speed regulation of governor (R)	0.05 pu
Inertia of synchronous generator (H)	4 s

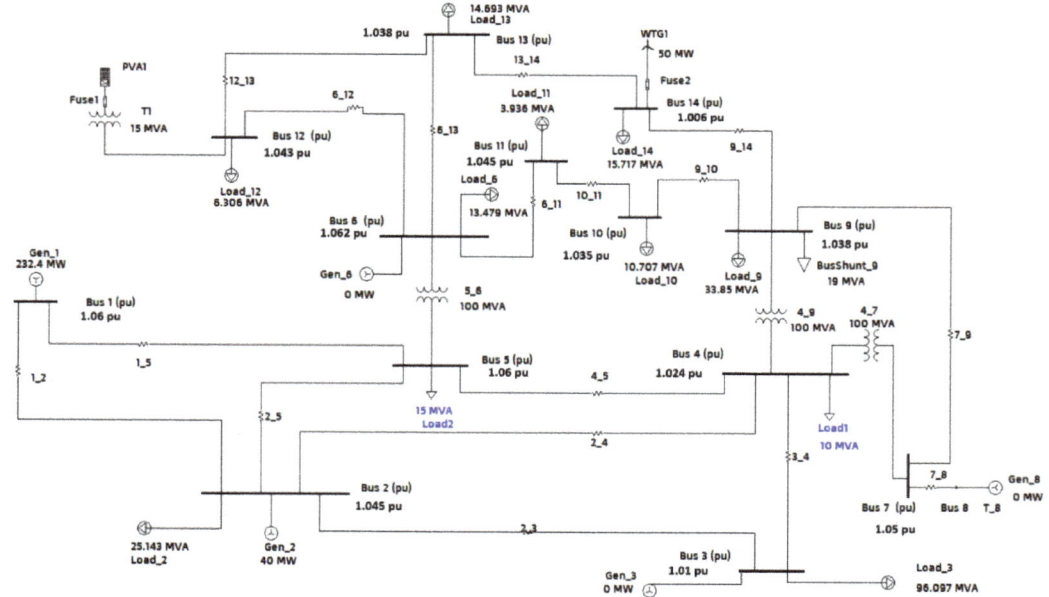

Figure 4. IEEE 14 bus test system (bus voltages in pu after power flow analysis).

4. Simulation Results and Performance Evaluation

As per the available literature [23–25], four main consecutive control steps have been utilized to restore the deviated frequency to the nominal value. 1. Inertial control 2. Primary control 3. Secondary control 4. Tertiary control. The fastest control is generally exhibited by SG, which instantaneously responds to supply–demand disturbance before the primary controller is activated. However, in recent years, SGs have been replaced by IIDGs to a large extent, and thus, the overall system damping and inertia are reduced significantly. Therefore, during load throw-off or numerous kinds of faults, primary control is not an effective solution to many problems, due to their sluggish response, which in the worst conditions causes blackouts or a complete failure of the system. The above discussion is shown in Figure 5.

4.1. Sudden Load Variation in Grid-Connected VSG

In this scenario, the capability of the proposed STO-based STC has been investigated. The load changes at bus-12 from 500 kW (for simulation purposes, the load connected at bus-12 was taken to 500 kW instead of 6.306 MVA for simplicity in designing the MATLAB model) to 650 kW at t = 2 s, and subsequently, the frequency droop has been observed from the nominal value f_n; consequently, a load has been decreased by up to 100 kW at bus-12 at t = 15 s (a sudden load variation through variable AC resistors in MATLAB) and a frequency overshoot, along with RoCoF, has been investigated for the selected state of the arts.

The frequency dynamic for different control schemes has been portrayed in Figure 6. An IIDG with zero or negligible inertia suffers from an impulsive overshoot; nonetheless, the oscillating time has been drastically reduced, as compared to the constant parameter, viz. the CP-VSG scheme (a constant inertia fixed up to H = 4.5 s).

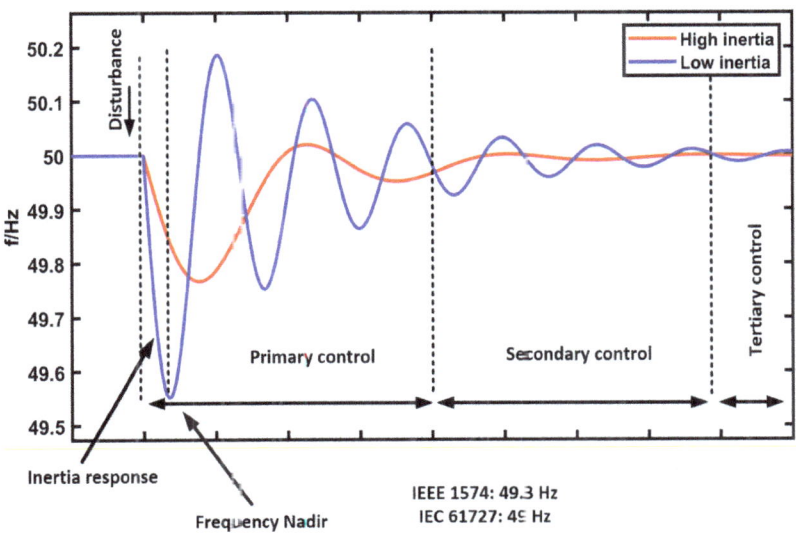

Figure 5. Frequency dynamics over time with various control steps.

The obvious disadvantages of a fixed high-inertia system viz. CP-VSG are sustained periodic oscillations and a subsequently inferior dynamic performance. Furthermore, fuzzy-based VSG illustrates a superior inertia emulation during the disturbances, but the settling time of oscillation is high, as compared to the proposed STO-based STC. The proposed STO-based STC demonstrates RoCoF = −0.25 Hz/s at t = 2 s, and 0.75 Hz/s at t = 15 s, with a 49.9 Hz frequency nadir and a 50.5 Hz frequency overshoot. The investigated results recommend a superior dynamic performance of STO-based STC over numerous control schemes, as shown in Figure 6. The inertia emulation for various schemes has been examined, based on the swing equation [23], and is given as:

$$J = \frac{\Delta P_m - \Delta P_e - D(\omega_m - \omega_0)}{\dot{\omega}_m} \quad (19)$$

Now, the inertia J, and consequently, the inertia constant $H(s)$ adaptively increased during the disturbances. STO-based STC efficiently emulates and regulates the inertia during disturbance, which is constant in the CP-VSM scheme. An encapsulation of the time–domain performance evaluation of popular controllers and STO-based STC on sudden load variation has been compiled in Table 3.

Table 3. Performance evaluation and comparison on sudden load change.

Control Scheme Attributes	No VI	CP-VSG	STO-Based STC
Frequency nadir [Hz]	49.2 (*Violates IEEE 1574*)	49.4	49.8
Frequency overshoot [Hz]	52 (*Violates IEEE 1574*)	51.9	50.9
RoCoF [Hz/s]	−0.8	−0.7	−0.3
Steady state error (%)	1.45	0.7	0.13
Settling time [ts/s]	5	10	3
Inertia H [s]	0	4.5	*Adaptive (4.4–4.9)*

Control Scheme Attributes	Fuzzy-Based VSG	STALG	MPC Scheme
Frequency nadir [Hz]	49.8	49.7	49.6
RoCoF [Hz/s]	−0.4	−0.5	−0.3
Settling time [ts]	4	3.7	4.2

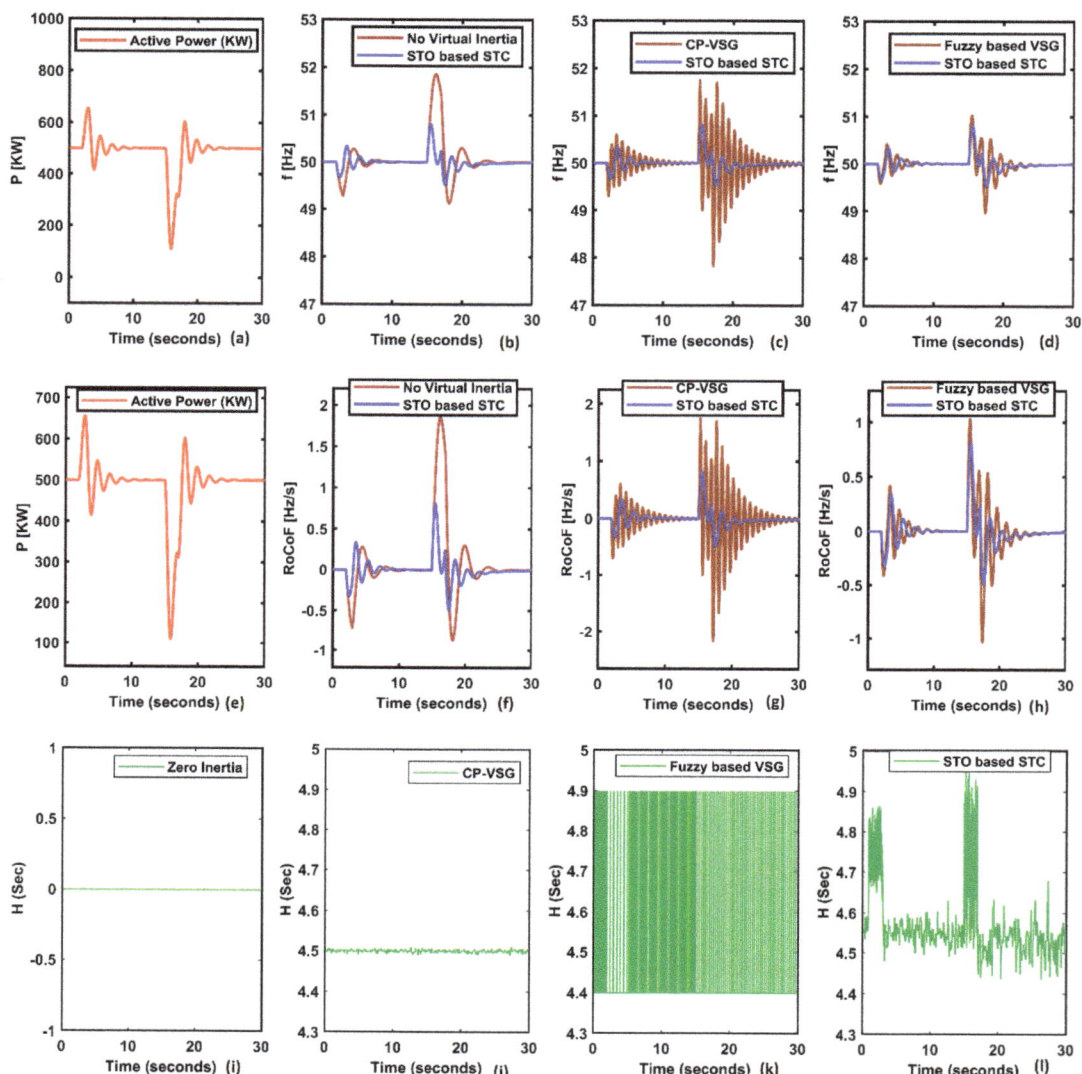

Figure 6. Grid-connected mode. (**a**) Active power variation, (**b**–**d**) dynamic frequency response of various control schemes, (**e**–**h**) rate of frequency variation for various control schemes, (**i**–**l**) inertia emulation.

4.2. Inertia Response on AC Fault

A 3 − Φ fault has been simulated and created at 1 s, and subsequently cleared at 1.5 s at bus-13. Further, the bus-12 voltage falls as per fault impedance, which has been portrayed in Figure 7a. Subsequently, the bus-12 frequency has been estimated after the occurrence of the 3 − Φ fault, and the frequency plots for zero inertia, CP-VSG, and STO-STC have been shown in Figure 7b. The frequency response illustrated the superior damping capability of STO-STC, with a settling time of below 5 s after fault clearance. No VI exhibited an impulsive frequency rise, due to zero inertia emulation, and consequently, no variation in the DC-link voltage, as shown in Figure 7c. Furthermore, to observe the off-shore power fluctuation owing to fault occurrence at the same location, the bus-13 and SG-1 (bus-1) power oscillations have been examined, as shown in Figure 7d. This characterizes superior damping for the STO-STC-based IIDG at bus-12. Other schemes have also been compared and compiled in Tables 4 and 5, respectively.

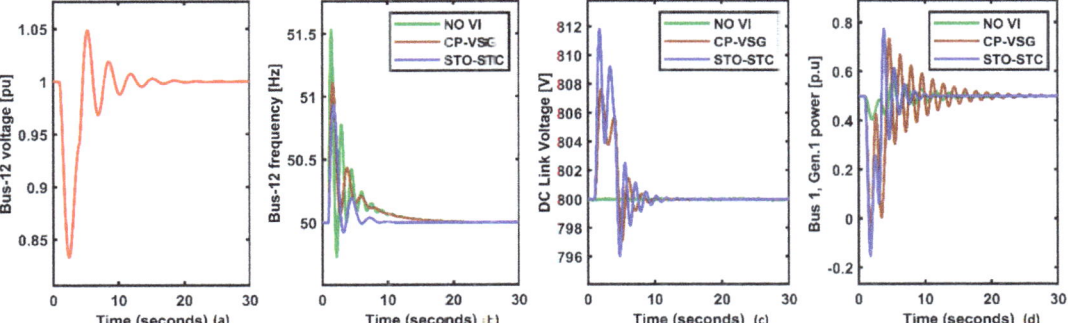

Figure 7. (a) Bus-12 voltage sag during fault (pu), (b) bus-12 frequency overshoot (Hz), (c) DC-link variation on fault (V), (d) off-shore SG power fluctuation (pu) comparison for various schemes.

Table 4. Time–domain specifications on a three phase fault (bus-12).

Attribute	Fuzzy Based VSG	STC-STC	MPC Scheme
Settling time (s)	6	6.3 oscillatory responses	5.8
DC link variation (V)	5	7	13
Bus-1, Gen.1 settling time (s)	12	9	8

Table 5. Time–domain specifications with off-shore performance assessment.

Attribute	No VI	CP-VSG	AI Based VSG
Settling time (s)	Unspecified	20 oscillatory responses	6.5
DC link variation (V)	No variation	5	9
Bus-1, Gen.1 settling time (s)	20	14	8

4.3. Normalized Active Power Response under Sudden Load Variation

Figure 8 exhibits normalized power extracted from IIDG on a sudden load change (ΔP_L = 50 kW) on bus-13, next to bus-12. Numerous controllers have been taken to examine the power response at bus-12 required to mitigate the frequency variation and the frequency nadir.

It has been found that the virtual inertia response is impulsive and instantly injects active power to improve the dynamic frequency stability. Initially, IIDG was operating at 0.2 pu. Further, the injected active power dies out soon as the frequency normalizes to nominal values. Primarily, IIDG at bus-12 was delivering 0.2 normalized power, and at t = 2 s, sudden load changes to 50 kW and power responses have been captured in MATLAB, as shown in Figure 8. Further, in Figure 8 is a performance evaluation and comparison of various schemes on injected normalized power during a sudden load change (pu): Initially IIDG was operating at 0.2 pu.

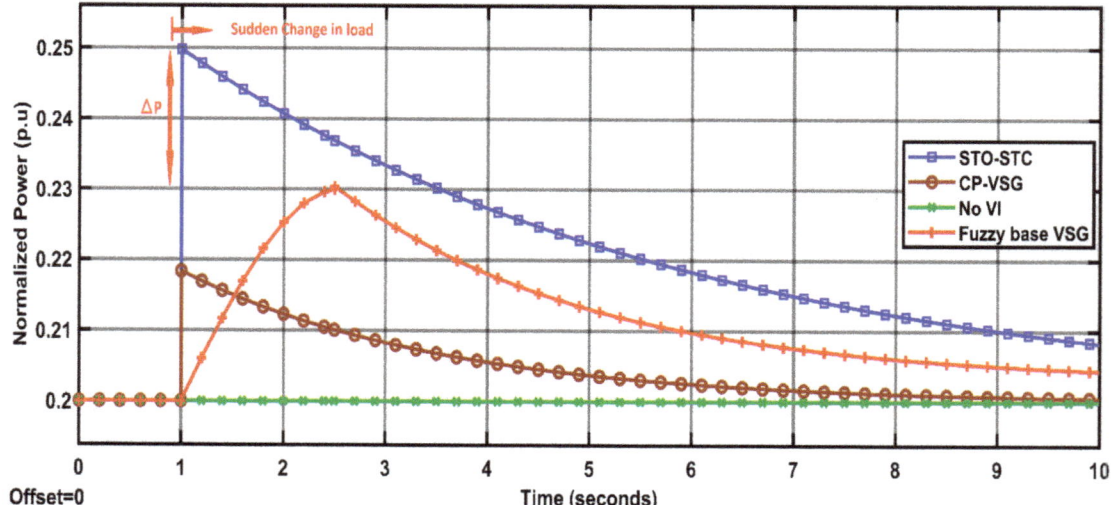

Figure 8. Performance evaluation and comparison of various schemes on injected normalized power during a sudden load change (pu).

STO-based STC estimates an accurate ΔP which is required based on the nominal frequency deviation and RoCoF ($\omega_m, \dot{\omega}_m$) owing to the sudden load change ΔP_L, which is the bounded disturbance d_1. Further, unlike fuzzy-based VSG, STO-STC exhibits an instantaneous power response and sustains supplementary injected power until the deviated frequency has not reached the nominal frequency. The time–domain simulation results illustrate the superior performance of the STO-STC scheme over recent popular controllers viz. model prediction, fuzzy-based VSG, CP-VSG, or self-tuning controllers. It is noteworthy that the emulated inertia by the STO-STC scheme increased adaptively during the disturbance, as shown in Figure 6l. Furthermore, optimal real power ΔP has been evaluated and injected during a sudden load change, as discussed in Figure 8. The proposed STO-STC scheme observes the various states using STO, and regulates the real power flow using STC during the disturbance. This will further enhance the overall dynamic frequency stability of the system.

5. Conclusions

In this work, a novel STO-tuned STC scheme has been efficiently explored and verified using time–domain simulation results. Firstly, the inertial response has been examined in grid-connected IIDG. The virtual inertia emulation has been proven to be the most promising scheme to enhance the dynamic frequency responses of low-inertia systems. Based on the mathematical modeling of VSG, it exhibits a non-linear behavior of VSG. Therefore, a non-linear controller such as sliding mode control (SMC) is required to handle these non-linearities, which enhances the performance of VSG-IIDG. Further, STO is based on high-order SMC (HOSMC). For STC design, STO evaluates numerous system states in finite time $t < T_0$. The proposed STO-STC scheme evaluates the exact corrective active power ΔP, based on the optimal selection of system inertia, which reduces the frequency pulsations. The STO-STC scheme exhibits superior damping in a multi-machine system as compared to the popular schemes viz. the self-tuning STALG scheme, CP-VSG, fuzzy-based VSG, and the MPC scheme. This is accomplished by the optimal selection of inertia and corrective power ΔP, which regulate both the frequency and power variations. This scheme also works well in fault occurrence, which has been revealed through numerous case studies.

Author Contributions: Investigation, S.K.S. (Sudhir Kumar Singh); Methodology, R.S.; Resources, H.A.; Supervision, S.K.S. (Sanjeev Kumar Sharma); Writing—review & editing, G.S. and P.N.B. All of the authors contributed to designing and developing the mathematical model of IIDG, and the collection of numerous information. All authors have read and agreed to the published version of the manuscript.

Funding: This research received no external funding.

Institutional Review Board Statement: Not applicable.

Informed Consent Statement: Not applicable.

Data Availability Statement: Not applicable.

Conflicts of Interest: The authors declare no conflict of interest.

References

1. Majumder, R. Some aspects of stability in microgrids. *IEEE Trans. Power Syst.* **2013**, *28*, 3243–3252. [CrossRef]
2. Torres, L.M.A.; Lopes, L.A.C.; Morán, T.L.A.; Espinoza, C.J.R. Self-tuning virtual synchronous machine: A control strategy for energy storage systems to support dynamic frequency control. *IEEE Trans. Energy Convers.* **2014**, *29*, 833–840. [CrossRef]
3. Alipoor, J.; Miura, Y.; Ise, T. Power system stabilization using a virtual synchronous generator with alternating moments of inertia. *IEEE J. Emerg. Sel. Top. Power Electron.* **2015**, *3*, 451–458. [CrossRef]
4. Ravanji, M.H.; Parniani, M. Modified virtual inertial controller for prudential participation of DFIG-based wind turbines in power system frequency regulation. *IET Renew. Power Gener.* **2017**, *13*, 155–164. [CrossRef]
5. Van de Vyver, J.; De Kooning, J.D.; Meersman, B.; Vandevelde, L.; Vandoorn, T.L. Droop control as an alternative inertial response strategy for the synthetic inertia on wind turbines. *IEEE Trans. Power Syst.* **2015**, *31*, 1129–1138. [CrossRef]
6. Ofir, R.; Markovic, U.; Aristidou, P.; Hug, G. Droop vs. virtual inertia: Comparison from the perspective of converter operation mode. In Proceedings of the IEEE International Energy Conference (ENERGYCON), Limassol, Cyprus, 3–7 June 2018; IEEE: Piscataway, NJ, USA, 2018.
7. Borsche, T.; Dorfler, F. On placement of synthetic inertia with explicit time-domain constraints. *arXiv* **2017**, arXiv:1705.03244.
8. Wang, F.; Zhang, L.; Feng, X.; Guo, H. An adaptive control strategy for virtual synchronous generator. *IEEE Trans. Ind. Appl.* **2018**, *54*, 5124–5133. [CrossRef]
9. Khajehoddin, S.A.; Karimi-Ghartemani, M.; Ebrahimi, M. Grid-supporting inverters with improved dynamics. *IEEE Trans. Ind. Electron.* **2019**, *66*, 3655–3667. [CrossRef]
10. Wu, H.; Ruan, X.; Yang, D.; Chen, X.; Zhao, W.; Lv, Z.; Zhong, Q.-C. Small-signal modeling and parameters design for virtual synchronous generators. *IEEE Trans. Ind. Electron.* **2016**, *63*, 4292–4303. [CrossRef]
11. Markovic, U.; Chu, Z.; Aristidou, P.; Hug-Glanzmann, G. Lqr-based adaptive virtual synchronous machine for power systems with high inverter penetration. *IEEE Trans. Sustain. Energy* **2018**, *10*, 1501–1512. [CrossRef]
12. Singh, S.K.; Singh, R.; Diwania, S.; Singhal, A.; Saway, S. Impact of Inverter Interfaced DG Control Schemes on Distributed Network Protection Recent Advances in Power Electronics and Drives. In *Lecture Notes in Electrical Engineering*; Springer: Singapore, 2021; Volume 707.
13. Andalib-Bin-Karim, C.; Liang, X.; Zhang, H. Fuzzy-secondary controller-based virtual synchronous generator control scheme for interfacing inverters of renewable distributed generation in microgrids. *IEEE Trans. Ind. Appl.* **2018**, *54*, 1047–1061. [CrossRef]
14. Lakshmi, V.S.; Purushotham, P. ANFIS controller with virtual synchronous generator control for parallel inverters in microgrids. *Int. J. Innov. Technol.* **2017**, *5*, 2155–2161.
15. Wang, S.; Dehghanian, P.; Alhazmi, M.; Nazemi, M. Advanced control solutions for enhanced resilience of modern power-electronic-interfaced distribution systems. *J. Mod. Power Syst. Clean Energy* **2019**, *7*, 716–730. [CrossRef]
16. Ademola-Idowu, A.; Zhang, B. Frequency Stability Using MPC-Based Inverter Power Control in Low-Inertia Power Systems. *IEEE Trans. Power Syst.* **2021**, *36*, 1628–1637. [CrossRef]
17. Karimi, A.; Khayat, Y.; Naderi, M.; Dragičević, T.; Mirzaei, R.; Blaabjerg, F. Inertia Response Improvement in AC Microgrids: A Fuzzy-Based Virtual Synchronous Generator Control. *IEEE Trans. Power Electron.* **2020**, *35*, 4321–4331. [CrossRef]
18. Yap, K.Y.; Sarimuthu, C.R.; Lim, J.M.-Y. Virtual inertia-based inverters for mitigating frequency instability in grid-connected renewable energy system: A review. *Appl. Sci.* **2019**, *9*, 5300. [CrossRef]
19. Rosales, A.; Yu, Z.; Ponce, P.; Molina, A.; Ayyanar, R. VSG scheme under unbalanced conditions controlled by SMC. *IET Renew. Power Gener.* **2019**, *13*, 3043–3049. [CrossRef]
20. Akagi, H.; Ogasawara, S.; Kim, H. The theory of instantaneous power in three-phase four-wire systems: A comprehensive approach. In Proceedings of the Conference Record of the 1999 IEEE Industry Applications Conference 34th IAS Annual Meeting (Cat. No. 99CH36370), Phoenix, AZ, USA, 3–7 October 1999; Volume 1, pp. 431–439.
21. Schiffer, J.; Zonetti, D.; Ortega, R.; Stanković, A.M.; Sezi, T.; Raisch, J. A survey on modeling of microgrids: From fundamental physics to phasors and voltage sources. *Automatica* **2016**, *74*, 135–150. [CrossRef]

22. Vekić, M.; Rapaić, M.R.; Šekara, T.B.; Grabić, S.; Adžić, E. Multi–Resonant observer PLL with real-time estimation of grid unbalances. *Int. J. Electr. Power Energy Syst.* **2019**, *108*, 52–60. [CrossRef]
23. Singh, S.K.; Singh, R.; Ashfaq, H.; Kumar, R. Virtual Inertia Emulation of Inverter Interfaced Distributed Generation (IIDG) for Dynamic Frequency Stability & Damping Enhancement Through BFOA Tuned Optimal Controller. *Arab. J. Sci. Eng.* **2021**, *47*, 3293–3310.
24. Zheng, X.; Wang, C.; Pang, S. Injecting positive-sequence current virtual synchronous generator control under unbalanced grid. *IET Renew. Power Gener.* **2019**, *13*, 165–170. [CrossRef]
25. Kumari, K.; Chalanga, A.; Bandyopadhyay, B. Implementation of Super-Twisting Control on Higher Order Perturbed Integrator System using Higher Order Sliding Mode Observer. *IFAC-PapersOnLine* **2016**, *49*, 873–878. [CrossRef]
26. Southall, B.; Bernard, F.B.; John, A.M. Controllability and Observability: Tools for Kalman Filter Design. In Proceedings of the British Machine Vision Conference, Southampton, UK; 1998.
27. Xue, Y.; Pequito, S.; Coelho, J.R.; Bogdan, P.; Pappas, G.J. Minimum number of sensors to ensure observability of physiological systems: A case study. In Proceedings of the 2016 54th Annual Allerton Conference on Communication, Control, and Computing (Allerton), Monticello, IL, USA, 27–30 September 2016; pp. 1181–1188. [CrossRef]

Article

Real-Time Peak Valley Pricing Based Multi-Objective Optimal Scheduling of a Virtual Power Plant Considering Renewable Resources

Anubhav Kumar Pandey, Vinay Kumar Jadoun * and Jayalakshmi N. Sabhahit

Department of Electrical & Electronics, Engineering, Manipal Institute of Technology, Manipal Academy of Higher Education, Manipal 576104, India
* Correspondence: vjadounmnit@gmail.com

Abstract: In the era of aiming toward reaching a sustainable ecosystem, the primary focus is to curb the emissions generated by non-conventional resources. One way to achieve this goal is to find an alternative to traditional power plants (TPP) by integrating various distributed energy resources (DERs) via a Virtual Power Plant (VPP) in modern power systems. Apart from reducing emissions, a VPP enhances the monetary benefits to all its participants, including the DER owners, participants, and utility personnel. In this paper, the multi-objective optimal scheduling of the VPP problem considering multiple renewable energy resources has been solved using the multi-objective black widow optimization (MOBWO) algorithm. Renewable resources consist of solar PV modules, wind turbines, fuel cells, electric loads, heat-only units, and CHP units. The weighting factor method was adopted to handle the multi-objective optimal scheduling (MOOS) problem by simultaneously maximizing profit and minimizing emission while satisfying the related constraints. In this research, a peak valley power pricing strategy is introduced and the optimal scheduling of the VPP is attained by performing a multi-objective scheduling strategy (MOSS), which is day-ahead (on an hourly basis) and 15-min based (for a one-day profile), to observe the behavior of the anticipated system with a better constraint handling method. This algorithm is capable of dealing with a complex problem in a reduced computational time, ensuring the attainment of the considered objective functions. The numerical results obtained by the MOBWO algorithm after 100 independent trials were compared with the latest published work showing the effectiveness and suitability of the developed system.

Keywords: virtual power plant; renewable energy resources; black widow optimization; multi-objective optimal scheduling; peak valley pricing

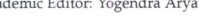

Citation: Pandey, A.K.; Jadoun, V.K.; Sabhahit, J.N. Real-Time Peak Valley Pricing Based Multi-Objective Optimal Scheduling of a Virtual Power Plant Considering Renewable Resources. *Energies* **2022**, *15*, 5970. https://doi.org/10.3390/en15165970

Academic Editor: Yogendra Arya

Received: 15 July 2022
Accepted: 2 August 2022
Published: 18 August 2022

Publisher's Note: MDPI stays neutral with regard to jurisdictional claims in published maps and institutional affiliations.

Copyright: © 2022 by the authors. Licensee MDPI, Basel, Switzerland. This article is an open access article distributed under the terms and conditions of the Creative Commons Attribution (CC BY) license (https://creativecommons.org/licenses/by/4.0/).

1. Introduction

A virtual power plant (VPP) is an assembly of energy resources that are owned by a private entity and can be interconnected for combined operation. The essence of a VPP is that it is owned and operated independently, it can be controlled centrally and monitored with the help of advanced software. This helps the distributed resources which are coming from disparate locations to respond quickly to the energy supply and demand Refs. [1–4]. The goal of the VPP is to handle the energy demand of consumers communally and to resolve the future failure of networks. A VPP consists of remote software that helps to standardize specific energy use by linking, organizing, and controlling decentralized and controlled charging generators.

A VPP acts as an energy hub and is capable of becoming the future of power systems. It offers variable generation at reduced inertia levels and at the same time provides ancillary services, including controlling the voltage imbalances, frequency regulation, and congestion management and helps in the black start. The VPP also resolves some of the most prominent grid-related issues, viz., improved forecasting, real-time mitigation of power quality-related concerns, and proper balancing of the demand and supply gap. Apart from offering technical services, a virtual power plant offers commercial services by participating in wholesale and energy reserve markets. By enhancing the visibility of small, distributed energy resources (DERs), they are capable of offering pecuniary services by managing spot pricing (SP) and time of use (TOU) pricing schemes. The simple characterization of DER is any energy resource that is connected to the grid at the distribution level, viz., fuel cells, captive power plants, natural gas turbines, electric vehicles (EVs), and energy storage (ES).

Virtual power plants can certainly be a supporting system for implementing 5-min bidding and can respond to various rapid and fast-moving markets through various DERs. Some of the diverse challenges which can be answered by a VPP are listed below.

(a) A VPP diminishes the need for the conventional generation to provide the provision of dynamic ancillary services.
(b) It controls a cluster of heterogeneous renewable energy resources (RERs).
(c) The intermittency and uncertainty caused by renewables such as solar and wind power, which are highly weather dependent, can be reduced to a certain extent.
(d) It maintains favorable grid conditions for real-time management and supervision in emergencies.

There are various other requirements to make the operation of VPPs more robust including ensuring cyber security which is a very important concern due to the involvement of highly advanced software. Information and communication technologies (ICTs) are also a prerequisite for reliable communication. Integrated resource planning (IRP), peer-to-peer (P2P) transactions, net-metering policy (NM), and behind the meter (BTM) technology are some of the regulatory mechanisms which play a key role in deciding the future of this technology.

A case study was carried out by P. Pal et al., Ref. [5] in which a PV panel, fuel cell, wind turbine, micro-turbine, and battery-connected energy storage system were connected to analyze the optimal scheduling of a VPP. The setup was made in such a way that three kinds of scenarios were considered using a beetle search antenna algorithm for optimal dispatching. The authors compared the performance of this method with algorithms such as particle swarm optimization (PSO) and genetic algorithm (GA). Some of the important parameters such as load analysis on an hourly basis and dynamic pricing of the grid were utilized and implemented for the day-ahead market strategy. S. Han et al., Ref. [6] highlighted the benefits of a VPP considering incentive prices and load data on an annual basis to assess the revenue generated. The profit evaluation of a VPP involves load filling as well as load shaving. A VPP requires the lowest investment cost to achieve the same load shaving effect when compared to gas power, coal power, pumped storage, and energy storage as well. The operating cost for the demand response (DR) is very low compared to an old-style power plant. The overall feasibility of the development of VPPs is verified by conducting simulation studies from the perspective of the power grid. P. Lombardi et al., Ref. [7] presented an article in which a synchronized measurement was implemented in the state estimation algorithm which could turn out to be a solution of equations strictly related to state estimation. They selected a CHP, photovoltaic plant, and wind park to supply energy and therefore can be called energy suppliers. An external boiler was also included in the analysis to manage the surplus electricity. A CHP unit was included in the study as it is a useful source by which the power load can be managed efficiently. To ensure the optimal operation of the VPP, forecasting errors that arise due to the intermittent nature of RERs and unpredictable weather need to be reduced as much as possible. S. M. M. Saabit et al., Ref. [8] explored why the concept of Vehicle-to-Home (V2H) has been given more emphasis instead of V2G and G2V. Three approaches were well-thought-out and include the grid-connected PV with battery storage system (BSS), grid-connected battery storage systems without PV, and just a battery storage system. The intention was to highlight the capabilities of batteries when they are installed at retail sites and eventually set up a VPP. Two types of load profiles were analyzed: are flat consumption and varying load profile. The Perturb and Observe technique was employed to track the maximum power point. The settlement period of the proposed system was estimated to be 9 years to overcome the installation cost. A hybrid energy generation system was used to resolve the problem of optimal scheduling which consisted of battery-connected solar PV modules, wind farms, and thermal generators in S. S. Reddy et al., Ref. [9]. A two-point estimation system and genetic algorithm were used to test the efficacy of their proposed strategy. Based on the simulation studies carried out, it could be observed that there was a marginal enhancement in the generation scheduling of the day-ahead strategy. In Ref. [10], Y. Zhang et al., attempted to propose the scheduling strategies of a VPP that can accommodate various development phases in the electricity market. A bi-directional context was proposed in the paper so the VPP participates in the electricity markets and acts only as a price-taker. Apart from raising the overall revenue, it also focused on the objective of social welfare which can be achieved by the effective allocation of resources.

To solve the unbalancing problem, T. Zhang et al., Ref. [11] proposed a VPP optimal scheduling model. They studied the energy cost model of the VPP and developed an optimal dispatch strategy in which the uncertainty of the energy prices and variable output of RESs were included. The VPP structure comprised small-scale wind power plants, solar PV systems, and gas turbine power plants along with energy storage systems. Few types of VPPs, when participating in the electricity market, are briefly discussed, namely the Joint-Venture model, Bi-lateral Transaction model, and medium-long term contract model. They considered the overall power balance without any network constraints. A small-scale VPP model was comprised of two DG sources and two controllable loads, and the same was verified in MATLAB/ Simulink by Naina et al., Ref. [12]. An algorithm of an energy management system that acts as a centralized controller operates in three modes, i.e., grid import (operated in an off-peak hour), grid export (operated in peak hours), and no power exchange mode (power exchange does not take place between the grid and VPP). Three different models were included in this analysis, the main grid, dynamic load, and distribution generation model. In this study, consecutive energy management (CEM) is proposed to make the regular energy management system more economic by satisfying the electric constraints of the power system. The control objective was to develop a framework so that the VPP can contribute to the energy market and at the same time cater to ancillary services.

A. Zahedmanesh et al., Ref. [13] discussed two hierarchies; i.e., the first involves a daily scheduling approach while the second is reactive power compensation which is a very crucial requirement of a power system. As per the claim, both voltage quality and energy cost were improved by employing CEM in the analysis without violating any electric constraints. The power of the conventional generators relies on the quantity of energy produced by renewable energy sources. The storage facility plays a prominent role in case the load shaving factor increases. The optimal design of the storage facility is highly recommended by Lombardi P. et al., Ref. [14] so that the system can balance the intermittent generation. The total demand for analysis is taken as 1 MW which is fulfilled by conventional generators and partially by renewable energy resources such as solar and wind. It was observed that the shaving factor has a very minimal impact on the optimum storage capacity only if the quantity of energy generated by the RERs is high.

S. Mishra et al., Ref. [15] offered an innovative business model in which consumers could be given full control over how much power they require, giving them the liberty to utilize power per their requirements and needs, and is known as energy-as-a-service (EAAS). This model is empowered by peer-to-peer (P2P) energy exchange especially designed for the local power markets. They also proposed a novel computation strategy in which a comparison was made between profit-based ordering and random selection to conclude the most preferred strategies for an energy-related transaction. A design that is termed a smart contract was discussed which allows a particular system to be independent and reliable for handling exchanges when multiple sectors are involved. This study aims at a multi-stage PSO method to improve energy penetration along with small-scale signal stability by T. K. Renuka et al., Ref. [16].

The proposed technique was tested on the IEEE 14-bus system and the results were validated. This bus system contained three synchronous generators and turbine governors. Real-time coordination, as well as a controller for the active supervision of real and reactive power in all the grids, are necessary for addressing the intermittency issues and increased penetration of renewable energy. It has been emphasized that realistic control strategies can be obtained by optimization studies to perform an integrated operation O. H. Mohammed et al., Ref. [17] defined an economic problem that consists of the optimal sizing of the system, state of charge (SOC) of the battery, and high reliability of the system. The characteristics and advantages of the PSO are highlighted over various conventional algorithms. The objective function consists of the total net present cost with the sole intention to optimize the generated power of a hybridized renewable energy system comprised of PV modules, batteries, tidal, and wind turbines. The important aspects of the batteries such as the floating charge voltage and their maintenance are considered to achieve the desired optimization. The problem that is being explored in this study is more economical and converted into a multi-objective problem in which the purpose is to minimize the cost of energy and the total net present cost (TNPC) of the system without compromising the flexibility and versatility of the hybrid energy system.

Pio Lombardi Ref. [18] optimized a multi-criterion-based VPP that was autonomous. This study involved three main criteria, i.e., service reliability, the cost associated with the system, and the quantity of pollution. In the desired load management program, the VPP comprised a wind farm, PV plant, and CHP in which the prime movers were a gas turbine, active consumers, and a battery switch station. In addition, the authors performed a sensitivity analysis to verify that the optimal solution is robust. The autonomous VPP is economically competitive as the total cost generation is less than that compared to traditional power systems. M. F. Dehghanniri et al., Ref. [19] examined the involvement of a VPP in the real-time market, day-ahead market, and reserves. The sources considered were wind turbines, combined heat and power, diesel generators, and electric vehicles, with electrical and thermal as their two storages. To maximize the profit, two-stage planning was used. In the first stage, optimization of DA and reserves was accomplished, followed by optimizing the real-time market. The simulation was carried out on an IEEE 21 bus network to assess the performance evaluation of the VPP. They emphasized two important parameters, namely, the inclusion of EVs and the price sensitivity of the load. An artificial neural network was used for forecasting the data and planning the required number of EVs required for day-ahead scheduling.

M. Khandelwal et al., Ref. [20] dealt with the impact of locational marginal prices on a VPP in terms of resource allocation by satisfying the constraints related to network flow within the limits. To enhance the profit by a considerable margin, technical and market aspects were discussed. The problem formulation of profit was created in such a way that it was the difference between the cost of energy and revenue from market trading. To analyze the impact of aggregation, 24-h scheduling was considered. M. Gough et al., Ref. [21] focused on the technical and grid-related constraints. They developed a VPP that dealt with technical details rather than commercial or financial outcomes. The impact of voltage profile, power losses, and network congestion was analyzed along with the thermal relief of the consumers. The obtained profit was split into two parts, namely, revenue from the electricity sold to commercial clients and the cost of functioning the technical VPP in consideration of both economic and technical constraints. S. Hadayeghparast et al., Ref. [22] projected a typical model for optimizing a VPP's day-ahead scheduling, consisting of power dispatching and unit commitment (UC). The multi-objective approach deals with capitalizing on the day-ahead net daily turnover of the VPP and curtailing the pollutants and daily emissions. In this study, the VPP was assumed to control all resources in the local network, including loads and DERs. The scenario-based approach was used for modeling the uncertainty of the market price, solar radiation, and electrical load.

Based on the critical review carried out in the literature survey, it was found that the problem statement requires an advanced meta-heuristic technique that can handle the objective function in a simpler yet efficient way by reducing the complexities which are unavoidable due to the non-linear nature of the problem when solved by established techniques such as PSO, GA, etc. The below-mentioned points are anticipated as research gaps that were not given the attention they require in the explored literature.

- To ensure efficient management of the grid, sources such as fuel cells and CHP can be considered for optimal scheduling to reduce the cost of power generation along with emissions in a VPP system.
- To handle a non-convex problem such as VPP efficiently, advanced, and recently developed soft computing (SC) techniques can be implemented or modified by choosing related constraints.
- To incentivize the participants, peak-valley pricing mechanisms with the incorporation of (15 min) interval scheduling is introduced and compared with day-ahead scheduling.

In Table 1, various control methods, mostly used in the work reported in the literature, to carry out optimization are listed. MOBWO is the only technique, to the best of our knowledge, that is not being employed to carry out the intended VPP problem. The control method is mentioned for the existing studies along with its characteristics and the nature of the problem. The limitations of the existing work and advantages of the anticipated MOBWO algorithm, along with the improvements, are highlighted over other techniques.

Table 1. Control methods employed for optimal scheduling purposes.

Refs. No.	Nature of Problem	Control Method	Features of Control Method
[22,23]	Heuristic	PSO/MOPSO	Fewer parameters. Ease of implementation. Local entrapment.
[24,25]	Stochastic	ABC	Poor in exploitation stage. Limited population diversity.
[26,27]	Computational	ANN	More precise predictions. Good computational efficiency.
[28,29]	Heuristic	GA	Can determine multiple solutions simultaneously.
[30,31]	Meta-heuristic	ACO	Can discover good solutions rapidly.
[32,33]	Mathematical	Fuzzy Logic	Improved prediction accuracy. Use of Fuzzy sets.
[34,35]	Mathematical	Game Theory	Computational load increases as the no. of participants increases.
This paper	Meta-heuristic	BWO/MOBWO	High searching accuracy. Better updating strategy. Converge to the global optimum in lesser iterations.

Virtual power plants are gaining interest very rapidly in the new era of energy management to have better management of the associated resources in modern electrical systems. Even though a lot of researchers are showing interest in exploring the domain of VPPs to know their feasibility and economic viability, all the studies related to optimal scheduling, which were carried out previously, were only day-ahead types, and emission was not given the kind of attention it requires, especially in today's scenario in which a considerable reduction in emissions is needed to contribute towards sustainable living. This study deals specifically with the optimal scheduling of a VPP considering both day-ahead scheduling and 15 min scheduling which shows close resemblance to the real-time scenario.

In this article, an attempt has been made to carry out multi-objective optimal scheduling and the key features of this research work are highlighted below:

- One centrally controlled VPP system comprised of multiple resources including solar PV modules, WT, fuel cells, electric loads, heat-only units, and CHP units has been attempted to solve the multi-objective optimal scheduling problem.

- The multi-objective optimal scheduling of the VPP considering renewable resources has been solved using the weighting factor method to simultaneously maximize profit and minimize emissions.
- Peak valley's power pricing strategy is introduced in the multi-objective optimal scheduling of the VPP problem.
- The new price-based multi-objective black widow optimization (MOBWO) is presented and implemented by considering constraint handling.
- Statistical analysis was performed for both single and multi-objective optimal scheduling of the VPP problems and quality solution sets were obtained from the MOBWO algorithm after 100 different independent trials.
- Pareto optimal solutions were obtained specifically for multi-objective optimal scheduling of the VPP problem for the maximization of profit along with simultaneously minimizing the emissions for both scenarios I and II, respectively.
- Results obtained by the proposed MOBWO algorithm were also compared with the latest published works.

The organization of this research article is as follows:

Section 2 deals with the basic concepts, challenges, and framework of VPP. Section 3 presents problem formulation which comprises the objective function, followed by the constraint handling. The optimization algorithm-related explanation is then discussed in Section 4. The implementation part of the case studies along with simulation outcomes is presented in Section 5, and, lastly, concluding notes followed by future work highlights are given in Section 6.

2. VPP Concept

A VPP facilitates the synchronization of power generation and uses more efficiently. It provides a sustainable supply and demand adjustment mechanism with a high level of precision and encourages the storage of electricity that generates the capacity for the usage of renewable energy in the power division. As a result, there is a significant potential for the VPP to link operating technologies with communications infrastructure and external data properties, thus collecting the forecasting data from scattered entities. In addition, the VPP can provide deep insights into the results by offering smoother and quicker decision-making and taking real-time action to enhance performance. The framework of the VPP followed is depicted in Figure 1.

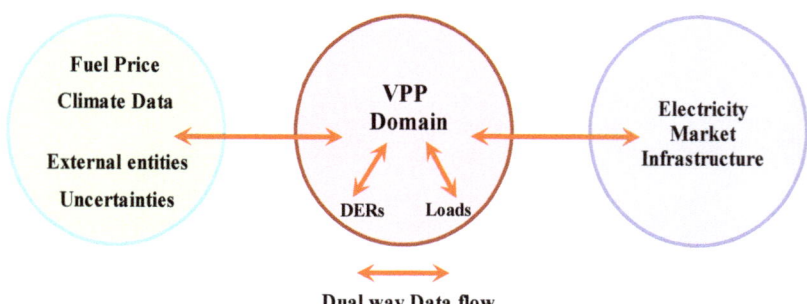

Figure 1. The Framework of a VPP.

Several challenges involved when considering a VPP can be categorized in terms of technical, commercial, and regulatory restrictions, followed by environmental concerns which are often ignored and not given the importance that it deserves. In Figure 2 some of the most prominent challenges of the VPP are highlighted.

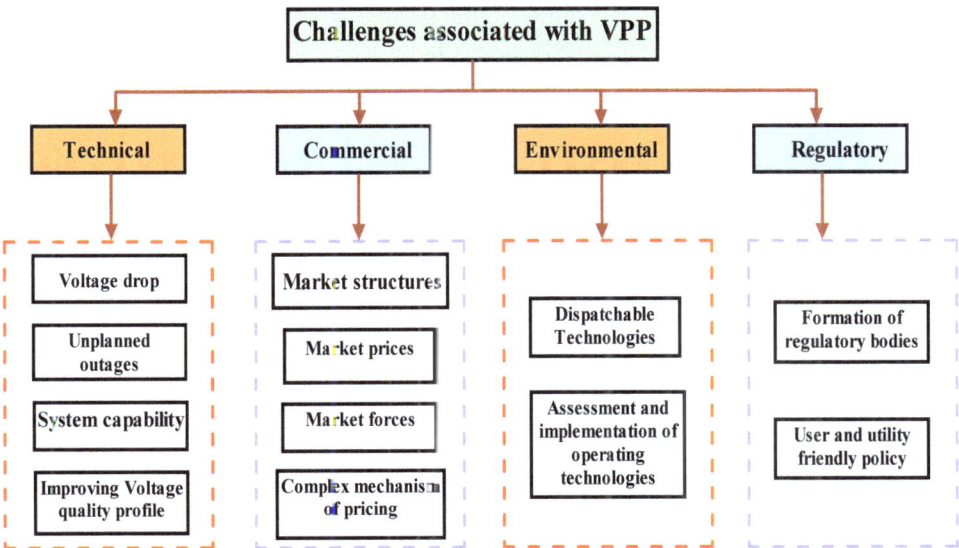

Figure 2. Different aspects of VPP.

The crux of a VPP is that it is not restricted geographically, unlike a Microgrid. A VPP is more concerned with flexible resources and indulges in power trading in the energy market and their mode of interconnection is always grid-connected Ref. [36].

Methodology

The process flow followed is demonstrated in Figure 3 in which the detailed organization is sequenced to make it easy for the readers looking for optimum operation and scheduling of a virtual power plant considering multi-objective profit maximization and emission minimization. The optimization is carried out with respect to the constraints handling which also includes the technical constraints of the associated resources. The methodology starts with the selection of resources, i.e., solar PV, wind, and combined heat and power, which comprise heat-only units, followed by the fuel cell. The next step after selecting appropriate resources for the optimal scheduling of the virtual power plant is the collection of data in which raw data are obtained for renewable sources such as wind and solar power and a few secondary data taken from the literature. The next step is to set up the computational framework in which a suitable selection of an advanced meta-heuristic technique is performed, keeping the requirements into consideration. In the next stage, a scheduling strategy is adopted in which two scenarios are considered.

In scenario I, day-ahead scheduling, i.e., 24 h, is performed followed by a 15-min interval which is scenario II. The multi-objective case is performed for both scenarios I and II. Based on the optimization algorithm, a detailed statistical analysis has been accomplished and highlighted in tabular form and the same is compared with another optimization technique. The superiority and effectiveness of the same are evident in the form of numerical results. The convergence characteristics are displayed for all three cases, respectively, and the two scenarios are followed by the Pareto graph which helps the VPP operator in deciding the best available trade-off in terms of monetary advantages as well as emission considerations that helps in achieving the sustainable development goals which is one of the sole purposes of a VPP.

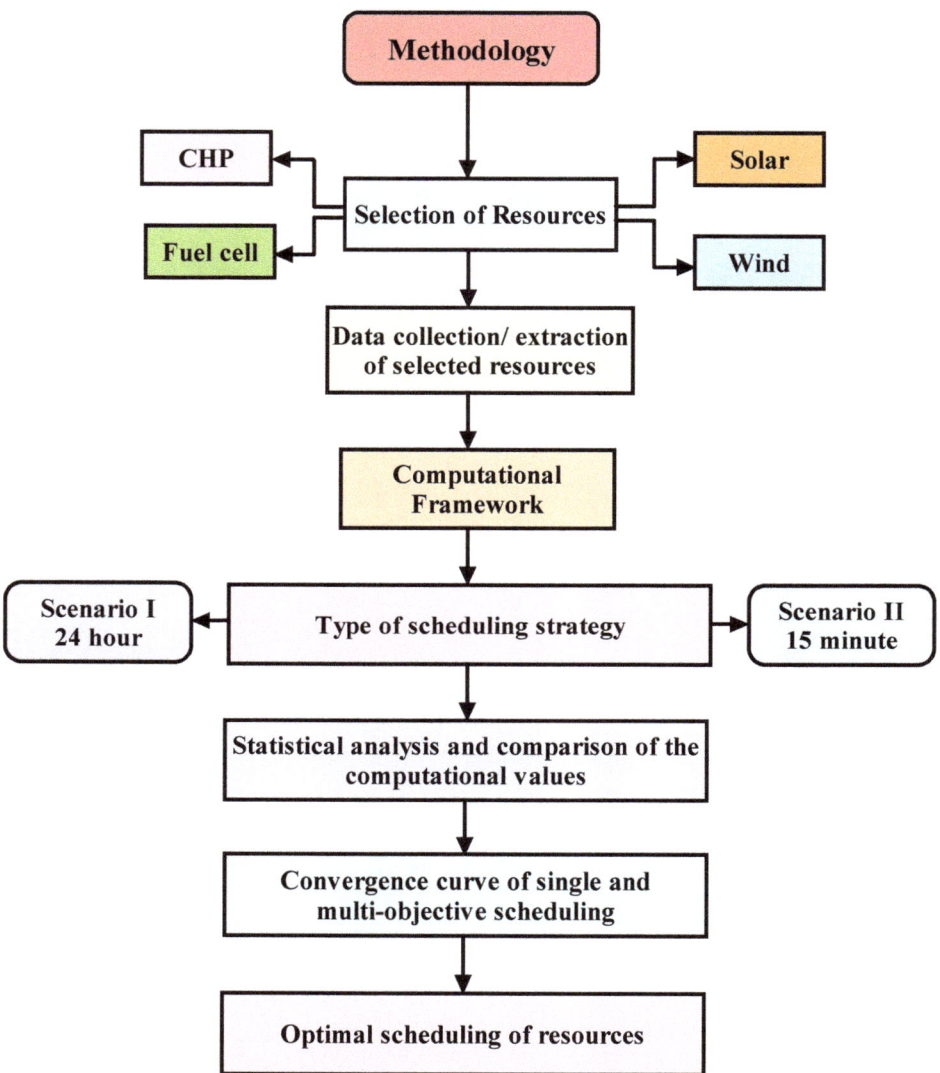

Figure 3. Methodology and process flow of the proposed strategy.

In Figure 4, a pictorial representation of the envisioned system under study is represented in which renewables and co-generation units are considered. The resources associated with the current VPP system under study are solar photovoltaics (PV), wind turbine (WT), fuel cell (FC), combined heat and power (CHP), and electric load (EL), followed by connecting the energy market (EM) and electricity price (peak valley) to the VPP operator. These resources are used to supply the electricity to the consumers of VPP based on the transactions between the energy market and the VPP operator. The energy market, which is indeed an integral feature of the VPP, is also associated to make the delivery of peak valley electricity pricing when applicable.

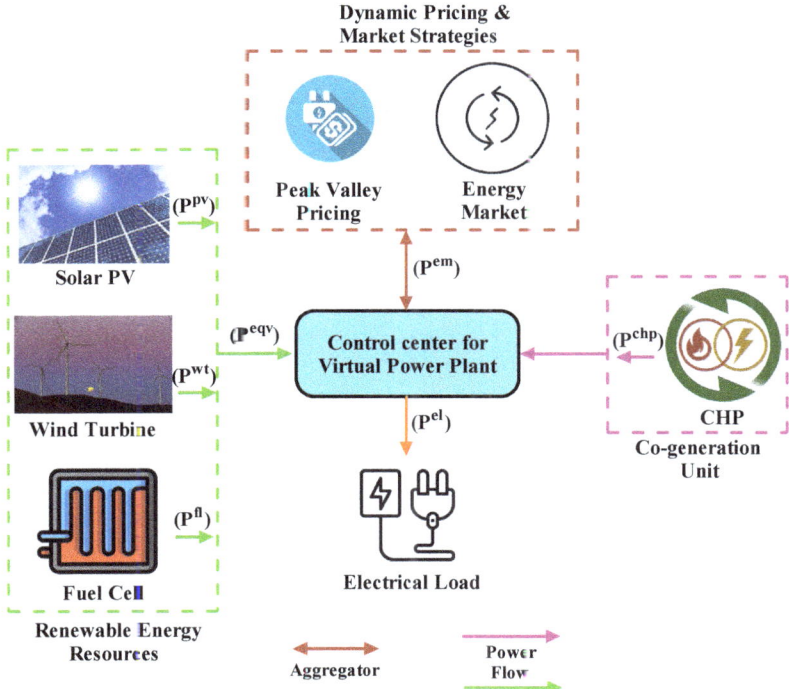

Figure 4. The proposed VPP system under study.

3. Problem Formulation
3.1. Objective Function

The objective function is segmented into two categories. The first deals with the maximization of the net profit and is followed by the minimization of total emissions generated by the associated resources, specifically with CHP and heat-only units.

3.1.1. Net Profit

The running costs of the system consists of power acquisition costs from the main grid, power selling costs to the main grid, and costs associated with wind generators, solar power, and fuel cells. The installation cost of the wind turbine and the photovoltaic array is not considered. The objective function is categorized into two segments, i.e., the main objective function deals with the maximization of net profit followed by the other objective function which is emission minimization.

Net Profit =

$$\text{maximization} \sum_{s=1}^{Ns} \pi s \times \Sigma_{t=1}^{t} \left\{ -\sum_{p=1}^{Np} \left\{ c_{chp} + c_{hou} \right\} \left\{ \rho_{em} \times c_{ph} + c_{se} \right\} \sum_{p=1}^{Np} \left\{ (1 \times c_{ph}) + (1 \times p_{hou}) - c_{ens} \right\} \right\} \quad (1)$$

where:

p and s are set of plants and scenarios, t is time ranges from 1 to 24 in Day-ahead scheduling, followed by 1 to 96 in 15-min scheduling.

c_{ph} and c_{se} are the tariffs for purchasing and selling power from the grid system.

π_s is the probability of scenarios.

p_{em} is the energy market price; c_{ens} is the cost of energy not served.

c_{chp} and c_{hou} are the cost function of CHP and heat-only units.

p_{hou} is the price of heat-only units.

3.1.2. Emission

Along with the main objective function of net profit represented in the above Equation (1), the emission is also considered as a supportive objective function in which minimization of the day-ahead emission is carried out to see the effect of the statistical analysis performed in Section 5.

$$\text{Emission} = \text{minimization} \sum_{s=1}^{Ns} \pi s \times \Sigma_{t=1}^{t} \left\{ \sum_{p=1}^{Np} \left\{ e_{chp} + e_{hou} \right\} \times \left\{ e_{ph} + e_{se} \right\} \right\} \quad (2)$$

where:

e_{chp} and e_{hou} are the emissions by CHP and heat-only units, respectively.

e_{ph} and e_{se} are the emissions by the grid system.

The total emissions by CHP and heat-only units, and the purchase and selling power coming from the main grid system can be evaluated by the below-mentioned Equations (3)–(5).

$$e\_chp = (NO_x^{chp} + SO_2^{chp} + CO_2^{chp}) \times P_{t,p}^{chp} \quad (3)$$

$$e\_hou = (NO_x^{hou} + SO_2^{hou} + CO_2^{hou}) \times H_{t,p}^{hou} \quad (4)$$

$$e\_ph + e\_se = (NO_x^{ph+se} + SO_2^{ph+se} + CO_2^{ph+se}) \times max\{-P_{t,p}^{ph+se}, 0\} \quad (5)$$

In the aforementioned Equation (5), the measure of emissions related to the grid system is considered only when the electricity is purchased from the power market.

3.1.3. Multi-Objective Framework

The objective function of the multi-objective optimal scheduling of the VPP problem is to handle objectives, namely, the maximization of profit and simultaneous minimization of the emission in such a way as to obtain the best compromise solution. Multi-objective optimal scheduling of the VPP problem is solved by using the weighting factor method and the mathematical expression is given by:

$$Fitness = w \times Net\ Profit + (1 - w) \times Emission \quad (6)$$

where:

w is considered as 0.5 for giving equal weightage to both objectives.

3.2. Constraints Handling

3.2.1. Power Balancing

The electrical power balance is represented in Equation (7) in which $P_{s,t,p}^{eqv}$ depicts the equivalent electrical output power of each plant.

$$P_{s,t,p}^{eqv} = P_{ex}^i + P_{fl}^i + P_{wt}^i + P_{pv}^i - P_{el}^i \quad (7)$$

where:

$P_{s,t,p}^{eqv}$ is equivalent to power scenario s, time t, and plant p.

P_{ex}^i is the exchanging power between the main grid and the CHP system at interval i (MW).

P_{fl}^i is the power of the fuel cell at interval i (MW).

P_{wt}^i is the power of the wind turbine at interval i (MW).

P_{pv}^i is the power of the solar photovoltaic at interval i (MW).

P_{el}^i is the electrical load at interval i (MW).

3.2.2. Heat Balancing

Constrictions related to the heat balancing equation must satisfy the waste heat and gas boiler.

$$P_{fl}^i * r_{fl}^i * \eta_{hr\_bl} + P_{gb}^i - P_{th}^i = 0 \tag{8}$$

where:

r_{fl}^i is the ratio of heat to the electricity of the fuel cell at interval i (MW).

η_{hr_bl} is the efficiency of the heat rate boiler (MW).

P_{gb}^i is the power of a gas boiler at interval i (MW).

P_{th}^i is thermal power balance at interval i (MW).

3.3. Power Switching between Main Grid and CHP Units

The operational constraints between the CHP and main grid are expressed in the following equation. The switching power which takings place between these two is within the permitted limits.

$$P_{ex}^{min} \leq \left| P_{ex}^i \right| \leq P_{ex}^{max} \tag{9}$$

where:

P_{ex}^{min} is the minimum exchange of power between the main grid and the CHP system.

P_{ex}^{max} is the maximum exchange of power between the main grid and the CHP system.

3.4. Constraints of Waste Heat and Gas Boiler

The waste heat and gas boiler are able to generate power in their precise electrical capacity.

$$P_{bl}^{min} \leq P_{fl}^i * r_{fl}^i * \eta_{hr\_bl} \leq P_{bl}^{max} \tag{10}$$

$$P_{gb}^{min} \leq P_{gb}^i \leq P_{gb}^{max} \tag{11}$$

where:

P_{bl}^{min} is the minimum limit of the waste heat boiler.

P_{bl}^{max} is the maximum limit of the waste heat boiler.

P_{gb}^{min} is the minimum limit of the gas boiler.

P_{gb}^{max} is the maximum limit of the gas boiler.

3.5. Fuel Cells

The efficacy of fuel cells can be expressed in the part-load ratio (PLR). The modeling is adapted from Ref. [37]. The mathematical formulations associated with the function of PLR are defined in Equations (12)–(14). The fuel cell units are capable of supplying part of the demand load in the form of electrical energy.

When PLR_i < 0.05

$$\eta_{fl}^i = 0.2716; \; r_{fl}^i = 0.6816 \tag{12}$$

When PLR_i ≥ 0.05

$$\eta_{fl}^i = 0.9033 PLR_i^5 - 2.9996 PLR_i^4 + 3.6503 PLR_i^3 - 2.0704 PLR_i^2 + 0.4623 PLR_i^1 + 0.37 \tag{13}$$

$$r_{fl}^i = 1.0785 PLR_i^4 - 1.9739 PLR_i^3 + 1.5005 PLR_i^2 - 0.2817 PLR_i^1 + 0.6838 \tag{14}$$

where:

η_{fl}^i is the fuel cell efficiency at interval i (p.u.) and

r_{fl}^i is the ratio of heat to the electricity of the fuel cell at interval i (MW).

Ramp Rate Limit of Fuel Cell

The power generated by the P^i_{fl} unit is not allowed to exceed the power generated in the previous interval P^{i-1}_{fl} by more than a specific amount.

$$\Delta P_{fl\_down} \cdot T \leq P^i_{fl} - P^{i-1}_{fl} \leq \Delta P_{fl\_up} \cdot T \tag{15}$$

where:

$\Delta P_{fl_{up}}$ T is the up-ramp limit and ΔP_{fl_down} is the down-ramp limit of the fuel cell.
P^i_{fl} is the power generated by the fuel cell at interval i (kW).
P^{i-1}_{fl} is the power generated by the fuel cell at the previous interval (kW).

3.6. CHP Units

To achieve optimal scheduling, two categories of CHP units having diverse feasible regions of operation (FOR) have been considered Ref. [38]. Equations (16) and (17) represent the maximum electric and thermal output power constrictions of the CHP unit.

$$0 \leq P^{chp}_{t,p} \leq P^{chp}_{p,A} \times V^{chp}_{t,p} \tag{16}$$

$$0 \leq H^{chp}_{t,p} \leq P^{chp}_{p,B} \times V^{chp}_{t,p} \tag{17}$$

where:

$P^{chp}_{t,p}$ is the electrical output power of the CHP and
$V^{chp}_{t,p}$ is the commitment status of the CHP.

The feasible regions of operation (FOR) are shown in Figure 5. In the Type 1 CHP unit FOR is a convex type which is followed by a non-convex FOR as shown in Type 2.

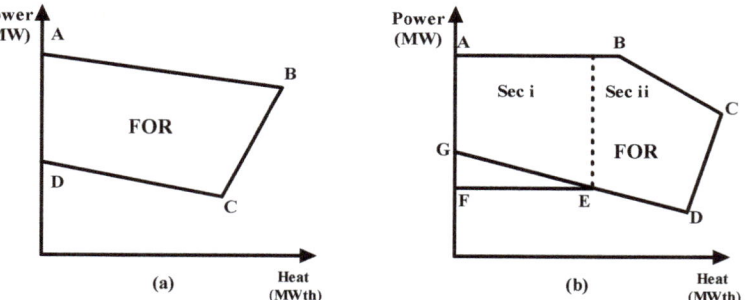

Figure 5. Feasible operating region (**a**) Type 1, (**b**) Type 2.

3.7. Solar PV Modules

The output of the photovoltaic modules is affected largely by solar radiation, characteristic of the module, followed by the ambient temperature of the specific location Ref. [39]. The availability of solar power is plentiful in the daytime, which can be seen in Figure 6 in which the peak ranges from 11 a.m. to 3 p.m. Usually, it follows the beta probability distribution function (PDF).

Certain parameters influence the output power of PV modules. The fill factor (FF) is the measure of the efficiency of a photovoltaic module. The ideal FF of a solar cell is around 0.7. It is defined as the maximum power obtainable from the solar module to the actual power obtained and it is expressed in Equation (18).

$$FF = \frac{V_{MPP} \times I_{MPP}}{V_{OC} \times I_{SC}} \tag{18}$$

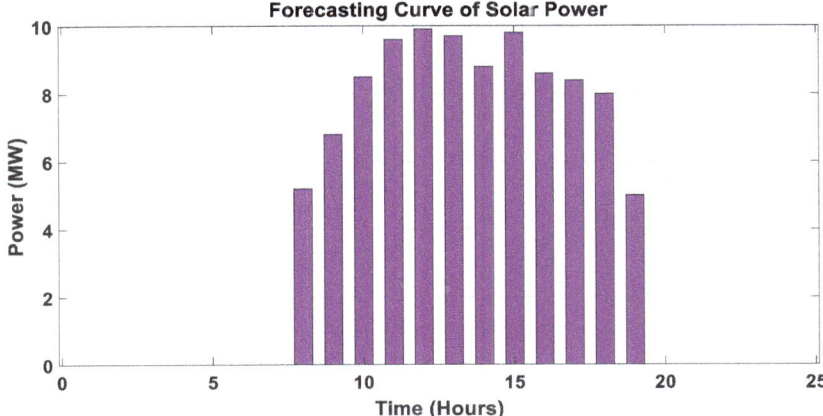

Figure 6. Forecasting curve of solar PV modules.

The characteristics of the module are represented below:

$$P_{s,t,p}^{pv}(sor_{s,t}) = N_p^{pv} \times FF \times V_{s,t} \times I_{s,t} \quad (19)$$

$$V_{s,t} = V_{OC} - K_V \times T_{Cs,t} \quad (20)$$

$$I_{s,t} = sor_{s,t} \times [I_{sc} + K_i \times (T_{Cs,t} - 25)] \quad (21)$$

where:

$P_{s,t,p}^{pv}$ is output power, N_p^{pv} is the number of PV modules, and $sor_{s,t}$ is the solar radiation (KW/m^2).

V_{OC} is the open-circuit voltage and I_{sc} is the short circuit current.

K_i is the current temperature and K_V is the voltage temperature coefficient.

$T_{Cs,t}$ is the solar cell temperature and FF denotes the Fill Factor of the PV module.

3.8. Wind Turbine

Wind power follows the Weibull PDF due to its ability to represent the variation in wind speeds. The forecasting curve of wind power generation with respect to time is shown in Figure 7. The primary problem with wind power is the rapid fluctuation due to varying climate conditions.

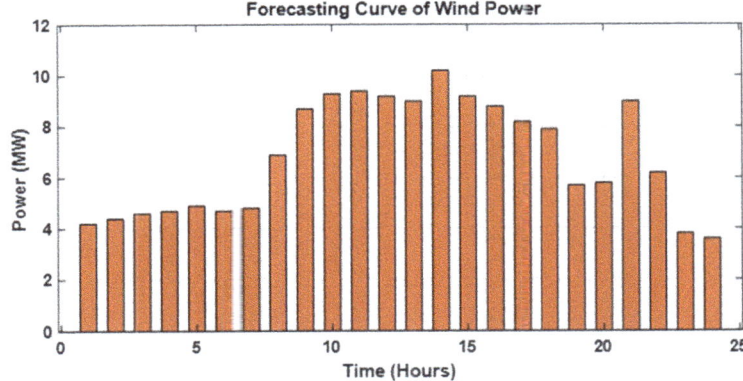

Figure 7. The forecasting curve of wind power generation.

The power from the wind turbine is calculated from Equations (22)–(25). The modeling of the wind turbine is referred to from Ref. [40].

$$P_{s,t,p}^{wt}(v_{s,t}) = N_p^{wt} \times \{0, v_{s,t} < v_{in}^c\}\{0, v_{s,t} > v_{out}^c\} \quad (22)$$

$$P_{s,t,p}^{wt}(v_{s,t}) = N_p^{wt} \times \left\{ P_{rated}^{wt} \times \left(\frac{v_{s,t} - v_{in}^c}{v_{rated} - v_{in}^c} \right)^3 \right\} \quad (23)$$

$$P_{s,t,p}^{wt}(v_{s,t}) = N_p^{wt} \times \{v_{in}^c \leq v_{s,t} \leq v_{rated}\} \quad (24)$$

$$P_{s,t,p}^{wt}(v_{s,t}) = N_p^{wt} \times \{v_{rated} \leq v_{s,t} \leq v_{out}^c\} \quad (25)$$

where:

$P_{s,t,p}^{wt}$ is the output power of the wind turbine (MW).
P_{rated}^{wt} is the nominal power of the wind turbine (MW).
N_p^{wt} is the number of wind turbines.

The crucial parameters considered to evaluate the intended objective function are listed in Tables 2–6 and referred from [22,40–43].

Table 2. Emission factors related to SO_2, NO_X, and CO_2.

Emissions	Heat-Only Unit	CHP Unit
SO_2	0.0027	0.0036
NO_X	0.3145	0.1995
CO_2	401.43	723.94

Table 3. Parameters for the wind turbine.

P_{rated}^{wt} (MW)	v_{in}^c (m/s)	v_{out}^c (m/s)	v_{rated} (m/s)	N_p^{wt}
150	3.5	25	13.5	3

Table 4. Parameters for solar PV.

V_{OC} (V)	I_{SC} (A)	K_i (I/°C)	K_V (V/°C)	N_{OT} (°C)	I_{MPPT} (A)	V_{MPPT} (V)	N_p^{pv}
21.98	5.32	0.003	0.0144	43	4.76	17.32	2240

Table 5. Parameters for heat-only units.

$H_{max,p}^{hou}$ (MWth)	a_p ($/MWth^2)	b_p ($/MWth)	c_p ($)
1.2	0.052	3.0651	4.8

Where a_p, b_p, c_p are the cost coefficients of heat-only units.

Table 6. Parameters of the CHP unit.

g_p ($/MW^2)	h_p ($/MW)	i_p ($)	j_p ($/MWth^2)	k_p ($/MWth)	i_p ($/MW.MWth)	C_p^{su} ($)	C_p^{sd} ($)
0.0345	44.5	26.5	0.03	4.2	0.031	20	20

Where $g_p, h_p, i_p, j_p, k_p, i_p$ are the cost coefficients of CHP units and C_p^{su} and C_p^{sd} are the startup and shutdown costs.

The complete procedure adopted for solving the optimization problem is mentioned in Figure 8. All the mathematical equations involved in this study are highlighted in the flowchart.

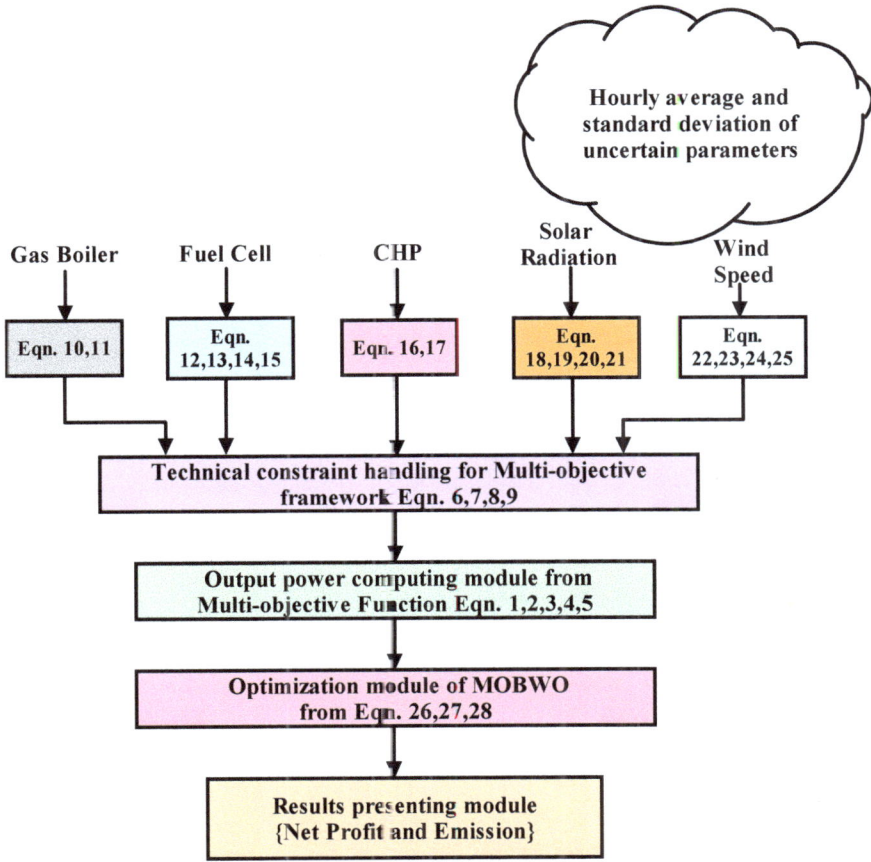

Figure 8. Process flow for the adopted methodology.

There are uncertain factors that affect the stability of the VPP system due to the uncertain nature of renewable energy generation, fluctuation in market prices, and varying load demand. These three aspects are crucial in overcoming the barrier caused by these uncertainties and can be managed by incorporating methods, viz., the Monte Carlo simulation (MCS), robust optimization (RO), and auto-regressive integrated moving average (ARIMA) probabilistic and possibilistic methods to name a few Ref. [44]. The focus on carrying out the uncertainty aspect is out of scope for the current work, and it is left for future work.

4. Optimization Algorithm

The nature-inspired algorithms are the best way to solve the selected non-convex type problems that come with lesser mathematical complexities and an efficient way to ensure the reachability of a global optimum value. To carry out the intended optimization, the MOBWO was selected over other algorithms due to its unique ability to overcome the local optima trap. It offers numerous search agents to estimate the global optimum, which is remarkable in this recently developed highly advanced optimization technique. The two most important aspects of every optimization are exploration and exploitation. Producing numerous offspring enhances the exploration of search space, followed by omitting the unfeasible solutions to move toward the best possible solution. The resulting early convergence is certainly a trait of selected optimization which is missing in most of the well-established nature-inspired techniques such as PSO and GA.

The system is initialized with a population of random spiders and searches for global optima by updating the population. The BWO algorithm attempts to solve an objective function by generating a mutation population (mute pop) and mutation variables (mute vars) Ref. [45]. The BWO concept consists of terminologies in which the search agents in the form of the widow are assumed to solve any specific problem in the same way that the crossover and mutation operator does in GA, followed by the population of particles in PSO.

4.1. Population Initialization

The array of the search agent is $1 \times M_{var}$ and the dimensional value is M_{var} in the optimization problem expressed in Equation (26).

$$Searchagents = [k_1, k_2, k_3 \ldots k_M \, var] \qquad (26)$$

where:

$k_1, k_2 \ldots k_m$ are the floating numbers in the form of variables.

A fitness search agent in the form of the widow is determined using the fitness function M expressed in Equation (27).

$$Fitness = f(Searchagents) \qquad (27)$$

4.2. Procreation, Cannibalism, Mutation

The BWO algorithm starts with selecting the parents which is known as the procreate stage. This step is very much necessary to start the exploration in the search area to avoid the local optima trap. To reproduce in BWO algorithms, the array referred to α is created and consists of random numbers which contain the offspring produced by Equation (28) in which x_1, x_2 denote parents and O_1, O_2 refer to the offspring.

$$O_1 = \alpha \times x_1 + (1-\alpha) \times x_2 \text{ and } O_2 = \alpha \times x_2 + (1-\alpha) \times x_1 \qquad (28)$$

where:

O_1 and O_2 are the offspring.
x_1 and x_2 are the parents.
α is the array matrix.

The process is repeated for M_{Var} two times, in which duplication should be avoided to enhance the accuracy of the fitness value among the pairs. The next step is the cannibalism rate which ensures better performance for the exploitation and guarantees faster convergence at the same time for BWO. Every operator corresponds to a contender solution to the given problem. The mutation stage is considered to bring the balance between both exploration and exploitation.

The performance of a BWO depends on its parameters such as reproduction rate (RP), cannibalism rate (CR), mutation rate (MR), and, of course, the lower bound (LB) and upper bound (UB). Appropriate selection of the controlling parameter, which is the cannibalism operator, may ensure superior performance of the exploitation stage by disregarding the unfitting individuals from the population, resulting in a lesser number of iterations to reach the optimum solution. The flowchart of the proposed MOBWO algorithm is shown in Figure 9 and an effort has been made to simplify the process of obtaining the optimum value by minimizing the complications associated with this algorithm. The MOBWO is carefully chosen to carry out the numerical analysis in this research as it is a recently developed meta-heuristic technique with a unique ability to handle the multi-objective problem with reduced complications, improved convergence characteristics, and better computational efficiency. Every possible effort has been made while developing the code to avoid local entrapment to reach the global optimum reasonably.

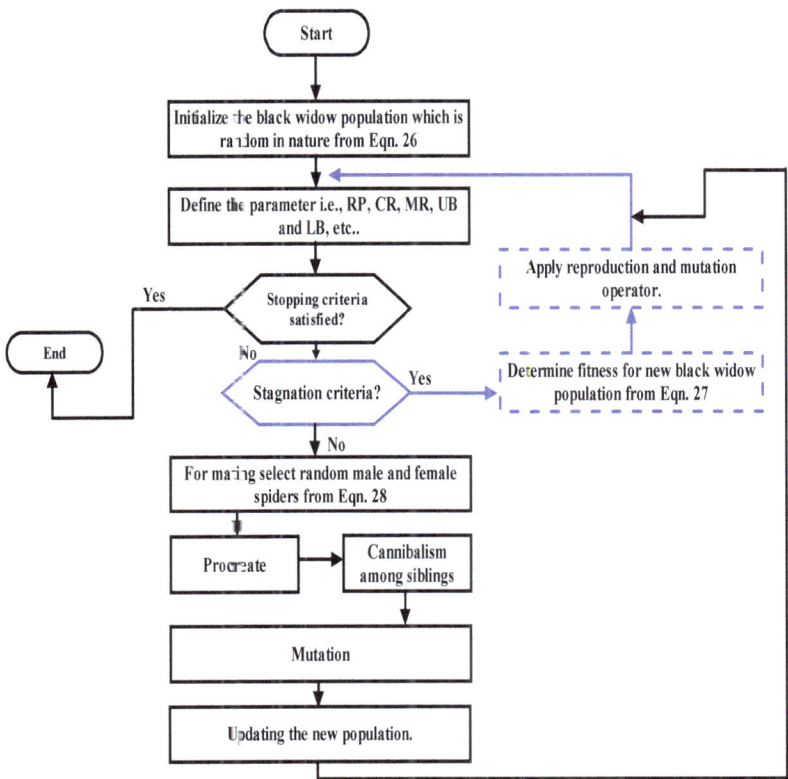

Figure 9. Flowchart of the proposed MOBWO algorithm.

5. Results

During the off-peak periods, the VPP obtains power from the energy market as a purchaser when electricity prices are on the lower side; when prices are high, the VPP trades electricity to the energy market. When electricity costs are cheaper, buying energy from the wholesale market is more profitable than making electricity which is also an attractive feature of CHP units. By enabling these units, the demand and supply gap can be balanced with great ease, especially when the power from solar and wind units cannot be extracted at night-time.

In this study, three cases have been considered to evaluate the feasibility of the virtual power plant.

1. Case I: Single-objective scheduling for profit maximization.
2. Case II: Single-objective scheduling for emission minimization.
3. Case III: Multi-objective scheduling for profit/emission, i.e., maximization followed by minimization.

To ensure security and continuity of power supply, the balance between load and power generation must be synchronized. The large presence of renewables, especially intermittent and highly volatile sources, i.e., solar and wind energy, makes this more complex. The surplus power in the network can be utilized in peak times when the electricity price is higher. This excess power can be stored in energy storage systems (ESSs), viz., battery storage or electric vehicles, and can be made available as and when needed through the VPP operator at a higher price that opens the option of attaining a higher economic benefit, which is beneficial to all the participants involved in the VPP system.

In the case of day-ahead scheduling, the electrical power balance for both Cases I and II are presented on an hourly basis and shown in Figures 10 and 11. Power is represented on the primary Y-axis which is on the left-hand side expressed in MW. To reduce the complexity, the electricity price is plotted on the secondary Y-axis which is on the right-hand side and is expressed in ($/MWhr).

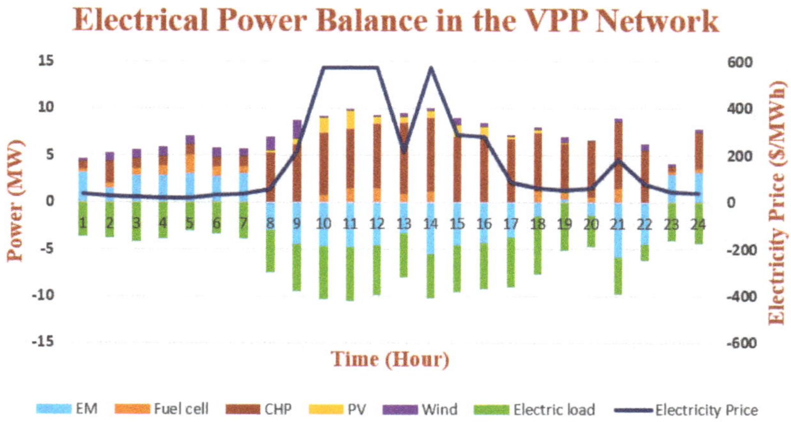

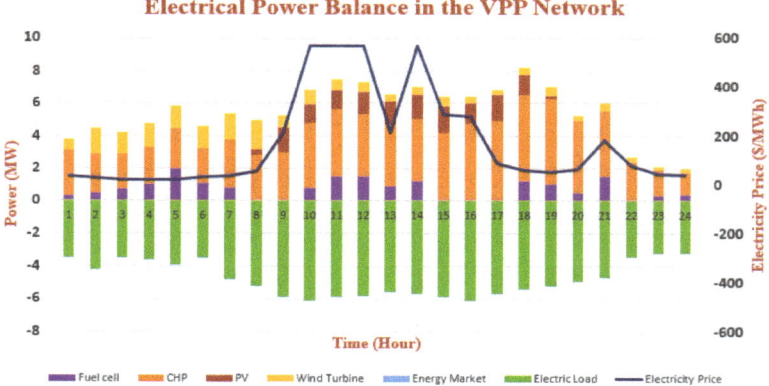

Figure 10. Case I Electrical power balance in the VPP network for 24 h.

Figure 11. Case II Electrical power balance in the VPP network for 24 h.

When the prices of electricity are low, the CHP units can be changed to a not committed (NC) status as the acquisition of electricity from the electricity market is considered more cost-effective and the VPP buys the electricity from the trading market. On the other hand, when the electricity prices are high, it is not economical to purchase the power from the wholesale market and in those instances, the VPP similarly trades power similar to any other conventional power plant.

5.1. Scenario Generation

Two scenarios have been considered to examine the performance of the projected technique. Scenario I is considered for day-ahead scheduling, followed by scenario II which is 15-min scheduling. All the results obtained in Scenario I are based on hourly scheduling and the performance is quite satisfactory and improved when compared with the other different techniques.

The numerical studies were carried out by using the multi-objective black widow optimization (MOBWO) algorithm with a code developed with utmost care and keeping in mind that the computation time should be as low as possible. In the BWO algorithm, the number of search agents and iterations is assumed to be 100 and 200, respectively. The behavior of the algorithm seems to be quite satisfactory, and it is capable of avoiding any premature convergence as well as local entrapment which is one of the most unfortunate traits of any nature-inspired algorithm. Along with BWO, other popular techniques, i.e., artificial bee colony (ABC) and ant colony optimization (ACO), are also selected to see their performance and the obtained results are discussed in the subsequent section. The case studies and scenario generation are executed and carried out in MATLAB software on a DELL laptop with a processor having an Intel(R) Core (TM) i7-6600U CPU @ 2.60 GHz.

5.1.1. Scenario I (Day-ahead Scheduling)

For demonstration purposes, the program was developed for day-ahead scheduling (DA), i.e., a 24-h period. The obtained results of the convergence for profit and emission by the proposed method are discussed and displayed in Case I and Case II, respectively, and compared with published results. In this paper, two scenarios are considered, i.e., day-ahead (hourly-based scheduling), followed by a 15 min-based scheduling. All the numerical results obtained from the proposed MOBWO algorithm in scenario I are compared with the existing published results in which the MOPSO technique is used. For the selected problem statement of the proposed system, only one relevant paper is reported in the literature, i.e., Ref. [22], to the best of our knowledge. Another comparison has been made, i.e., Ref. [46], and the obtained value of net profit from the proposed technique was higher. Similarly, the emission value was compared with Ref. [47] and found that the obtained values of both net profit and emission in a single objective, as well as multi-objective optimal day-ahead scheduling, are better in addition to the reduced computational time of the proposed MOBWO algorithm over all the other control methods as evident in Tables 7 and 8 for each case, respectively.

Table 7. Comparison of net profit with different techniques for Case I.

Output	Ref. [46]	MOPSO [22]	ABC	ACO	Proposed MOBWO
Maximum Profit ($)	19,737	23,302.8271	24,191.8221	24,950.7372	27,785.6723
Minimum Profit ($)	-	22,600.1679	19,636.7483	20,190.8183	21,400.3254
Mean Profit ($)	-	22,955.3462	21,914.2852	22,570.7776	24,592.9985
Computational time (Seconds)	-	148.095 (For 20 runs)	139.3737 (For 100 runs)	135.4932 (For 100 runs)	123.058 (For 100 runs)

Table 8. Comparison of emissions with different techniques for Case II.

Output	Ref. [47]	MOPSO [22]	ABC	ACO	Proposed MOBWO
Minimum Emission (Kg)	56,270	64,432.3217	62,467.8291	61,346.4838	57,532.2738
Maximum Emission (Kg)	77,430	67,077.3937	72,383.7292	71.463.2612	67,342.3798
Mean Emission (Kg)	-	66,070.1682	67,425.7792	66,404.8725	62,437.3268
Computational Time (Sec)	-	171.4826 (For 20 runs)	153.4826 (For 100 runs)	131.3633 (For 100 runs)	81.3745 (For 100 runs)

Case I: The statistical results obtained after 100 independent runs from MOBWO converge to an optimum value of 27,785.6723 $ which outperforms the existing value of 23,302.8271 $ obtained from multi-objective particle swarm optimization (MOPSO) Ref. [22]. Additionally, the numerical results for other techniques are mentioned in Table 7.

In Case I, the computational time for the proposed MOBWO of 123.0548 (for 100 runs) is less when compared with the multi-objective particle swarm optimization (MOPSO) of 148.0945 (for 20 runs). The convergence curve of the proposed algorithm for Case I is presented in Figure 12 and it is evident that the proposed technique produces the optimum value.

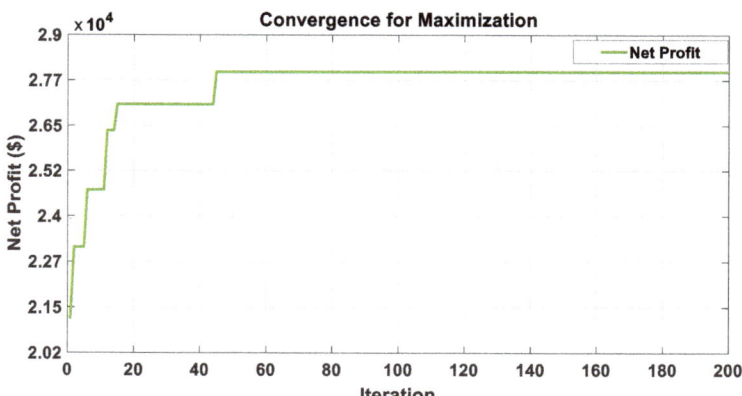

Figure 12. Convergence curve of MOBWO algorithm for Case I.

Case II: On the other hand, in Case II, the purpose is to reduce emissions and the level of emission is reduced significantly by a considerable margin. The unit of the emission is in Kg and it is the combination of three emission factors, namely, NO_X, CO_2, and SO_2. The statistical results obtained after 100 independent runs from MOBWO are shown in Table 8 as converging to an optimum value of 57,532.2738 Kg which outperforms the existing value of 64,432.3217 Kg obtained from MOPSO Ref. [22]. In Case II, the computational time for the proposed MOBWO is 81.3745 (for 100 runs), much less when compared with 171.4826 (for 20 runs) from MOPSO. In addition, the convergence characteristics of the proposed algorithm for Case II are shown in Figure 13.

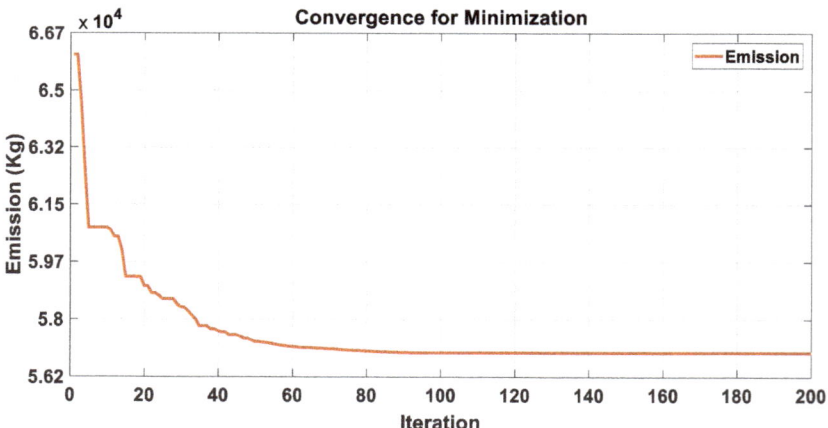

Figure 13. Convergence characteristics of the MOBWO algorithm for Case II.

Computation time plays a significant role in selecting a suitable optimization technique which is usually problem specific. Cumulative factors are involved in solving real-world problems, i.e., selection of parameters, constraints handling, computation time, etc. Based on any particular optimization technique, these factors vary to some extent and the best trade-off is selected for the chosen technique. The nature-inspired algorithms are the best way to solve these non-convex type problems and come with lesser mathematical complexities and an efficient way to reach a global optimum value. Black widow optimization (BWO) is selected to carry out the numerical analysis in this research as it is a recently developed meta-heuristic technique and it has a unique ability to handle the multi-objective problem with reduced complications, improved convergence characteristics, and better computational efficiency. The cases which have been considered for single, as well as multi-objective, are compared with PSO and the resulting computation time is better with BWO.

Case III: The multi-objective optimal scheduling is carried out and the convergence of both objectives is displayed together in Figure 14, followed by Figures 15 and 16 with other techniques. It is worth mentioning that including emission as an objective function has an immense effect on the overall working of the system since the CHP unit is considered one of the resources. Here, the peak valley electricity pricing is considered which is very essential in the peak load shifting to enhance the economic benefit. For every time period, the sale and purchase prices are given in Table 9. The effect of this pricing scheme on the operating costs has been given proper consideration in this paper.

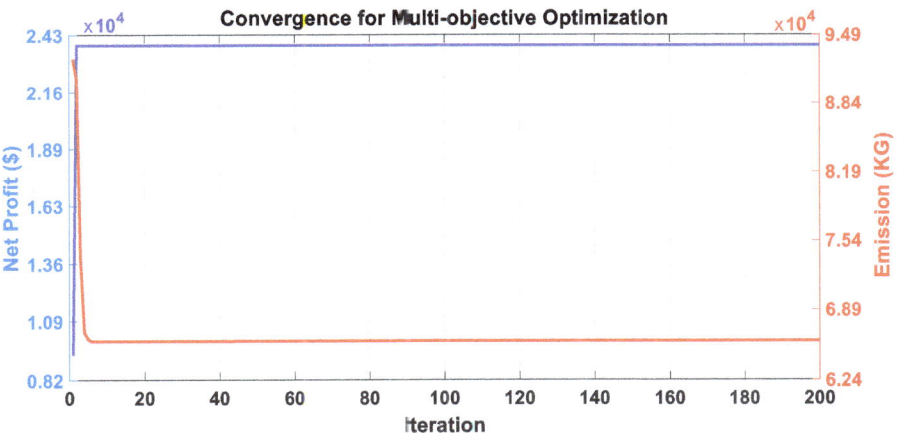

Figure 14. Convergence characteristics of the ABC algorithm for Case III.

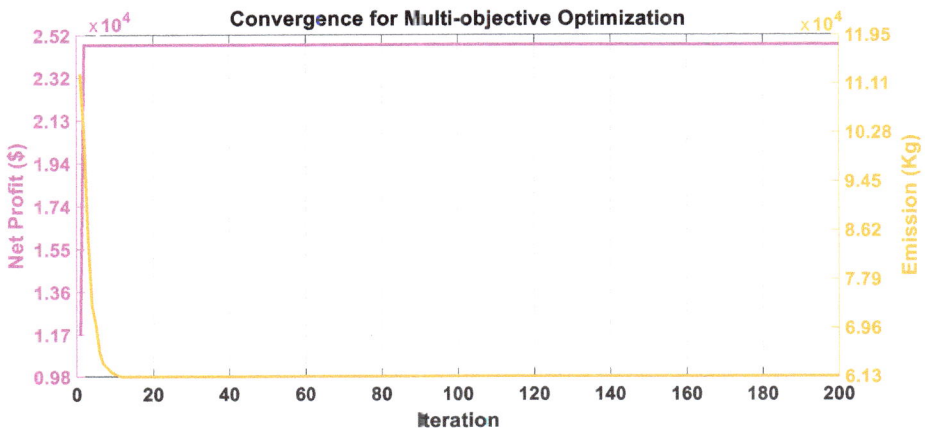

Figure 15. Convergence characteristics of the ACO algorithm for Case III.

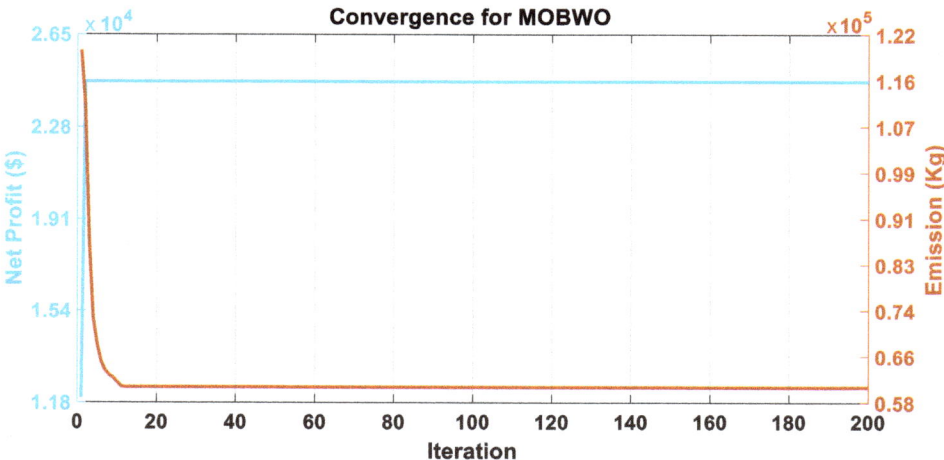

Figure 16. Convergence characteristics of the MOBWO algorithm for Case III.

Table 9. Comparison of emissions with different techniques.

Period	Detail Time (Hr.)	Purchase Price ($/MWh)	Sale Price ($/MWh)
Peak	9,12,17,22	0.0079	0.0044
Intermediate	13,16	0.0070	0.0035
Valley	1,8,23,24	0.0062	0.0026

A comparison of the statistical analysis is also presented in Table 10 and obtained after 100 independent runs. It can be observed that by adopting the peak valley power pricing concept in this paper, better compromise solutions are given which show enhancement in the performance and behavior of the price-based-MOBWO algorithm.

Table 10. Comparison of multi-objective with published and proposed techniques.

Objective Functions	Profit ($)				Emissions (Kg)			
Parameters	MOPSO [22]	ABC	ACO	MOBWO	MOPSO [22]	ABC	ACO	MOBWO
MaxF^{profit}	23,302.83	24,286.82	25,183.74	26,167.78	122,963.46	119,789.29	119,432.37	116,400.85
Min$F^{emission}$	9883.69	10,320.38	11,723.47	11,808.47	64,432.32	62,467.83	61,346.48	58,785.34

The main purpose of obtaining the Pareto graph is to provide the VPP operator a chance to select a trade-off solution that is in line with the environmental restrictions and satisfies the economic constraints at the same time. To this end, Case III has been implemented with multi-objective scheduling and obtaining the Pareto optimal solutions, respectively, as shown in Figure 17.

In the existing program, a pricing strategy was also incorporated to see the behavior of the convergence obtained from the proposed MOBWO. The code was developed very carefully, and the program was run for multiple trials. All the displayed results were obtained after running the program for 100 trials to ensure the precision of the algorithm.

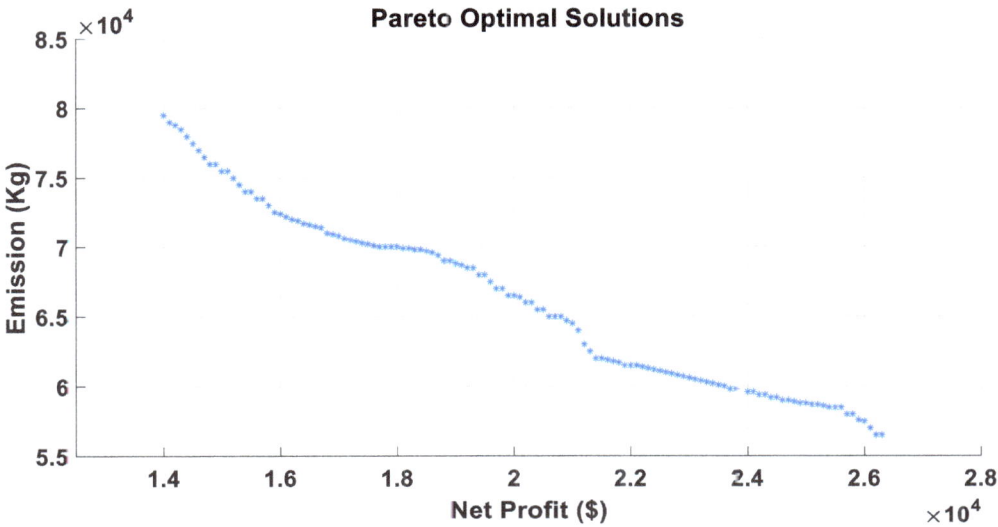

Figure 17. Pareto optimal solutions for Case III of Scenario I (proposed MOBWO technique).

5.1.2. Scenario II (15-min Interval Scheduling)

For further analysis, scenario II is presented in which the type of scheduling is transformed to a 15-min basis and the available data are extrapolated for carrying out the real-time analysis. The overall power balance in the case of a 15 min schedule for a one-day profile, is represented in Figures 18 and 19, and the unit of time is taken in hours. All the associated resources are represented in a vertical bar and the electricity price is projected on the right-hand side of the secondary Y-axis.

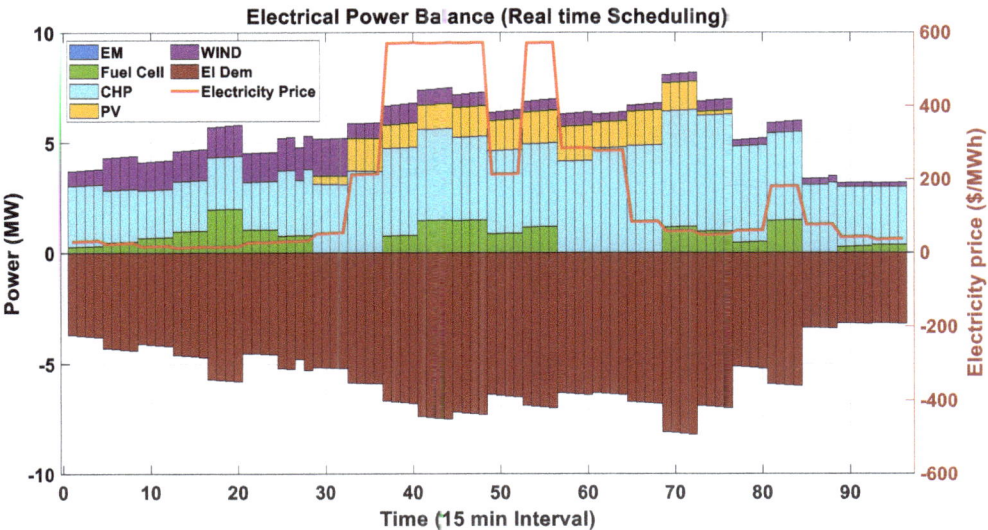

Figure 18. Case I power balance (electrical) in the VPP system for 15-min.

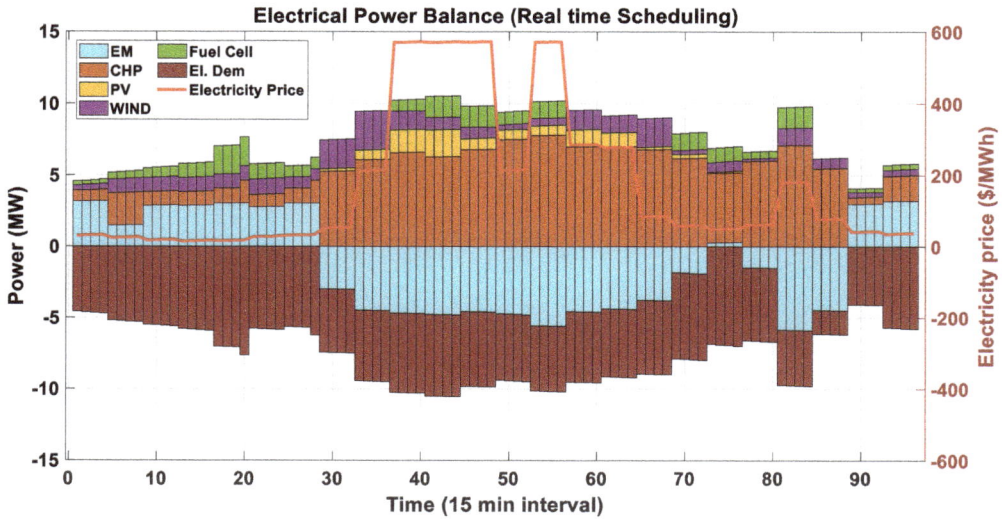

Figure 19. Case II power balance (electrical) in the VPP network for 15-min.

Likewise, in scenario II, i.e., 15-min scheduling, the proposed scenario has no comparison available in the existing literature. The pay-off table is mentioned below, and it can be observed that the results are quite reasonable in real-time scheduling, especially in terms of profit as compared to the day-ahead scheduling which can be seen in Table 11, followed by Figures 20–22 obtained by ABC, ACO, and MOBWO after 100 independent trials.

Table 11. The pay-off table for the proposed MOBWO (15-min scheduling).

Objective Functions	Profit ($)			Emissions (Kg)		
Parameters	ABC	ACO	MOBWO	ABC	ACO	MOBWO
MaxF^{profit}	27,392.5631	26,312.3523	28,415.3525	120,913.4532	120,325.463	119,843.4532
Min$F^{emission}$	10,404.7262	11,123.4253	11,929.7262	56,402.4216	57,342.4235	59,921.3248

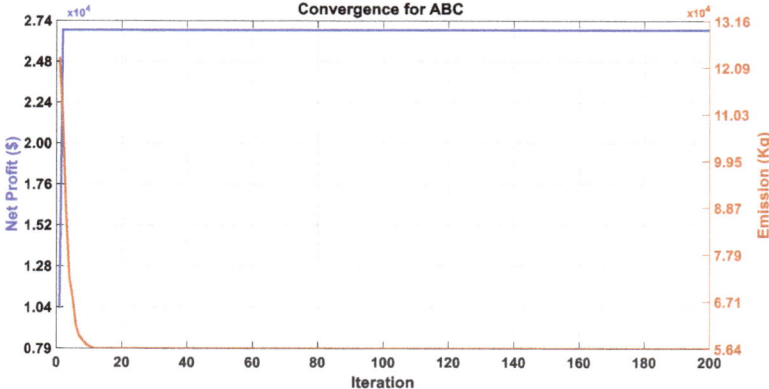

Figure 20. Convergence curve of the ABC algorithm.

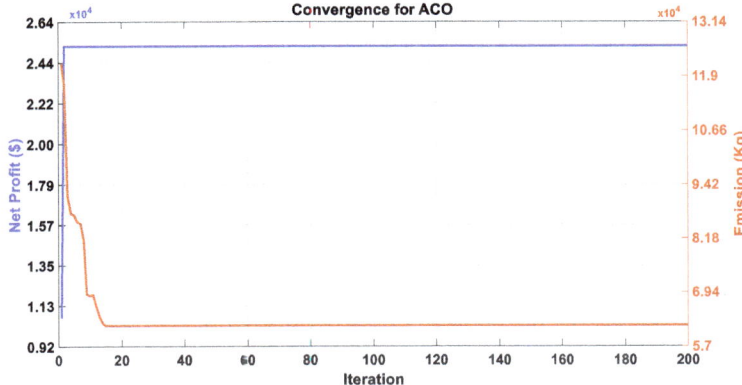

Figure 21. Convergence curve of the ACO algorithm.

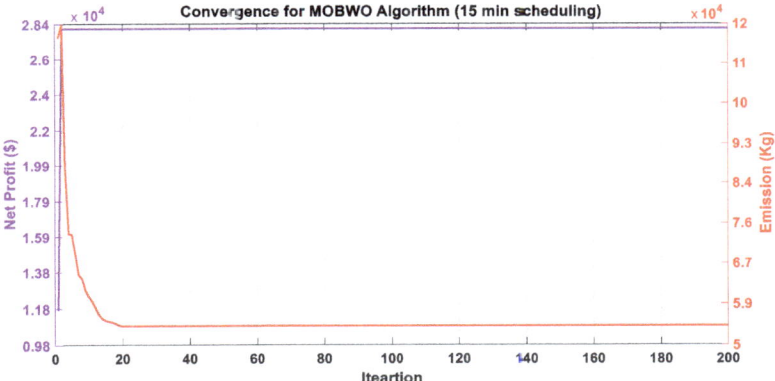

Figure 22. Convergence curve of the proposed MOBWO algorithm.

Figure 23 shows the Pareto graph for scenario II, the 15-min scheduling. In generating the graph, the operator of the VPP can select an appropriate solution that deals with the possibilities of both technical constraints as well as the related economical limitations.

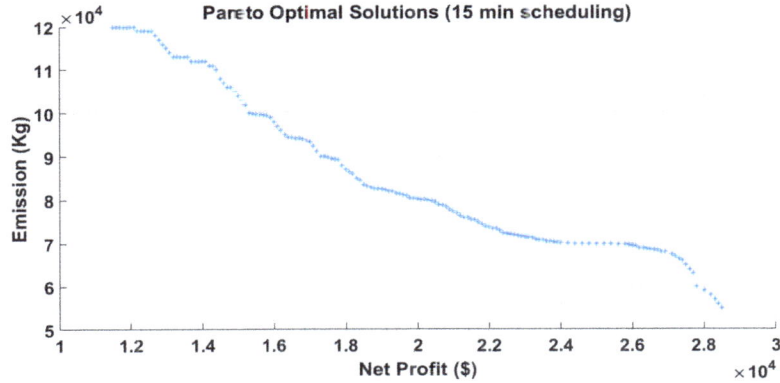

Figure 23. Pareto optimal solutions for Case III of Scenario II (proposed MOBWO technique).

Marcos Tostado-Veliz et al. Ref. [48] proposed a MILP optimization framework that is segmented into two parts, i.e., based on historical consumption, the most suitable tariff is determined followed by the optimal hours which are referred to as 'Happy hours' tariff plans. Three different tariffs are discussed, viz., the fixed tariff, time-variable tariff, and happy hours tariff. The Spanish retail market is considered for the developed framework case study in which the price of selling electricity is set to zero during the happy hours time to avert any unrealistic transactions among the home and utility grid. In a deregulated electricity market, a framework for the decision for tariff selection is presented along with some useful results in which the Happy hours, i.e., 7:00–8:00 a.m., are stressed.

To balance the price volatility, scheduling strategies play a vital role. They are categorized as day-ahead scheduling, for a 24 h profile, and 15 min schedule. Merits of day-ahead scheduling are highlighted:

(a) Buying and selling of electricity one day before the following day.
(b) The VPP acts as a price taker in the day-ahead market.
(c) Ease of unit commitment and power dispatching.

The advantages of the 15-minute scheduling are mentioned below:

(a) This scheduling helps to determine the imbalance in settlement prices.
(b) Offers the purchase and selling of electricity during the functioning day.
(c) Real-time scheduling stabilizes the differences between day-ahead and real-time demand and production of electricity.
(d) Operating systems that work in real-time can execute quickly without any delay, resulting in a nearly immediate output.

For a practical case, real-time scheduling (a few seconds to 5-min intervals) is preferable due to its inherent characteristics of reliability and performance-orientation, and the requirement of only minimal, latest, and most relevant data followed by a rapid response for given circumstances.

The proposed approach is more suitable for advancing the existing system by adding a few more resources that can act as a spinning reserve, i.e., an energy storage system, and ensure the continuity of electricity supply in case of any power deficit. The capacity of the power supply system considered in the current VPP system is 10.48 MW. Since the main concept of a VPP is that it is not restricted by any geographical location, it is always connected to the grid which is not the case when discussing a Microgrid that can be islanded and constrained in a confined region of operation. The selected VPP system under study is assumed to be a national power system.

The multi-objective optimal scheduling of VPP considering various renewable resources has been solved using the weighting factor method to simultaneously maximize profit and minimize emission. Peak valley's power pricing (PVPP) strategy is introduced in the multi-objective optimal scheduling of the VPP problem. Moreover, a new price based MOBWO is presented and implemented, satisfying all the related constraints. For both scenarios I and II, Pareto optimal solutions are obtained specifically for the multi-objective optimal scheduling of the VPP problem for maximization of profit along with minimization of emissions. The statistical results for all three cases show the effectiveness and suitability of the proposed approach and are compared with various other techniques along with the published results.

6. Conclusions

The optimal scheduling of a VPP consisting of various resources is carried out and the performance of the BWO algorithm shows superiority in terms of the numerical results obtained. It can be inferred that taking emission as an objective function leads to a considerable reduction in the amount of day-ahead emission. In scenario I, the net profit for hourly-based scheduling leads to 26,167.7823 ($), followed by the daily emission which is 116,400.8473 (Kg). Although, when the same process is followed for scenario II, which is a 15-min interval, the numerical results are reasonably better, i.e., net profit leads to

28,415.3525 ($). However, in the case of emission, there is no significant improvement as the obtained value is 119,843.4532 (Kg). The single objective scheduling is performed for both net profits along with emission as an objective function, followed by multi-objective scheduling; Pareto optimal solutions are presented for both scenarios. Statistical analysis was performed for both single and multi-objective scheduling of the VPP problem. Optimal scheduling of the selected VPP problem was performed with three different techniques, i.e., ABC, ACO, and MOBWO, and it was observed that the developed MOBWO algorithm performance was superior in terms of reduction in computation time as well as the ability to escape from local entrapment by reaching a global optimum.

In the future, more resources such as hydro, thermal, or a combination of hydro-thermal will be incorporated into the VPP system. It will be very interesting to see how the performance of the system and the aspect of the reliability index will be explored. Similarly, adding some specific pricing and bidding strategies along with exploring the impact of market constraints on effective analysis is anticipated in a forthcoming study.

Author Contributions: Conceptualization, A.K.P. and V.K.J.; methodology, A.K.P. and V.K.J.; software, A.K.P. and V.K.J.; validation, A.K.P., V.K.J. and J.N.S.; formal analysis, A.K.P.; investigation, A.K.P., V.K.J. and J.N.S.; resources, A.K.P. and V.K.J.; data curation, A.K.P.; writing—original draft preparation, A.K.P.; writing—review and editing, V.K.J. and J.N.S.; visualization, A.K.P., V.K.J. and J.N.S.; supervision, V.K.J. and J.N.S.; project administration, V.K.J. and J.N.S.; and funding acquisition, A.K.P., V.K.J. and J.N.S. All authors have read and agreed to the published version of the manuscript.

Funding: This research is funded by DST sponsored INSPIRE Fellowship, Govt. of India, fellowship number [INSPIRE Code: IF 190938] and "The APC is funded by MANIPAL ACADEMY OF HIGHER EDUCATION (MAHE) MANIPAL".

Institutional Review Board Statement: Not applicable.

Informed Consent Statement: Not applicable.

Data Availability Statement: Not applicable.

Acknowledgments: The first author positively acknowledges the funding organization: DST sponsored INSPIRE Fellowship, Govt. of India, [INSPIRE Code: IF 190938] in carrying out the research work with their incessant funding. The authors also acknowledge gratefulness to the host University Manipal Institute of Technology, Manipal Academy of Higher Education, Manipal, Karnataka, India for providing a conducive research atmosphere to carry out the envisioned research work.

Conflicts of Interest: The authors declare no conflict of interest.

Nomenclature

All the abbreviations used in the manuscript are listed:

TPP	Traditional power plant
DERs	Distributed energy resources
VPP	Virtual power plant
MOBWO	Multi-objective black widow optimization
MOOS	Multi-objective optimal scheduling
MOSS	Multi-objective scheduling strategy
SP	Spot pricing
TOU	Time of use
EVs	Electric vehicles
ES	Energy storage
RERs	Renewable energy resources
ICT	Information communication technology
IRP	Integrated resource planning
P2P	Peer to peer
NM	Net metering
BTM	Behind the meter

PSO	Particle swarm optimization
GA	Genetic algorithm
DR	Demand response
V2H	Vehicle to home
BSS	Battery storage system
CEM	Consecutive energy management
SOC	State of charge
TNPC	Total net present cost
UC	Unit commitment
SC	Soft computing
PV	Photovoltaics
WT	Wind turbine
FC	Fuel cells
CHP	Combined heat and power
EL	Electric load
EM	Energy market
PLR	Part load ratios
FOR	Feasible regions of operation
PDF	Probability distribution function
FF	Fill factor
ACO	Ant colony optimization
ABC	Artificial bee colony
ANN	Artificial neural network
CR	Cannibalism rate
MR	Mutation rate
RP	Reproduction rate
PVPP	Peak valley power pricing

References

1. Saswat, S.S.; Patra, S.; Mishra, D.P.; Salkuti, S.R.; Senapati, R.N. Harnessing wind and solar PV system to build hybrid power system. *Int. J. Power Electron. Drive Syst.* **2021**, *12*, 2160. [CrossRef]
2. Salkuti, S.R. Short-term optimal hydro-thermal scheduling using clustered adaptive teaching learning based optimization. *Int. J. Electr. Comput. Eng.* **2019**, *9*, 3359. [CrossRef]
3. Reddy, S.S. Multi-objective based adaptive immune algorithm for solving the economic and environmental dispatch problem. *Int. J. Appl. Eng. Res.* **2017**, *12*, 1043–1048.
4. Sravanthi, P.; Vuddanti, S. Solving realistic reactive power market clearing problem of wind-thermal power system with system security. *Int. J. Emerg. Electr. Power Syst.* **2022**, *23*, 125–144.
5. Pal, P.; Krishnamoorthy, P.A.; Rukmani, D.K.; Antony, S.J.; Ocheme, S.; Subramanian, U.; Elavarasan, R.M.; Das, N.; Hasanien, H.M. Optimal Dispatch Strategy of Virtual Power Plant for Day-Ahead Market Framework. *Appl. Sci.* **2021**, *11*, 3814. [CrossRef]
6. Han, S.; Mao, T.; Guo, X.; Wu, W.; Sun, L.; Wang, T.; Zhou, B.; Zhao, W. Profit Evaluation for Virtual Power Plant in Power Load Response: From the Perspective of Power Grid. In Proceedings of the 2020 IEEE 4th Conference on Energy Internet and Energy System Integration (EI2), Wuhan, China, 30 October–1 November 2020; IEEE: Piscataway, NJ, USA, 2020; pp. 2874–2878.
7. Lombardi, P.; Powalko, M.; Rudion, K. Optimal operation of a virtual power plant. In Proceedings of the 2009 IEEE Power & Energy Society General Meeting, Calgary, AB, Canada, 26–30 July 2019; IEEE: Piscataway, NJ, USA, 2009; pp. 1–6.
8. Saabit, S.M.; Hasan, S.; Chowdhury, N.A. Virtual Power Plant Implementation and Cost Minimization for Retail Industry. In Proceedings of the 2021 2nd International Conference for Emerging Technology (INCET), Belagavi, India, 21–23 May 2021; IEEE: Piscataway, NJ, USA, 2021; pp. 1–6.
9. Reddy, S.S. Optimal scheduling of thermal-wind-solar power system with storage. *Renew. Energy* **2017**, *101*, 1357–1368. [CrossRef]
10. Zhang, Y.; Yang, K.; Zheng, B.; Zhu, G.; Wang, D.; Li, S.; Xu, K.; Tu, T. Optimal Scheduling Strategies of the Virtual Power Plant Considering Different Development Stages of the Electricity Market. In Proceedings of the 2021 6th Asia Conference on Power and Electrical Engineering (ACPEE), Chongqing, China, 8–11 April 2021; IEEE: Piscataway, NJ, USA, 2021; pp. 1010–1016.
11. Zhang, T.; Qin, Y.; Wu, W.; Zheng, M.; Huang, W.; Wang, L.; Xu, S.; Yan, X.; Ma, J.; Shao, Z. Research on Optimal Scheduling in Market Transaction for the Participation of Virtual Power Plants. In Proceedings of the 2019 6th International Conference on Information Science and Control Engineering (ICISCE), Shanghai, China, 20–22 December 2019; IEEE: Piscataway, NJ, USA, 2019; pp. 841–845.

2. Naina, P.M.; Rajamani, H.; Swarup, K.S. Modeling and simulation of virtual power plant in energy management system applications. In Proceedings of the 2017 7th International Conference on Power Systems (ICPS), Pune, India, 21–23 December 2017; IEEE: Piscataway, NJ, USA, 2017; pp. 392–397.
3. Zahedmanesh, A.; Muttaqi, K.M.; Sutanto, D. A consecutive energy management approach for a VPP comprising commercial loads and electric vehicle parking lots integrated with solar PV units and energy storage systems. In Proceedings of the 2019 1st Global Power, Energy and Communication Conference (GPECOM), Nevsehir, Turkey, 21–15 June 2019; IEEE: Piscataway, NJ, USA, 2019; pp. 242–247.
4. Lombardi, P.; Sokolnikova, T.; Styczynski, Z.; Voropai, N. Virtual power plant management considering energy storage systems. *IFAC Proc. Vol.* **2012**, *45*, 132–137.
5. Mishra, S.; Crasta, C.J.; Bordin, C.; Mateo-Fornés, J. Smart contract formation enabling energy-as-a-service in a virtual power plant. *Int. J. Energy Res.* **2022**, *46*, 3272–3294. [CrossRef]
6. Renuka, T.K.; Reji, P.; Sreedharan, S. An enhanced particle swarm optimization algorithm for improving the renewable energy penetration and small signal stability in power system. *Renew. Wind Water Sol.* **2018**, *5*, 6. [CrossRef]
7. Mohammed, O.H.; Amirat, Y.; Benbouzid, M. Particle swarm optimization of a hybrid wind/tidal/PV/battery energy system. Application to a remote area in Bretagne, France. *Energy Procedia* **2019**, *162*, 87–96. [CrossRef]
8. Lombardi, P. Multi criteria Optimization of an autonomous virtual power plant with high degree of renewable energy sources. In Proceedings of the 17th PSCC, Stockholm, Sweden, 22–26 August 2011; pp. 22–26.
9. Dehghanniri, M.F.; Golkar, M.A.; Jahangir, H. Virtual power plant performance strategy in the DA and RT market under uncertainties. In Proceedings of the 2021 11th Smart Grid Conference (SGC), Tabriz, Iran, 7–9 December 2021; IEEE: Piscataway, NJ, USA, 2021; pp. 1–5.
10. Khandelwal, M.; Mathuria, P.; Bhakar, R. Profit based self-scheduling of virtual power plant under multiple locational marginal prices. In Proceedings of the 2018 20th National Power Systems Conference (NPSC), Tiruchirappalli, India, 14–16 December 2018; IEEE: Piscataway, NJ, USA, 2018; pp. 1–6.
11. Gough, M.; Santos, S.F.; Lotfi, M.; Javadi, M.S.; Osorio, G.J.; Ashraf, P.; Castro, R.; Catalao, J.P.S. Operation of a Technical Virtual Power Plant Considering Diverse Distributed Energy Resources. *IEEE Trans. Ind. Appl.* **2022**, *58*, 2547–2558. [CrossRef]
12. Hadayeghparast, S.; Farsangi, A.S.; Shayanfar, H. Day-ahead stochastic multi-objective economic/ emission operational scheduling of a large scale virtual power plant. *Energy* **2019**, *172*, 630–646. [CrossRef]
13. Tascikaraoglu, A.; Erdinc, O.; Uzunoglu, M.; Karakas, A. An adaptive load dispatching and forecasting strategy for a virtual power plant including renewable energy conversion units. *Appl. Energy* **2014**, *119*, 445–453. [CrossRef]
14. Zhang, C.; Ouyang, D.; Ning, J. An artificial bee colony approach for clustering. *Expert Syst. Appl.* **2010**, *37*, 4761–4767. [CrossRef]
15. Aydin, D.; Özyön, S.; Yaşar, C.; Liao, T. Artificial bee colony algorithm with dynamic population size to combined economic and emission dispatch problem. *Int. J. Electr. Power Energy Syst.* **2014**, *54*, 144–153. [CrossRef]
16. Narkhede, M.S.; Chatterji, S.; Ghosh, S. Multi objective optimal dispatch in a virtual power plant using genetic algorithm. In Proceedings of the 2013 International Conference on Renewable Energy and Sustainable Energy (ICRESE), Coimbatore, India, 5–6 December 2013; IEEE: Piscataway, NJ, USA, 2013; pp. 238–242.
17. Eiben, Á.E.; Hinterding, R.; Michalewicz, Z. Parameter control in evolutionary algorithms. *IEEE Trans. Evol. Comput.* **1999**, *3*, 124–141. [CrossRef]
18. Kumar, S.; Pal, N.S. Ant colony optimization for less power consumption and fast charging of battery in solar grid system. In Proceedings of the 2017 4th IEEE Uttar Pradesh Section International Conference on Electrical, Computer and Electronics (UPCON), Mathura, India, 26–28 October 2017; IEEE: Piscataway, NJ, USA, 2017; pp. 244–249.
19. Rekik, M.; Chtourou, Z.; Mitton, N.; Atieh, A. Geographic routing protocol for the deployment of virtual power plant within the smart grid. *Sustain. Cities Soc.* **2016**, *25*, 39–48. [CrossRef]
20. Biswas, M.A.R.; Robinson, M.D.; Fumo, N. Prediction of residential building energy consumption: A neural network approach. *Energy* **2016**, *117*, 84–92. [CrossRef]
21. You, S. Developing Virtual Power Plant for Optimized Distributed Energy Resources Operation and Integration. Ph.D. Thesis, Technical University of Denmark, Lyngby, Denmark, 2010.
22. Simoes, M.; Bose, B.K.; Spiegel, R.J. Design and performance evaluation of a fuzzy-logic-based variable-speed wind generation system. *IEEE Trans. Ind. Appl.* **1997**, *33*, 956–965. [CrossRef]
23. Niknam, T.; Azizipanah-Abarghooee, R.; Narimani, M.R. An efficient scenario-based stochastic programming framework for multi-objective optimal micro-grid operation. *Appl. Energy* **2012**, *99*, 455–470. [CrossRef]
24. Wang, Y.; Ai, X.; Tan, Z.; Yan, L.; Liu, S. Interactive dispatch modes and bidding strategy of multiple virtual power plants based on demand response and game theory. *IEEE Trans. Smart Grid* **2015**, *7*, 510–519. [CrossRef]
25. Dabbagh, S.R.; Sheikh-El-Eslami, M.K. Risk-based profit allocation to DERs integrated with a virtual power plant using cooperative Game theory. *Electr. Power Syst. Res.* **2015**, *121*, 368–378. [CrossRef]
26. Pandey, A.K.; Jadoun, V.K.; Jayalakshmi, N.S. Virtual Power Plants: A New Era of Energy Management in Modern Power Systems. In Proceedings of the 2021 8th International Conference on Signal Processing and Integrated Networks (SPIN), Noida, India, 26–27 August 2021; IEEE: Piscataway, NJ, USA, 2021; pp. 538–543.

37. Wu, Z.; Gu, W.; Wang, R.; Yuan, X.; Liu, W. Economic optimal schedule of CHP microgrid system using chance constrained programming and particle swarm optimization. In Proceedings of the 2011 IEEE Power and Energy Society General Meeting, Detroit, MI, USA, 24–28 July 2011; IEEE: Piscataway, NJ, USA, 2011; pp. 1–11.
38. Nazari-Heris, M.; Abapour, S.; Mohammadi-Ivatloo, B. Optimal economic dispatch of FC-CHP based heat and power micro-grids. *Appl. Therm. Eng.* **2017**, *114*, 756–769. [CrossRef]
39. Zamani, A.G.; Zakariazadeh, A.; Jadid, S.; Kazemi, A. Stochastic operational scheduling of distributed energy resources in a large scale virtual power plant. *Int. J. Electr. Power Energy Syst.* **2016**, *82*, 608–620. [CrossRef]
40. Nojavan, S.; Zare, K.; Mohammadi-Ivatloo, B. Application of fuel cell and electrolyzer as hydrogen energy storage system in energy management of electricity energy retailer in the presence of the renewable energy sources and plug-in electric vehicles. *Energy Convers. Manag.* **2017**, *136*, 404–417. [CrossRef]
41. Ali, E.S.; El-Sehiemy, R.A.; El-Ela, A.A.A.; Tostado-Véliz, M.; Kamel, S. A proposed uncertainty reduction criterion of renewable energy sources for optimal operation of distribution systems. *Appl. Sci.* **2022**, *12*, 623. [CrossRef]
42. Available online: https://www.meteoblue.com/en/weather/archive/export/basel_switzerland_2661604 (accessed on 7 March 2022).
43. Available online: https://niwe.res.in/department_r&d,rdaf_time_series_data.php (accessed on 7 March 2022).
44. Hayyolalam, V.; Kazem, A.A.P. Black widow optimization algorithm: A novel meta-heuristic approach for solving engineering optimization problems. *Eng. Appl. Artif. Intell.* **2020**, *87*, 103249. [CrossRef]
45. Devi, M.R.; Jeya, I.J.S. Black Widow Optimization Algorithm and Similarity Index Based Adaptive Scheduled Partitioning Technique for Reliable Emergency Message Broadcasting in VANET. *Mapp. Intimacies* **2021**. [CrossRef]
46. Zamani, A.G.; Zakariazadeh, A.; Jadid, S. Day-ahead resource scheduling of a renewable energy based virtual power plant. *Appl. Energy* **2016**, *169*, 324–340. [CrossRef]
47. Zakariazadeh, A.; Jadid, S.; Siano, P. Stochastic multi-objective operational planning of smart distribution systems considering demand response programs. *Electr. Power Syst. Res.* **2014**, *111*, 156–168. [CrossRef]
48. Tostado-Véliz, M.; Mouassa, S.; Jurado, F. A MILP framework for electricity tariff-choosing decision process in smart homes considering 'Happy Hours' tariffs. *Int. J. Electr. Power Energy Syst.* **2021**, *131*, 107139. [CrossRef]

Article

Day-Ahead Load Demand Forecasting in Urban Community Cluster Microgrids Using Machine Learning Methods

Sivakavi Naga Venkata Bramareswara Rao [1], Venkata Pavan Kumar Yellapragada [2,*], Kottala Padma [3], Darsy John Pradeep [2], Challa Pradeep Reddy [4], Mohammad Amir [5] and Shady S. Refaat [6,7,*]

1. Department of Electrical and Electronics Engineering, Sir C. R. Reddy College of Engineering, Eluru 534007, India
2. School of Electronics Engineering, VIT-AP University, Amaravati 522237, India
3. Department of Electrical Engineering, Andhra University College of Engineering (A), Visakhapatnam 530003, India
4. School of Computer Science and Engineering, VIT-AP University, Amaravati 522237, India
5. Department of Electrical Engineering, Faculty of Engineering and Technology, Jamia Millia Islamia Central University, Delhi 243601, India
6. Department of Electrical Engineering, Texas A&M University, Doha P.O. Box 23874, Qatar
7. School of Physics, Engineering and Computer Science, University of Hertfordshire, Hatfield AL10 9AB, UK
* Correspondence: pavankumar.yv@vitap.ac.in (V.P.K.Y.); shady.khalil@qatar.tamu.edu (S.S.R.)

Abstract: The modern-day urban energy sector possesses the integrated operation of various microgrids located in a vicinity, named cluster microgrids, which helps to reduce the utility grid burden. However, these cluster microgrids require a precise electric load projection to manage the operations, as the integrated operation of multiple microgrids leads to dynamic load demand. Thus, load forecasting is a complicated operation that requires more than statistical methods. There are different machine learning methods available in the literature that are applied to single microgrid cases. In this line, the cluster microgrids concept is a new application, which is very limitedly discussed in the literature. Thus, to identify the best load forecasting method in cluster microgrids, this article implements a variety of machine learning algorithms, including linear regression (quadratic), support vector machines, long short-term memory, and artificial neural networks (ANN) to forecast the load demand in the short term. The effectiveness of these methods is analyzed by computing various factors such as root mean square error, R-square, mean square error, mean absolute error, mean absolute percentage error, and time of computation. From this, it is observed that the ANN provides effective forecasting results. In addition, three distinct optimization techniques are used to find the optimum ANN training algorithm: Levenberg–Marquardt, Bayesian Regularization, and Scaled Conjugate Gradient. The effectiveness of these optimization algorithms is verified in terms of training, test, validation, and error analysis. The proposed system simulation is carried out using the MATLAB/Simulink-2021a® software. From the results, it is found that the Levenberg–Marquardt optimization algorithm-based ANN model gives the best electrical load forecasting results.

Keywords: ANN training algorithms; cluster microgrids; load demand forecasting; machine learning methods; urban energy community

1. Introduction

Electricity is a need and a strategic asset for national economies. As a result, electric utilities strive to balance power generation and demand to provide a decent service at a reasonable cost. The microgrid is the integration of several renewable energy sources with adjustable or nonadjustable loads and storage systems, such as batteries/fly-wheels [1]. With more penetration of distribution generation (DG) sources, it is a big challenge for the service provider to supply reliable and consistent power to the customer premises due to time-varying weather conditions. Similarly, energy consumption also varies according to seasonal variations and human behavior [2]. As a result, reliable forecasting of generators and load needs is required to solve the unit commitment problem and schedule the energy sources and storage devices in a microgrid [3,4]. To obtain the total power of utilization from the energy sources, we first need to obtain how much power is extracted from the green energy sources with the use of forecasting methods. Due to the inconsistency of solar and wind power, it is very difficult to predict load demand accurately. This type of load prediction-based weather prediction leads to inaccuracy in the results. So, it is better to use historical data related to power instead of the numerical prediction of weather in a microgrid to develop short-term load forecasting. The basic goal of short-term load demand forecasting is to plan the electricity schedule to meet seasonal and periodical load demands. Energy demand consumption is influenced by so many factors, such as weather, special activities, and seasonal variations. In microgrid wind and solar energy, sources are driven by weather. Thus, the development of forecasting technologies to forecast the generations and load demands is significant to maintaining the power balance in a microgrid. Depending on the requirement of energy management, there are four types of forecasting methods available, such as extremely short, short, medium, and long-term forecasting. The following is the literature survey carried out so far for short-term (ST) load demand forecasting.

With the use of neural networks and the particle swarm optimization (PSO) algorithm, Reference [5] presents a method for assessing short-term electrical load. This study establishes a proper learning rate and the number of hidden layers in neural networks for forecasting the electrical load demand using the PSO algorithm. Using these altered parameters, the neural network is then utilized to forecast the short-term load demands. This methodology used a three-layer feed-forward neural network with back propagation and an updated global best PSO algorithm. However, beyond the simple and min-max scaling approaches, Reference [6] proposes two novel data pre-processing techniques for short-term load forecasting utilizing artificial neural networks. The two main pre-processing methodologies proposed to focus on the importance of specific neural network input variables in connection to output variables, yielding better prediction results than previous methods. This strategy offers better results in terms of "Mean Squared Error (MSE)", "Mean Absolute Error (MAE)", and "Mean Absolute Percentage Error" when compared to previous studies using data from the interconnected system in Greek. Later in [7], the use of an artificial neural network for load forecasting is discussed. Due to changes in the load profile on weekdays and weekends, neural network training for weekdays and weekends was performed independently for better forecasting performance. As a result, forecasting for weekdays and weekends is performed independently. However, in [8], an open-loop environment with actual load and weather data is used for training the ANN and then deployed in a closed-loop environment with the projected load as the feedback input to create a forecast. Unlike other artificial neural network-based forecasting methods, the own output of the proposed method is used as an input to increase the accuracy, essentially creating a load feedback loop and lowering the reliance on external inputs. A new approach for short-term load demand forecasting is proposed in [9]. For improved accuracy, this method was built by integrating a memory network with a convolutional neural network. Later in [10], the application, benefits, and limitations in power consumption of short-term forecasting approaches and electric energy consumption in microgrids are discussed. Two strategies are used to obtain the short-term load forecasts: artificial neural networks and a data management based group method are proposed. To predict short-term load demand in microgrids, Reference [11] suggested a dragonfly algorithm-based support vector machine technique. Empirical mode decomposition, particle swarm optimization (PSO), and adaptive network-based fuzzy inference systems are used in a hybrid approach for short-term load forecasting in microgrids. The proposed method decomposes the difficult load data series into a set of many intrinsic mode functions and a residue using the empirical mode decomposition algorithm and PSO algorithm to optimize an adaptive neuro-fuzzy inference system (ANFIS) model for each intrinsic mode function component and the residue [12].

A hybrid methodology for forecasting very short-term loads in microgrids, genetic algorithms, particle swarm optimization, and adaptive neural fuzzy inference systems are all used in the suggested method [13]. Later in [14], the authors used an adaptive fuzzy model to tackle the problem of short-term load forecasting for a day ahead to transfer information between different locations. The suggested solution divides daily load profile predictions into smaller, easier sub-problems, each of which is handled separately using a Takagi-Sugeno fuzzy model. This choice is made in order to solve smaller sub-problems more effectively, resulting in enhanced forecasting accuracy. A novel methodology is proposed in [15] for identifying and measuring the impact of important components in the energy demand forecasting model on the Mean Absolute Percentage Error (MAPE) criterion and error performance. The support vector machine approach, the random forest regression method, and the long short-term memory neural network method are three commonly used machine-learning methods for load forecasting discussed in [16], as well as their features and uses. These approaches' properties and applicability are studied and compared. A fusion forecasting strategy and a data preprocessing technique are proposed for improving forecasting accuracy by combining the advantages of these methods. On a building-by-building basis, eight techniques for day-ahead electrical load estimations at the grocery store, school, and home are compared in [17]. Machine learning and statistics were utilized to compare these methods, and a median ensemble was employed to combine the different forecasts. Reference [18] describes a short-term load forecasting system that employs a comparable day strategy to predict power demand 24 h in advance, as well as long short-term memory and wavelet transform to improve forecasting accuracy. A short-term power load forecasting approach based on the modified exponential smoothing grey model is presented in [19] to increase the prediction accuracy. Using grey correlation analysis first determines the major component affecting the power load. Second, the smoothed sequence is used to create a grey prediction model that agrees with the exponential trend and has an optimal background value. Later in [20], the load is divided into seven time periods using a regional load characteristic law, and a time-segment Back Propagation (BP) neural network model is developed. Moreover, in [21], a thorough examination of forecasting models is performed to determine which model is best suited for a specific instance or scenario. The comparison was based on 113 separate case studies described in 41 academic journals. Timeframe, inputs, outputs, scale, data sample size, error kind, and value have all been considered as comparison criteria. In [22], the mathematical model is constructed using a machine learning neural network intelligence algorithm in this study, and the optimization is enhanced from the perspectives of data preparation, network structure selection, and learning algorithm. Furthermore, short term load forecasting in a microgrid (MG) is performed using hybrid machine learning methods [23]. The suggested model combines Support Vector Regression (SVR) and Long Short-Term Memory (LSTM). The proposed method is applied to data from a rural MG in Africa to forecast the load demand. The input variables are factors that influence the MG load, such as varied household kinds and commercial entities, whereas the target variables are load profiles. To anticipate electric loads, an SVR model with an immunity algorithm (IA) is presented in [24] and the SVR model parameters are determined using the immunity algorithm. Later, different state-of-the-art ML techniques have been applied in [25] to examine their performance. These include logistic regression (LR), support vector machines (SVM), naïve Bayes (NB), decision tree classifier (DTC), K-nearest neighbor (KNN), and neural networks (NNs). The primary goal of this work is to provide a comparison of machine learning methods for short-term load forecasting (STLF) in terms of accuracy and forecast error. However, authors in [26] proposed the prophet model, which uses both linear and non-linear data to predict original load data, although there are still some residuals that are considered non-linear data. These residuals (non-linear data) are then trained using long short-term memory (LSTM), and both the Prophet predicted data and LSTM are then trained using Back Propagation Neural Network (BPNN) to improve prediction accuracy. Later, to predict daily energy consumption, Reference [27] investigates the performance of three machine learning models

(SVR, Random Forest, and Extreme Gradient Boosting (XGBoost)), three deep learning models (Recurrent Neural Networks (RNNs), Long Short-Term Memory (LSTM), and Gated Recurrent Unit (GRU)), and a classical time series model (Autoregressive Integrated Moving Average (ARIMA)).

In the above-discussed literature, the researchers used different machine learning forecasting methods to forecast the load demands limited to single microgrid cases. At the same time, many researchers suggested that machine learning algorithms are preferable for short-term and very short-term load forecasting. So, this article forecasts a 24-h day-ahead load demand using several machine learning methods, such as linear regression, support vector machines, long short-term memory, and artificial neural networks (ANN). To test the effectiveness of these methodologies in predicting short-term load demands in a microgrid cluster, several measures such as "root mean square error", "R-squared", "mean square error", "mean absolute error", "mean absolute percentage error", and calculation time are computed. The following are some key contributions of this article:

1. Cluster microgrids are proposed by interconnecting neighborhood microgrids.
2. Linear regression (quadratic), support vector machine, long short-term memory, and artificial neural networks machine learning algorithms are implemented for day-ahead load demand forecasting in cluster microgrids.
3. Levenberg–Marquardt optimization algorithm-based ANN model is proposed for effective load day-ahead load demand forecasting in the cluster microgrids.

Based on the objective discussed above, the rest of the article is structured as follows. Section 2 presents the layout of cluster microgrids, Section 3 provides various machine learning algorithms, Section 4 summarizes the results and findings, and Section 5 concludes with the article outcomes.

2. Description of the Proposed Cluster Microgrids

The cluster microgrids is also known as the interconnected microgrid system and is formed by interconnecting four neighborhood microgrids, as shown in Figure 1. It shows the layout of the "Cluster Microgrid" system considered for the study.

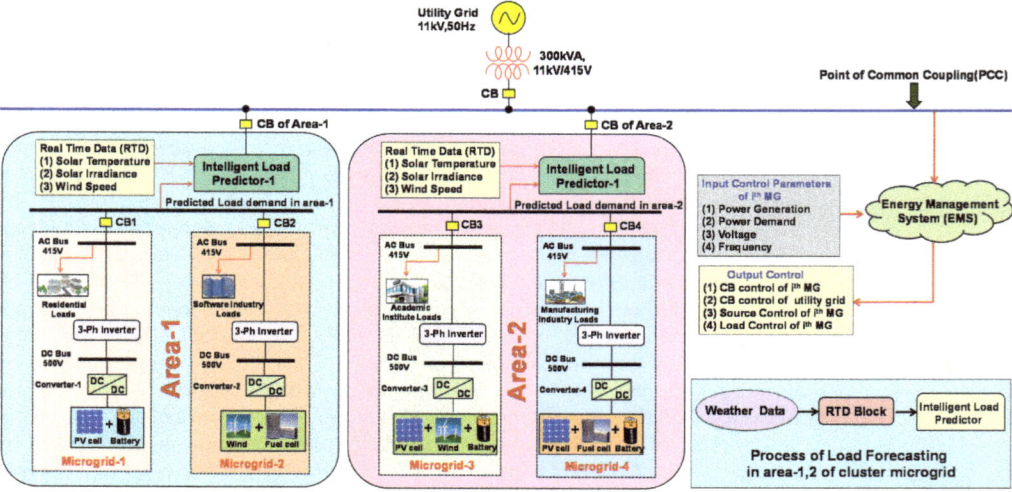

Figure 1. The architecture of cluster microgrid (single line diagram).

The proposed system is separated into two areas: area-1 and area-2, both of which are deemed interoperable inside the cluster. Area-1 is formed by connecting Microgrid-1 (Residential Building) and Microgrid-2 (Software Building) and area-2 is formed by connecting Microgrid-3 (Academic Institute Building) and Microgrid-4 (Manufacturing

Industry Building) in the selected location. These interconnected and interoperable areas are further integrated into the conventional grid through the Common Coupling Point. Each area in the cluster microgrids is equipped with renewable sources which are available freely in that location and also consists of two microgrids with local agent controllers of that building and an intelligent forecaster with a Real-Time Data (RTD) measurement block. The interrelated cluster microgrid is considered in such a way that it has to operate in both self-standing modes and also in grid-related modes. The RTD block collects the real-time information on solar temperature, irradiance, speed of the wind turbine, and predicted load demand from the selected location. The output data from the RTD block is applied to the corresponding intelligent predictor for forecasting. Forecasted load demand information of the cluster microgrid is applied as an input for the energy management system to function effectively. Correspondingly, the energy management system produces the control signals to the circuit breaker to import or export based upon the energy needs. The energy management system in this architecture is designed to make the energy transactions during excess/deficit power conditions to/from the neighborhood area or utility grid. However, the complete modelling of all renewable sources and constituents is given in [28]. Table 1 contains the information on various parameters considered for the simulation of the cluster microgrid system.

Table 1. Typical values of components used in the proposed system [29].

Parameter	Typical Ratings	Units
Electric Charge	1.6×10^{-19}	Coulomb
Boltzmann's Constant	1.3805×10^{-23}	Joule/Kelvin
Energy Gap	1.11	eV
Base power wind turbine	1100	VA
Rotor Efficiency	0.45	–
Battery voltage	90	volt
Battery capacity	7.5	Ampere hour
Battery state of charge	100%	–
Battery response time	50	sec
Temperature of fuel cell stack	342	kelvin
Faradays Constant	96,484,600	–
Gas Constant	8314.656	–
EMF of fuel cell (No load)	0.85	volt
No. of Cells in fuel cell stack	85	–
Utilization factor	0.85	–
H_2-O_2 flow ratio	1.268	–
Duty cycle of dc/dc converter	0.74	–
Inductor value of dc/dc converter	205	µH
Capacitor value of dc/dc converter	20	µF
Initial voltage across capacitor	220	volts
Switching frequency	75	kHz
Cutoff frequency of LPF	500	Hz
Damping factor of LPF	0.707	–
Length of transmission line	10	km
Power of a transformer	300	kVA
Frequency	50	Hz
Primary winding voltage (Line-Line)	420	volts
Resistance connected in primary winding	0.016	Ω
Secondary winding voltage (Line-Line)	420	volts
Resistance connected in secondary winding	0.016	Ω
Voltage of conventional grid	11,000	volts
Frequency of conventional grid	50	Hz
Source resistance of conventional grid	0.8929	Ω
Source Inductance of conventional grid	16.58	mH

Energy Management System (EMS)

An energy management system (EMS) is an information system on a software platform that supports the functionality of generating, transmitting, and distributing electrical energy at a low cost, according to the international standard IEC 61970. Energy management in microgrids is a computer-based control method that ensures the best functioning of the system. In a variety of ways, a microgrid must optimize the usage of renewable energy sources.

Machine interfaces and supervisory control and data acquisition systems (SCADA) are two EMS components that carry out decision-making procedures [28,29]. This EMS is implemented in the proposed cluster microgrid system to manage the resources connected to all microgrids. This ensures that energy transactions are seamless and that grid frequency is maintained under dynamic load conditions. The flowchart for implementing the EMS is shown in Figure 2.

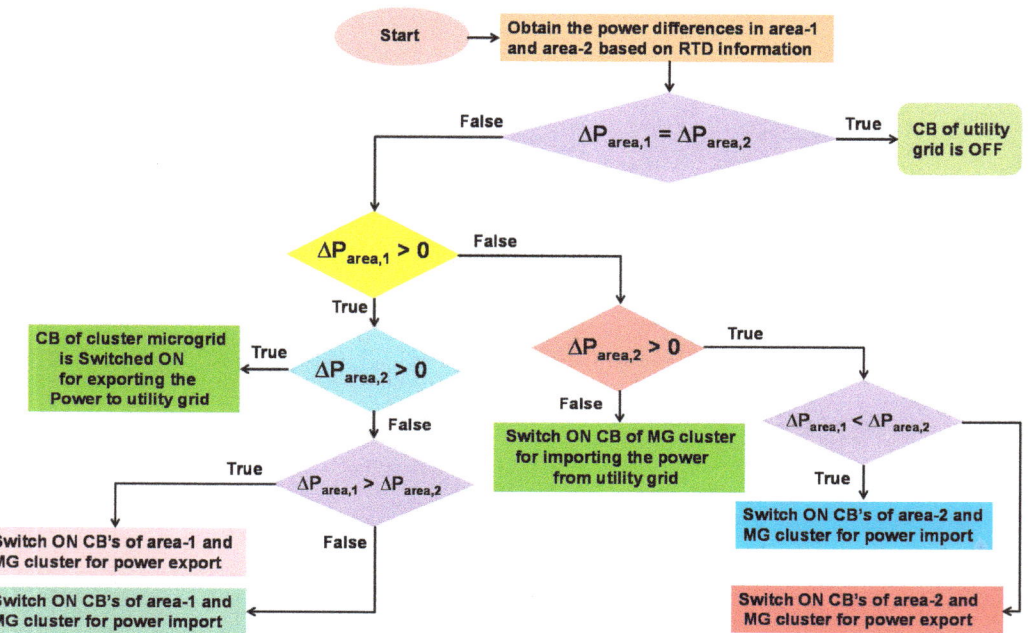

Figure 2. Flowchart of energy management system developed for cluster microgrids.

The economic benefit to the consumer can be enhanced by utilizing on-site distributed generators and lowering reliance on the main grid. The EMS consists of a central controller with direct commands to each distributed energy resource in each microgrid, data acquisition of microgrid operation characteristics and parameters, and information acquisition from the forecasting system, all of which are used to optimize appropriate unit commitment and resource dispatch in relation to the preset objectives. Three layers are incorporated into the cluster microgrid system, namely an external layer that is dedicated to data collection (live weather data, electricity consumption data, etc.), a prediction layer that is used to predict weather conditions and local demand, and an operational layer that consists of energy management algorithms, which are implemented to dynamically regulate energy flow among the devices based on prediction data. The goals of this centralized control algorithm are to forecast energy and electricity load, govern dynamic energy management, and send commands to physical equipment to respond appropriately.

3. Machine Learning Techniques
3.1. Linear Regression (Quadratic)

The model is called linear regression, which optimizes the fit of functions to training data by utilizing the squared Euclidean distance metric. In the simplest model $y = \lambda_1 x + \lambda_0$, a straight line with gradient λ_1 is fitted to the data, and the intercept of y is λ_0. The depiction of a linear regression model is given in Equation (1) [30].

To map the relationship between terms x_i and x_j, interaction terms $x_i x_j$ might be used. If the effect x_i on y is dependent on the other factors $x_{j, j \neq i}$, this produces better results than a simple linear regression given in Equation (2). In a quadratic model, a quadratic function is fitted to the data and optimized using least squares. As a result, as shown in Equation (3), the model has an intercept, linear terms, interactions, and squared terms. Figure 3 shows the flowchart for implementing the linear regression model in the MATLAB/Simulink environment.

$$y = \lambda_0 + \lambda_1 x_1 + \lambda_2 x_2 + \ldots + \lambda_n x_n \tag{1}$$

$$y = \lambda_0 + \lambda_1 x_1 + \lambda_2 x_2 + \ldots + \lambda_n x_n + \lambda_{12} x_1 x_2 + \ldots + \lambda_{1n} x_1 x_n + \ldots + \lambda_{(k-1)k} x_{k-1} x_k \tag{2}$$

$$\left. \begin{array}{r} y = \lambda_0 - \lambda_1 x_1 + \lambda_2 x_2 + \ldots + \lambda_n x_n + \lambda_{12} x_1 x_2 + \ldots \\ + \lambda_{1n} x_1 x_n + \ldots + \lambda_{(k-1)k} x_{k-1} x_k + \lambda_{11} x_1^2 + \ldots + \lambda_{kk} x_k^2 \end{array} \right\} \tag{3}$$

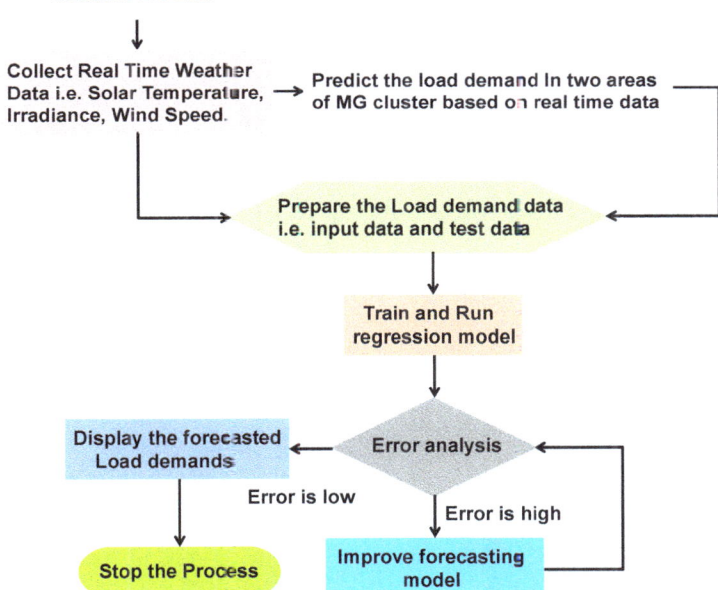

Figure 3. Flowchart for implementing linear regression (quadratic) method.

3.2. Support Vector Machine

Machine learning approaches for data classification and regression, such as support vector regression (SVR) and support vector machines (SVMs), have been used to forecast the electric load demand. Vapnik proposed the support vector machines (SVMs) in 1995. The SVR's main premise is to map the original data "α" nonlinearly into a higher dimensional feature space. Hence, the training dataset is given as $\{(\alpha_n, \beta_n)\}_{n=1}^{k}$, in which the input vector is α_n; the target vector is β_n, and k is the total number of data patterns of the training data. The target of SVM is to generate a decision function of SVM in Equation (4) by minimizing the risk function given in Equation (5) [31].

$$\mu = \tau(\alpha) = \omega \cdot \rho(\alpha) + \theta \tag{4}$$

$$R_R = \frac{M}{k} \sum_{i=1}^{k} L(\beta_i, \tau(\alpha)) + 0.5 \|\omega\|^2 \tag{5}$$

Therefore, the function value,

$$L(\beta_i, \tau(\alpha)) = \begin{cases} |\beta_i - \tau(\alpha)| - \varepsilon; & if\ |\beta_i - \tau(\alpha)| \geq \varepsilon \\ 0 & if\ |\beta_i - \tau(\alpha)| < \varepsilon \end{cases}$$

The following is the terminology used in Equations (4) and (5):
ω—Weight vector used to control the model smoothness;
θ—"Bias" parameter;
$\rho(\alpha)$—High-dimensional space which is mapped nonlinearly into input space α;
$\frac{M}{k}\sum_{i=1}^{k} L(\beta_i, \tau(\alpha))$—Empirical risk function;
$0.5\|\omega\|^2$—Regularization term used to determine function complexity;
ε—Tube size (user determined);
M—Regularized constant (user determined).

Two positive slack variables, namely ϕ, ϕ^*, are incorporated to signify the distance between original values and the ε-associated tube's edge values; then, Equation (5) is converted to the form given in Equation (6).

$$\min(R_R) = M \cdot \sum_{i=1}^{k} (\phi_i + \phi_i^*) + 0.5\|\omega\|^2 \tag{6}$$

Subjected to the constraints

$$\beta_i - \omega\alpha_i - \theta \leq \varepsilon + \phi_i; i = 1, 2, 3 \ldots k$$
$$\omega\alpha_i + \theta - \beta_i \leq \varepsilon + \phi_i^*; i = 1, 2, 3 \ldots k$$
$$\phi_i \phi_i^* \geq 0, i = 1, 2, 3 \ldots k$$

Using primal Lagrangian, the dual optimization problem of the above primal one is obtained as follows [31,32].

$$L(\omega, \theta, \phi_i, \phi_i^*, \mu_i, \mu_i^*, \Omega_i, \Omega_i^*) = 0.5\|\omega\|^2 + M\left(\sum_{i=1}^{k}(\phi_i + \phi_i^*)\right)$$
$$- \sum_{i=1}^{k} \Omega_i[\varepsilon + \phi_i + \omega\rho(\alpha_i) + \theta - \beta_i]$$
$$- \sum_{i=1}^{k} \Omega_i^*[\beta_i + \varepsilon + \phi_i^* - \omega\rho(\alpha_i) - \theta] \tag{7}$$

The above Equation (7) is minimized by using variables $\omega, \theta, \phi_i, \phi_i^*$ and is maximized with respect to $\mu_i, \mu_i^*, \Omega_i, \Omega_i^*$; so, the following Equations (8)–(11) will be obtained.

$$\frac{\partial L}{\partial \omega} = \omega - \sum_{i=1}^{k} \rho(\alpha_i)(\phi_i - \phi_i^*) = 0 \tag{8}$$

$$\frac{\partial L}{\partial \theta} = \sum_{i=1}^{k} (\Omega_i - \Omega_i^*) = 0 \tag{9}$$

$$\frac{\partial L}{\partial \phi_i} = M - \Omega_i - \mu_i = 0 \tag{10}$$

$$\frac{\partial L}{\partial \phi_i^*} = M - \Omega_i^* - \mu_i^* = 0 \tag{11}$$

Applying Kuhn–Tucker conditions to regression and Equation (6) gives dual Lagrangian by substituting Equations (8)–(11) into Equation (7). The following Equation (12) is the dual Lagrangian function obtained by considering the kernel function as $K(\alpha_i, \beta_i) = \rho(\alpha_i)\rho(\alpha_j)$

$$\psi(\Omega_i, \Omega_i^*) = \sum_{i=1}^{k} \beta_i(\Omega_i - \Omega_i^*) - \varepsilon \sum_{i=1}^{k}(\Omega_i + \Omega_i^*) - 0.5 \sum_{i=1}^{k}\sum_{j=1}^{k}(\Omega_i - \Omega_i^*)\left(\Omega_j - \Omega_j^*\right)K(\alpha_i\alpha_j) \qquad (12)$$

Subjected to the constraints

$$\sum_{i=1}^{k}(\Omega_i - \Omega_i^*) = 0; \text{ where } \begin{array}{l} 0 \leq \Omega_i \leq M; i = 1,2,3\ldots k \\ 0 \leq \Omega_i^* \leq M; i = 1,2,3\ldots k \end{array}$$

The Lagrangian multipliers defined in Equation (12) must satisfy the equality constraint $\Omega_i \Omega_i^* = 0$. Hence, the regression function is obtained as given in Equation (13).

$$\tau(\alpha, \Omega, \Omega^*) = \sum_{i=1}^{m}(\Omega_i - \Omega_i^*)K(\alpha, \alpha_i) + \theta \qquad (13)$$

Figure 4 shows the sequence of steps to be considered for implementing the support vector machine model in the MATLAB/Simulink environment.

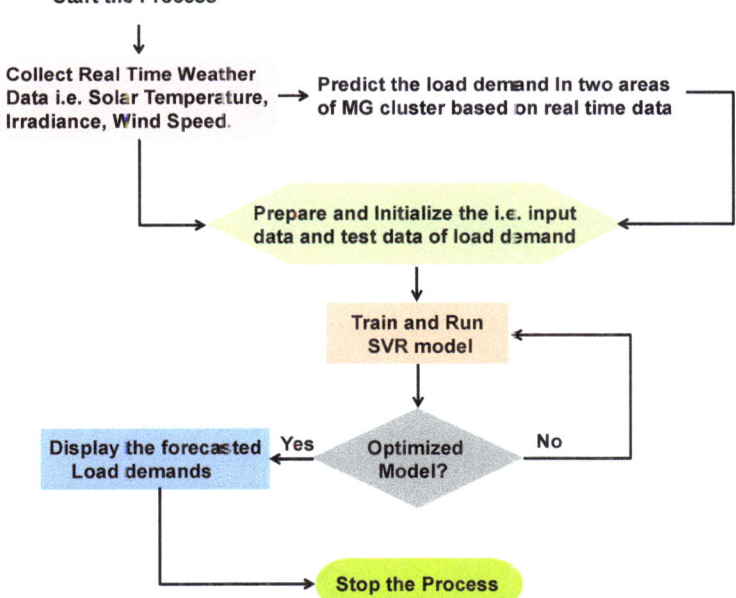

Figure 4. Flowchart for implementing support vector machine model.

3.3. Artificial Neural Networks (ANN)

In this article, as an application of artificial intelligence (AI), an artificial neural network (ANN) is employed as an intelligent predictor. The concept of ANN was introduced several years ago for different applications because of its capacity to forecast the data and also to control the system response effectively. It has been demonstrated that ANN is one of the effective solutions for all forms of real-time nonlinear issues. An artificial neural network (ANN) is designed based on the interconnection of processing elements that carries information. McCulloch et al. first introduced various neural network architectures, such as single layer and multilayer feed-forward networks, which are explained as follows [33].

3.3.1. Single Layer Feed Forward Network

In this schematic view, the network has two layers, namely the input and output layers. The primary function of the input layer is to transmit signals to other neurons. The neurons in the input layer receive the input signals in the input layer, and the output layer neurons send output signals. In this type of structure, the signals are transferred from the input layer to the output layer but vice versa is not possible, so it is named a feed-forward network. The general architecture is as shown in Figure 5a, where $x_1, x_2, x_3 \ldots x_n$ are the input layer elements and $y_1, y_2, y_3 \ldots y_m$ are the output layer elements, and w_{ji} are the weights associated between the input and output layer.

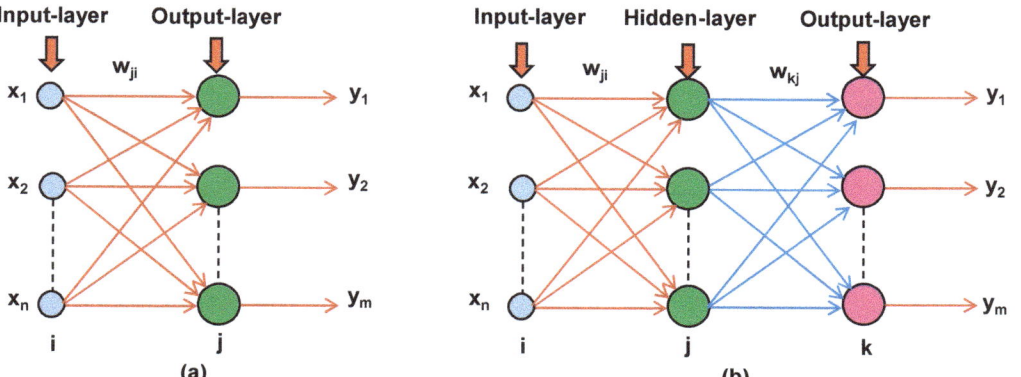

Figure 5. Schematic view of (**a**) Single-Layer Feed Forward (FF)-ANN, (**b**) Multi-Layer FF-ANN [33].

3.3.2. Multi-Layer Feed Forward Network

This network consists of one input layer, one output layer, and single or multi hidden layers. The processing elements for the hidden layer are hidden neurons only. Before sending the inputs to the output layer, computation is performed by hidden neurons in the hidden layer. The general schematic is shown in Figure 5b. Here, w_{ji} is the weight that links input and hidden layers and w_{kj} is the weight that links hidden and output layers.

Figure 6 depicts the flowchart of the intelligent predictor, i.e., artificial neural network (ANN), which is to be implemented in the Simulink environment. ANN is very flexible and can be easily adaptable to all complex nonlinear problems. The following are the steps to be taken for training the ANN with different weight updating algorithms:

- Select the input data, such as temperature, Diffuse Horizontal Irradiance (DHI), wind, and loads from selected locations;
- Select the number of hidden layers;
- Select proper active function for hidden and output layers.

For all feed-forward (FF) networks, the relationship between inputs, hidden, and output samples are obtained from Equations (14) and (15).

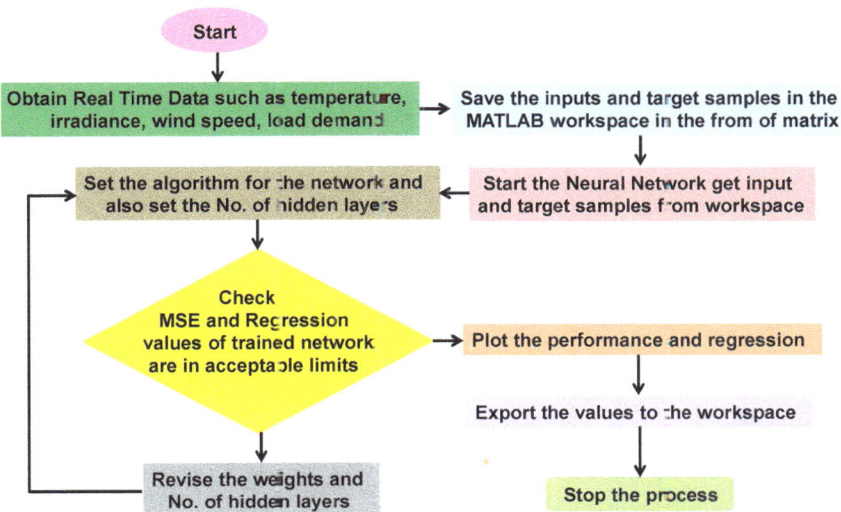

Figure 6. Flowchart for implementing artificial neural networks.

$$\psi_k = \zeta_n \left(\sum_{j=0}^{h} \omega_{(n)\theta j} \cdot \rho_j \right) \quad (14)$$

$$\text{where, } \rho_j = \zeta_{n-1} \left(\sum_{i=0}^{N} \omega_{(n-1)ji} \mu_i \right) \quad (15)$$

Here, N is the input layer dimension, h- is a hidden layer dimension, k is the output layer dimension, $\omega_{(n)\theta j}$ is the output layer weight, $\omega_{(n-1)ji}$ is the hidden layer weight, and ζ_n is the activation function used for the feed-forward neural network.

4. Results' Validation and Discussion

4.1. Day-Ahead Load Demand Forecasting Using Linear Regression, Support Vector Machine, and Artificial Neural Networks

The goal of this work is to determine the best day-ahead load demand forecasting solution in cluster microgrids. We gathered data on solar and wind factors from Vijayawada city in the state of A.P., India, with a "location ID" of 44665, a latitude of −16.65°, and a longitude of −80.65° [34,35]. Figure 7a–d shows the characteristics of the actual dataset, such as solar irradiance, temperature, wind speed, and the electric load consumption in a specified location for one month from 1 January 2019 to 31 January 2019. Test data are considered for the period from 10 January 2019 to 16 January 2019. Our job is to calculate daily, weekly, and monthly electricity usage by predicting consumption for each hour of the day. Machine learning algorithms anticipate the future value of a time series data collection by discovering correlations between historical data attributes and using the revealed associations to forecast the future value.

Pre-processing is essential for improving data quality and the effectiveness of machine learning algorithms. In each machine learning (ML) model, normalization and data transformation are two common pre-processing procedures. The variables in a cluster microgrid dataset are spread across various ranges, resulting in a bias favoring values with greater weights, lowering the effectiveness of the framework. A zero-mean normalization technique is employed in the study for data normalization on the load and temperature variables because attribute normalization improves the convergence rate and numerical stability of NN training

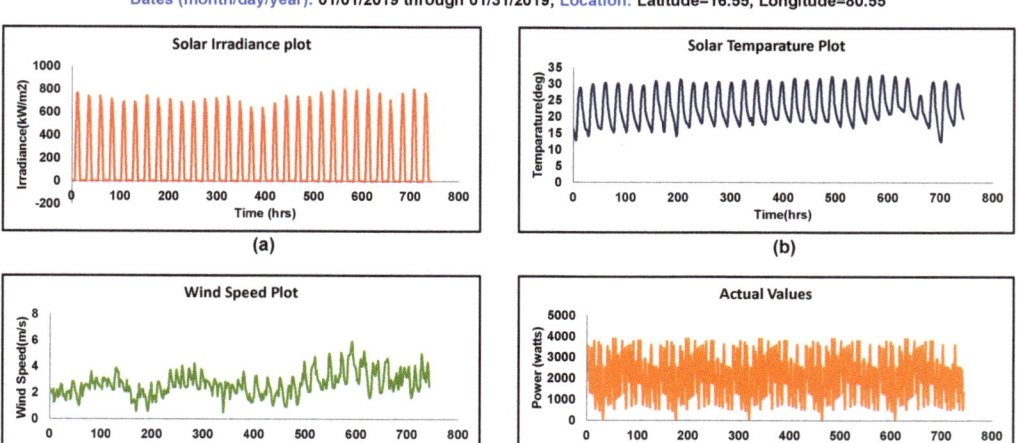

Figure 7. Characteristics of data considered in the specified location: (**a**) solar irradiance, (**b**) solar temperature, (**c**) wind speed, and (**d**) actual values of power.

Data must be quantifiable because machines process them using mathematical calculations. Data encryption is performed during data pre-processing, which converts quasi inputs to numeric inputs before giving them to ML frameworks because the majority of the dataset contains both category and numerical information. The data are then separated into training and testing datasets. The training dataset is used to create the machine learning algorithms, which are then tested against a new dataset to see how well they perform. In this study, 30% of the dataset is used to evaluate the performance of the developed ML algorithms, whereas 70% is used to comprehend the ML algorithms [25]. The "MATLAB/Simulink software 2021a" is used to model the proposed cluster microgrid and also to execute the machine learning algorithms.

In this work, we have obtained the real-time information of solar temperature, irradiance, and wind speed in the two interoperable areas at the abovementioned location and then obtained the loads in area1 and 2 concerning the real-time values, which are then applied as four inputs to the intelligent predictor to forecast the load demand in the next 30 days; then, the total estimated forecasted power at the PCC of the cluster microgrid is given to the EMS of the system. The EMS then generates control signals to export/import power to/from the central utility grid.

The stress on the electrical grid is lowered as a result of providing consumers with more reliable and efficient power. Figure 8 shows the performance plots of various machine learning algorithms used in this work for day-ahead load demand forecasting for the aforesaid period considered for the study. The plot is drawn by taking 120 data samples on the x-axis and the predicted load on the y-axis.

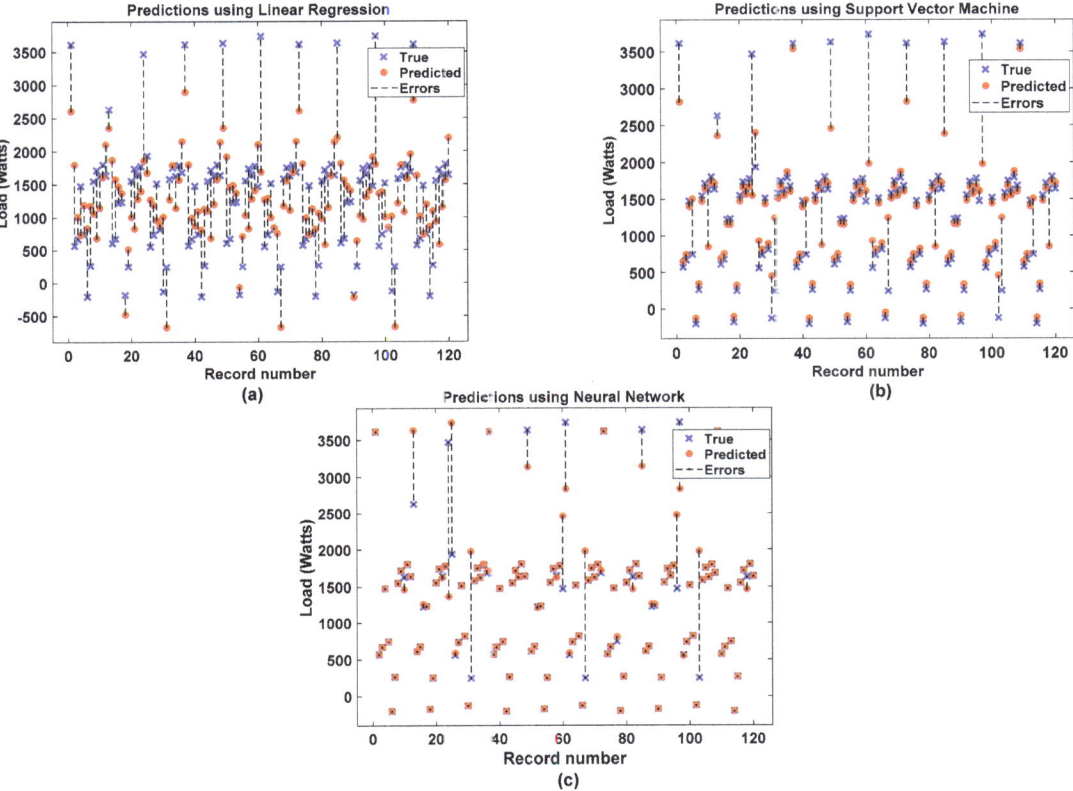

Figure 8. Performance plots showing the accuracy of predicted values with (**a**) Linear Regression (quadratic), (**b**) Support Vector Machine, and (**c**) Artificial Neural Network models.

The actual and predicted values are shown in the plot and also error values are marked with dashed lines. From the plot, it is observed that the performance of the neural network regression model gives more accurate load demand forecasted values when making the comparison with the remaining LR (quadratic) and SVM models. A residual measures how far a point is vertically from the regression line. To visually confirm the correctness of the machine learning model, we must use residual plots. Plotting all residual values across all independent variables can be tricky, so we can either make separate plots or use other validation statistics, such as adjusted R^2 or MAPE scores.

So, Figure 9 shows the typical residual plots of all the machine learning methods used. Curve fitting is described as the model that provides the greatest fit to the specific curves in one's dataset in regression analysis. Linear connections are easier to fit and interpret than curved variable relationships. We use root mean square error (RMSE), R-squared, mean square error (MSE), mean absolute error (MAE), mean absolute percentage error (MAPE), and time of computation metrics to assess the prediction accuracy of each machine learning model in all cases. The forecasting error metrics are obtained as given in Equations (16)–(20).

$$RMSE = \sqrt{\frac{\sum_{\tau=1}^{m}(\lambda_\tau - \hat{\lambda}_\tau)^2}{m}} \qquad (16)$$

$$MSE = \frac{1}{m}\sum_{\tau=1}^{m}\left(\lambda_\tau - \hat{\lambda}_\tau\right)^2 \qquad (17)$$

$$MAE = \frac{1}{m}\sum_{\tau=1}^{m}\left|\lambda_\tau - \hat{\lambda}_\tau\right| \qquad (18)$$

$$MAPE = \frac{1}{m}\sum_{\tau=1}^{m}\frac{\left|\lambda_\tau - \hat{\lambda}_\tau\right|}{\lambda_\tau} \qquad (19)$$

$$R^2 = 1 - \frac{\sum\left(\lambda_\tau - \hat{\lambda}_\tau\right)^2}{\sum\left(\lambda_\tau - \overline{\lambda}_\tau\right)^2} \qquad (20)$$

where m is the number of data points, λ_τ are the actual values, $\hat{\lambda}_\tau$ are the forecasted values, and $\overline{\lambda}_\tau$ are the mean values.

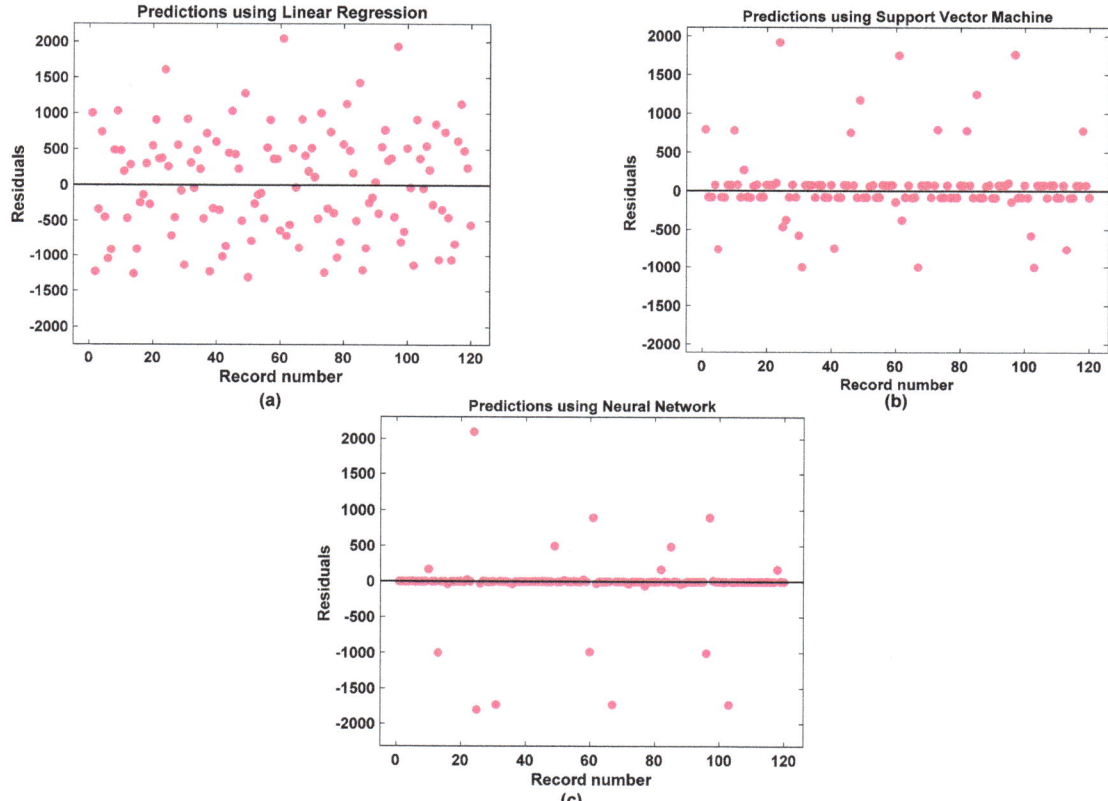

Figure 9. Residual plots of (**a**) Linear Regression (quadratic), (**b**) Support Vector Machine, and (**c**) Artificial Neural Network models.

Figure 10 shows how the data are fitted for perfect load forecasting using machine learning algorithms. In this, we observed how well the ANN is trained to give the best accurate results. Table 2 gives the comparison of different metrics obtained during the day-ahead load demand forecasting with the use of machine learning algorithms, such as linear regression [30], support vector machine [25,31,32], and artificial neural networks. Later, these results are compared and verified with the time series long short term memory

(LSTM) forecasting method [36]. Design parameters of the LSTM method are given in Appendix A. The results show that the artificial neural network regression model effectively forecasts day-ahead load demands in cluster microgrids. So, we propose the ANN is the best machine learning technique for forecasting the day-ahead load demands in cluster microgrids. The performance of various machine learning algorithms considered for the study can be viewed Figures 8–10 by considering performance metrics. However, the forecasted values are shown in Figure 11a, which are obtained with respect to the actual values. Similarly, the area plot is given in Figure 11b.

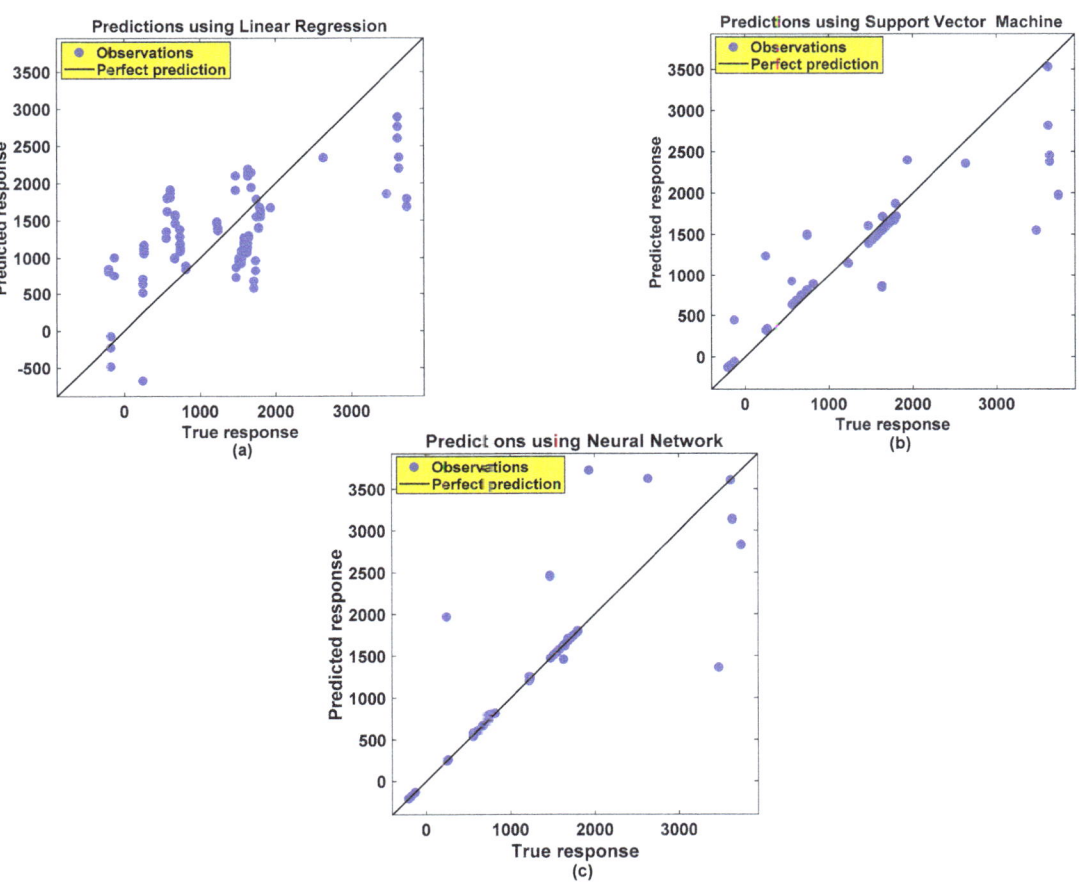

Figure 10. Plot fitting of predicted data and observations with (**a**) Linear Regression (quadratic), (**b**) Support Vector Machine, and (**c**) Artificial Neural Network models.

Table 2. Comparison of performance metrics computed for LR (quadratic), SVM, ANN, and LSTM.

Parameter	LR (Quadratic) [30]	SVM [25,31,32]	LSTM [36]	ANN (Proposed)
RMSE	736.68	438.54	1456.3	426.04
R-squared	0.37	0.78	0.85	0.79
MSE	5.427×10^5	1.9232×10^5	2.1208×10^6	1.8151×10^5
MAE	621.19	235.97	182.94	131.72
MAPE	26.34%	21.52%	42.35%	13.92%
Computation Time (s)	1.8124	0.9999	25	2.829

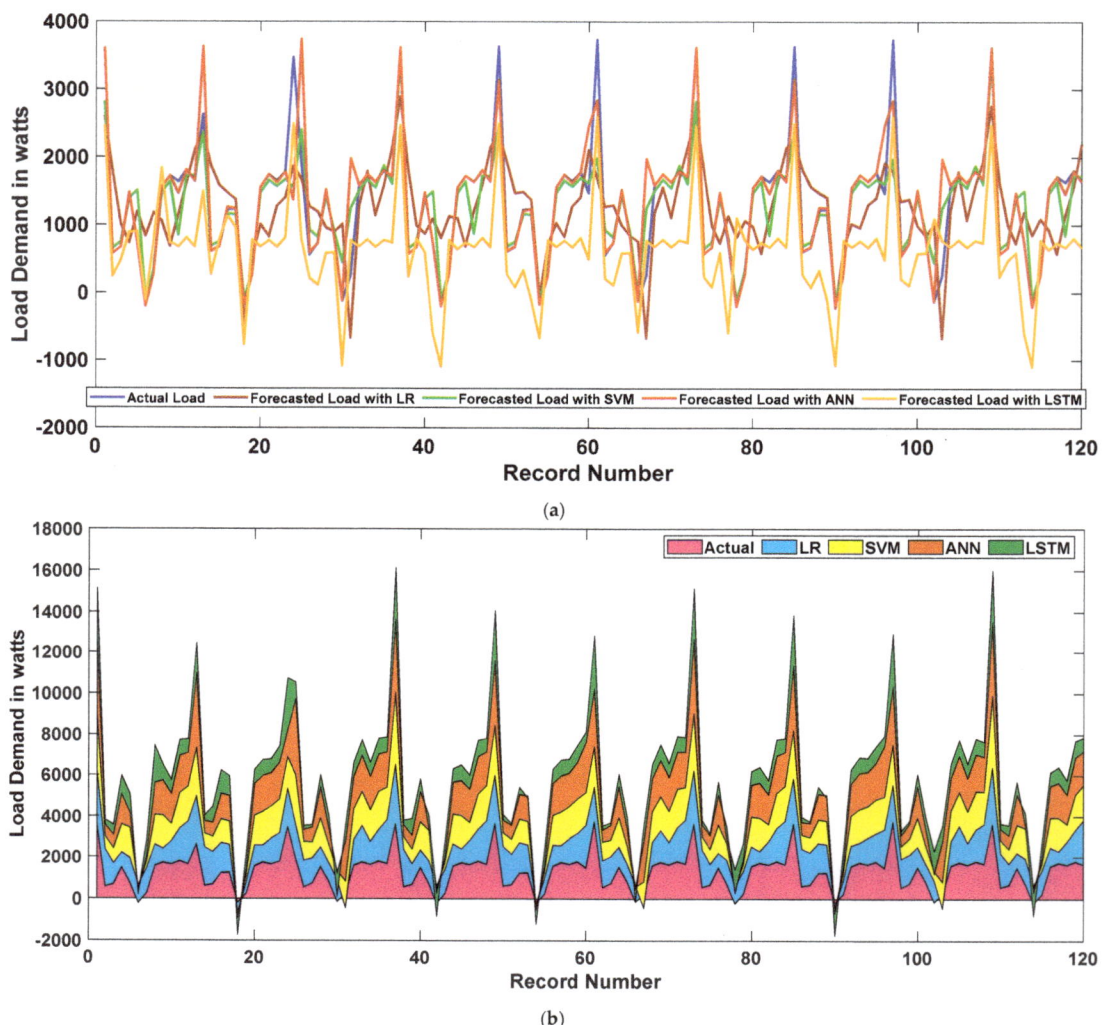

Figure 11. Cluster microgrid load profile plots. (**a**) Actual vs. forecasted load demands with LR (quadratic), SVM, ANN and LSTM. (**b**) Area plot with LR (quadratic), SVM, ANN, and LSTM.

4.2. Identification of Best Optimization Algorithm of Neural Networks for Effective Forecasting

The discussion in the previous Section 4.1 shows that the artificial neural network gives fruitful results in day-ahead load demand forecasting. Hence, in this section, we identified the best optimization algorithm to propose for neural networks for effective functioning. The three optimization algorithms, viz. (1) Levenberg–Marquardt (LM) algorithm, (2) Bayesian Regulation (BR) algorithm, and (3) Scaled Conjugate Gradient algorithm are considered for training the neural networks.

4.2.1. Levenberg–Marquardt (LM) Algorithm

The alternative name for the Levenberg–Marquardt (LM) algorithm is the "Damped Least Square" method. It is particularly designed to work based on the loss function, which is expressed as the sum of squared errors. The approximated hessian matrix is obtained by using the Jacobean matrix and gradient vectors, which are obtained using the following Equations (21) and (22). Jacobean matrix is obtained with the loss function and is given by [33].

$$J_{mn} = \frac{\partial E_m}{\partial W_n} \quad (21)$$

where $m = 1, 2, 3, \ldots, i$; $n = 1, 2, 3, \ldots, j$; the number of instances in the dataset is "i" and the number of parameters in the network is "j". The gradient vector is obtained as follows,

$$H_f \approx 2J^T \cdot J + \lambda I \quad (22)$$

The further approximated Hessian matrix is obtained from Equation (23),

$$W^{(r+1)} = W^{(r)} - \left(J^{(r)T} \cdot J^{(r)} + \lambda^{(r)} \cdot I\right)^{-1} \cdot \left(2 \cdot J^{(r)T} \cdot E^{(r)}\right); r = 0, 1, 2, 3 \ldots \quad (23)$$

4.2.2. Bayesian Regulation (BR) Algorithm

The Bayesian Regulation (BR) technique is more efficient than typical backpropagation methods and is based on Bayes' theorem. The nonlinear regression relations are translated to second-order linear regression-based mathematical equations during the BR process. The most difficult part is deciding on absolute fitting values for the function parameters. The BR framework in ANN works by interpreting provided network parameters probabilistically, which differs from typical training approaches. This chooses a set of weights based on the error function minimization. However, in the BR method, a performance function given in Equation (24) is utilized to find the error, or the difference between real and anticipated data, throughout the training phase. Regularization adds an extra term and function to a BR method to achieve smooth mapping, which uses a gradient-based optimization to minimize the objective and performance function as provided in Equation (25). To address the additional noise present in the targets, the posterior distribution of weights of the neural network will be modified as needed after the data are taken for training [33].

$$f_n = \mu_t(\tau|\omega, a) + \frac{1}{n}\sum_{x=1}^{p}(\alpha_x - \beta_x)^2 \quad (24)$$

$$f_n = \Omega \cdot \mu_t(\tau|\omega, a) + \Psi \mu_m(\omega/a) \quad (25)$$

Here, f_n is performance function, μ_t is network error values (sum of squares), τ is training set of input and output targets, a is an architecture of neural network which consists of information about the number of layers and their units, μ_m is network weights (sum of squares), $\Psi \mu_m(\omega/a)$ is weight decay, and Ψ is a rate of decay.

4.2.3. Scaled Conjugate Gradient (SCG) Algorithm

M. Hestenes and Eduard S. created scaled conjugate techniques. These are primarily used to solve linear equations. In conjugate gradient methods, there are numerous sub methods. One of these sub methods is the SCG algorithm. Constrained optimization, curve fitting, and more uses of the SCG algorithm can be found. It uses feed-forward artificial neural networks. These approaches solve when all errors are inside the range of anticipated values. The calculation of the direction of the weights, which is practically difficult, is the most important component of the conjugate methods. Equations (26) and (27), respectively, are the training data and the parameter vector functions. The main drawback of the SCG technique is that it does not supply any data for calculating and inverting the Hessian matrix [33].

$$S_{r+1} = G_{r+1} + S_r \cdot \alpha_x \quad (26)$$

$$W_{r+1} = W_{r+1} + S_r \cdot \beta_x; r = 1, 2, 3 \ldots \quad (27)$$

where α is SCG parameter, β is the training rate, S_o is initial direction vector, and W_o is the initial parameter vector. The performance in terms of best validation of the ANN connected in cluster microgrid is attained at epoch 53 with the LM algorithm, at epoch 259 with the BR algorithm, and at epoch 29 with the SCG algorithm, as shown in Figure 12. After training a feed-forward neural network, the error histogram is a histogram of errors between target and predicted values. These error figures can be negative because they represent how expected values depart from target values. The number of vertical bars that appear on the graph is referred to as bins. Here, the entire error range is broken down into 20 smaller bins. The number of samples from your dataset that fall into each category is shown on the Y-axis. The error histogram and also the training states of all optimization algorithms used to train ANN are shown in Figures 13 and 14.

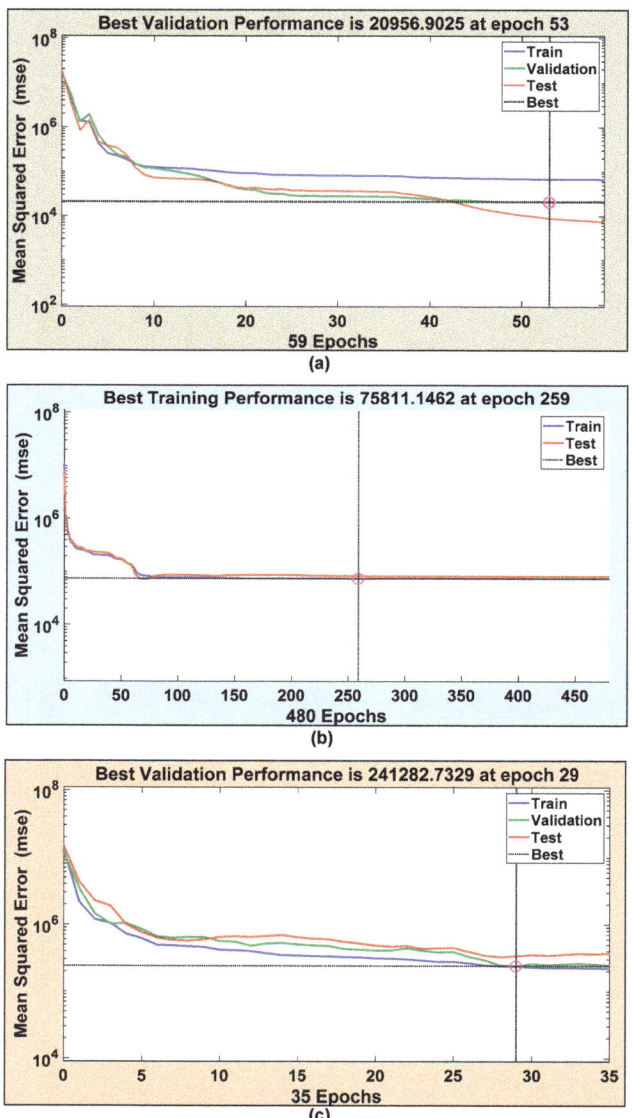

Figure 12. Performance plots of ANN with (**a**) Levenberg–Marquardt (LM) algorithm, (**b**) Bayesian Regulation (BR) algorithm, and (**c**) Scaled Conjugate Gradient (SCG) algorithm.

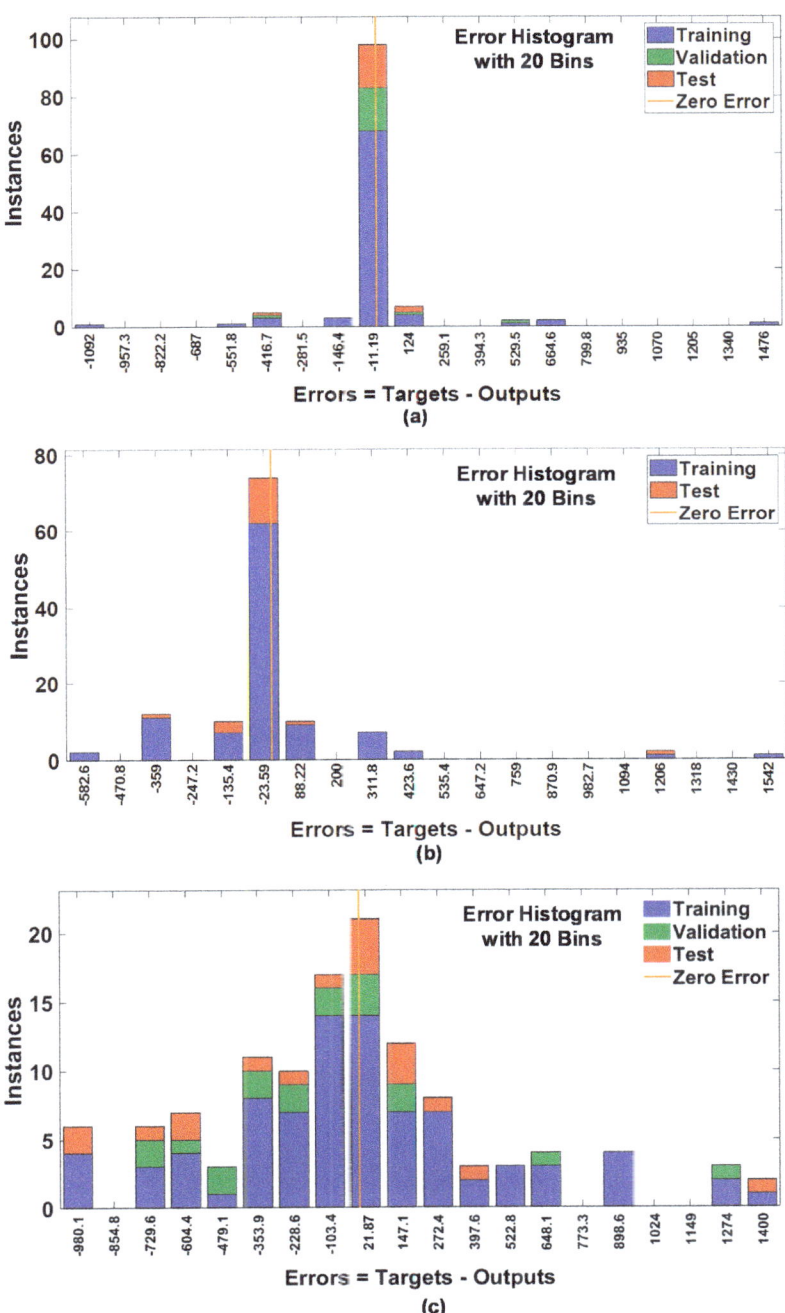

Figure 13. Error histogram plots of ANN with (**a**) Levenberg–Marquardt (LM) algorithm, (**b**) Bayesian Regulation (BR) algorithm, and (**c**) Scaled Conjugate Gradient (SCG) algorithm.

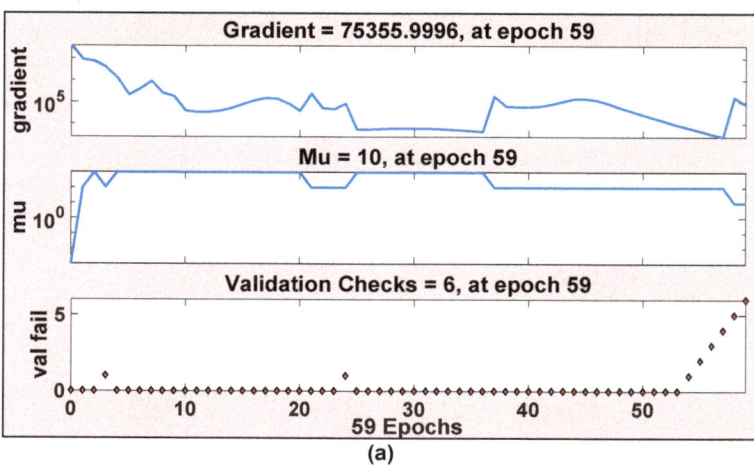

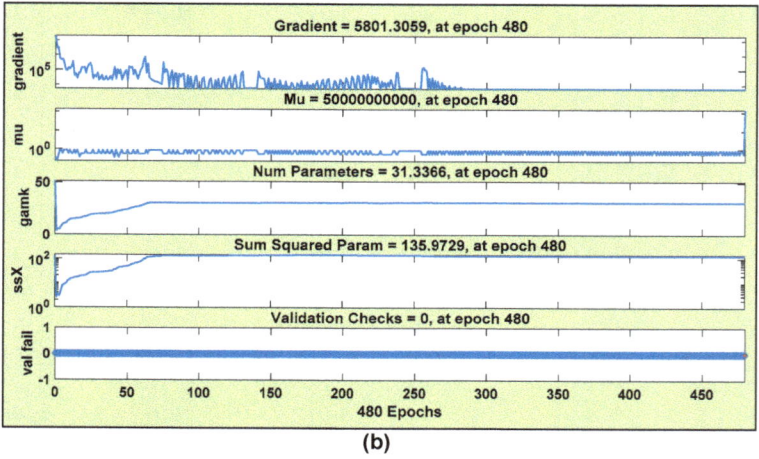

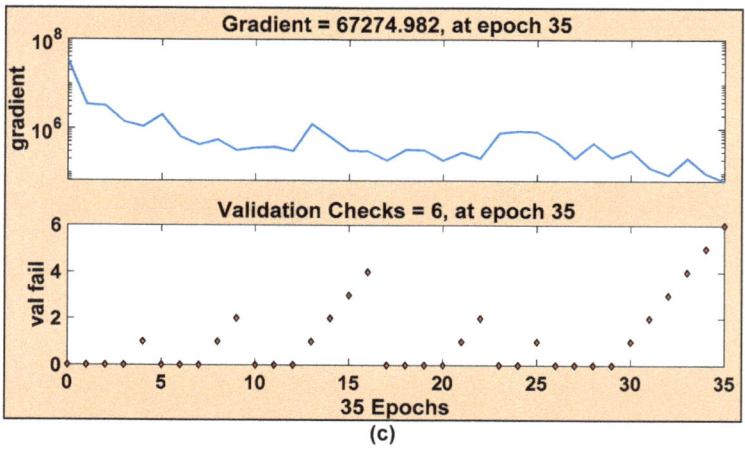

Figure 14. Training state plots of ANN with (**a**) Levenberg–Marquardt (LM) algorithm, (**b**) Bayesian Regulation (BR) algorithm, and (**c**) Scaled Conjugate Gradient (SCG) algorithm.

In terms of data training, validation, and testing the values generated by regression demonstrates the relationship between target samples and output samples. If R = 1 on the regression Figure 15, the line is angled at 45 degrees to the x-axis, suggesting that the target and output are the same. When the output sample and the target values are closely related, the value of "R" may be one. If the ANN regression values for all examples are greater than 0.95, the curve fitting is considered to be reasonably valid. From Table 3, it is observed that in all the cases of regression analysis, Levenberg–Marquardt optimization algorithm plays a dominant role. Hence, we proposed LM-based ANN for forecasting day-ahead load demands in cluster microgrids. Actual and forecasted values of day-ahead load demand in the cluster microgrid with different optimization algorithms based on ANN are shown in Figure 16. From the results, it is observed that ANN with the Levenberg–Marquardt optimization algorithm gives fruitful results; hence, we proposed LM algorithm-based ANN for day-ahead load demand forecasting.

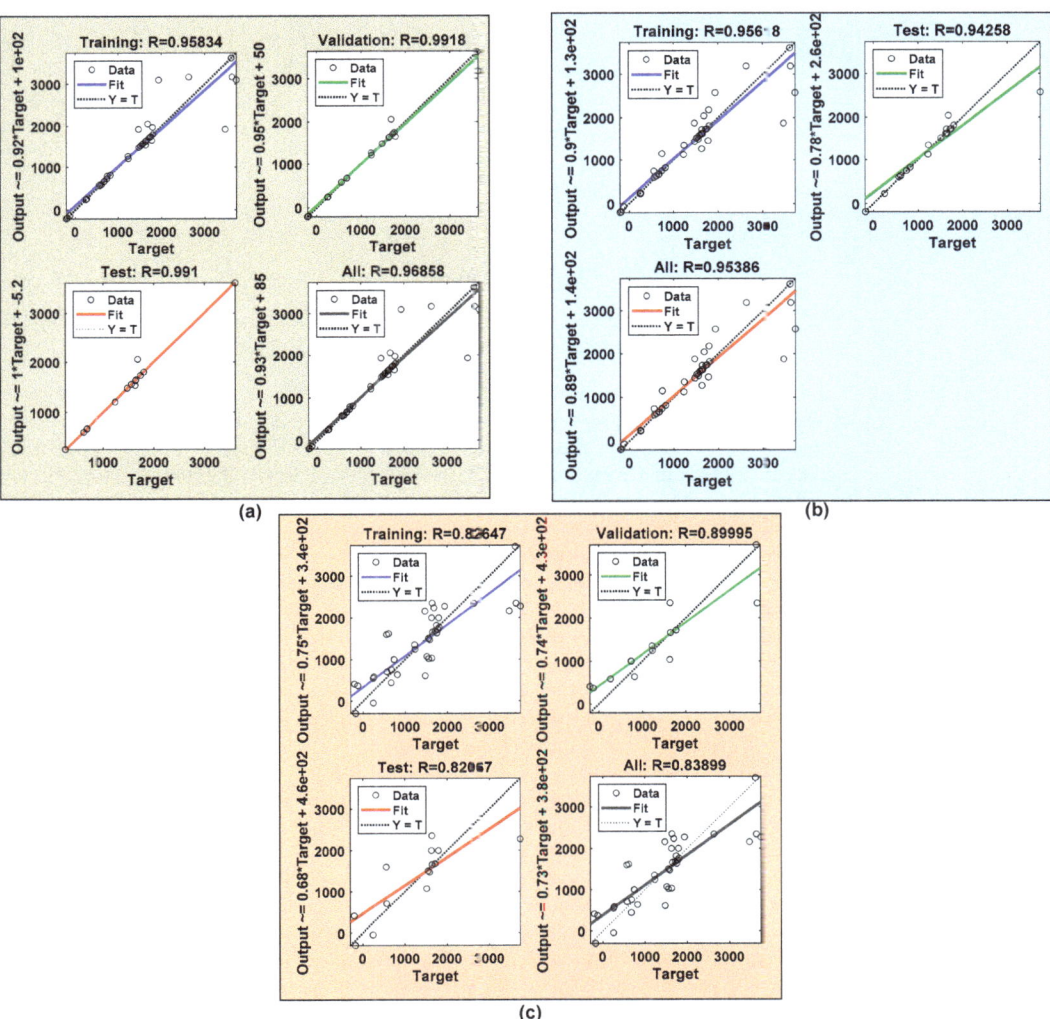

Figure 15. Regression plots of ANN with (**a**) Levenberg–Marquardt (LM) algorithm, (**b**) Bayesian Regulation (BR) algorithm, and (**c**) Scaled Conjugate Gradient (SCG) algorithm.

Table 3. Quantitative regression analysis of various optimization algorithms of ANN.

Name of the Algorithm	Training	Validation	Test	All	MSE
Levenberg–Marquardt	0.95834	0.9918	0.991	0.96858	68,722
Bayesian Regulation	0.95618	–	0.94258	0.95386	75,811
Scaled Conjugated Gradient	0.82647	0.89995	0.82067	0.83899	238,292

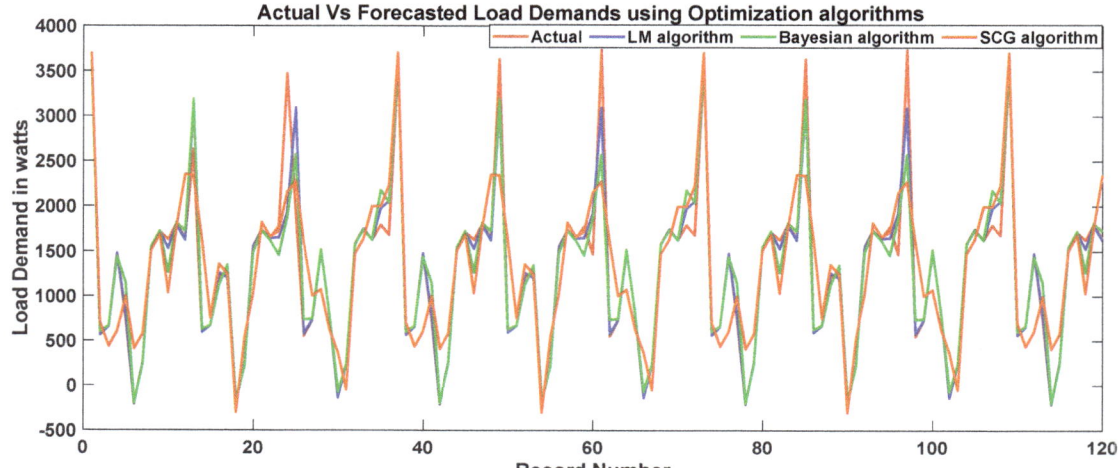

Figure 16. Forecasted load demand plots of ANN with Levenberg–Marquardt (LM) algorithm, Bayesian Regulation (BR) algorithm, and Scaled Conjugate Gradient (SCG) algorithm.

In terms of data training, validation, and testing, the values generated by regression demonstrate the relationship between target samples and output samples. If R = 1 on the regression in Figure 15, the line is angled at 45 degrees to the x-axis, suggesting that the target and output are the same. When the output sample and the target values are closely related, the value of "R" may be one. If the ANN regression values for all examples are greater than 0.95, the curve fitting is considered to be reasonably valid. From Table 3, it is observed that in all the cases of regression analysis, the Levenberg–Marquardt optimization algorithm plays a dominant role. Hence, we proposed LM-based ANN for forecasting day-ahead load demands in cluster microgrids. Actual and forecasted values of day-ahead load demand in the cluster microgrid with different optimization algorithms based on ANN are shown in Figure 16. From the results, it is observed that ANN with the Levenberg–Marquardt optimization algorithm gives fruitful results. Hence, we proposed LM algorithm-based ANN for day-ahead load demand forecasting.

5. Conclusions

With the rise of the smart grid, load forecasting is becoming more crucial. As a result, predicting the electrical load with high precision is a difficult assignment. The non-linearity and volatility of real-time energy consumption make it challenging to forecast load demand and consumption. To address this issue, multiple machine learning approaches, such as linear regression (LR), support vector machine (SVM), Long Short-Term Memory (LSTM), and artificial neural networks (ANN) are implemented in this article to estimate electric load demand forecasting in the cluster microgrid context. This work discovered the best models to perform day-ahead load demands by reviewing the validation results for the provided models. This encompasses both the accuracy of their forecasts and the low computational effort required to fit the models and make the predictions. The following are the salient outcomes of the proposed work in this article:

- All machine learning algorithms are compared in terms of performance by computing several factors, such as root mean square error (RMSE), mean square error (MSE), mean absolute error (MAE), and calculation time.
  - Based on the findings, it was identified that artificial neural networks are the best forecasting technique for day-ahead load demand forecasting. It outperforms SVM and LR in terms of RMSE (426.04), MAPE (0.79), MSE (1.815 × 10^5), and MAE (131.72), although the computation is high.
- Further, the ANN has also been evaluated using various optimization techniques, including Levenberg–Marquardt, Bayesian Regularization, and Scaled Conjugate Gradient algorithms, in order to determine the optimum algorithm for training ANN.
  - According to the findings, the Levenberg–Marquardt algorithm produces good results in terms of training, testing, validation, and error analysis.

Thus, this article concludes that the proposed ANN with the Levenberg–Marquardt algorithm is an optimum choice for forecasting day-ahead load demand in cluster microgrids.

Author Contributions: Conceptualization, V.P.K.Y.; Data curation, D.J.P.; Formal analysis, V.P.K.Y.; Funding acquisition, S.S.R.; Investigation, S.N.V.B.R.; Methodology, S.S.R.; Project administration, M.A.; Software, C.P.R.; Supervision, K.P. and S.S.R.; Writing—original draft, S.N.V.B.R.; Writing—review & editing, V.P.K.Y. All authors have read and agreed to the published version of the manuscript.

Funding: This research received no external funding.

Institutional Review Board Statement: Not applicable.

Informed Consent Statement: Not applicable.

Data Availability Statement: Not applicable.

Acknowledgments: This work was supported by the Qatar National Library. Open access funding provided by the Qatar National Library.

Conflicts of Interest: The authors declare no conflict of interest.

Appendix A

Table A1. Parameters considered for time series long short term memory (LSTM) Forecasting Method.

Parameter	Value
Maximum Epochs	200
Hidden Units	100
Gradient Threshold	1
Initial Learn rate	0.005
Learn rate Droop period	125
Learn rate Droop factor	0.2
No. of Records	120
Samples considered for training	96 (80%)
Samples considered for testing	24 (20%)

References

1. Saeed, M.H.; Fangzong, W.; Kalwar, B.A.; Iqbal, S. A Review on Microgrids' Challenges & Perspectives. *IEEE Access* **2021**, *9*, 166502–166517. [CrossRef]
2. Jacob, M.; Neves, C.; Vukadinović Greetham, D. Short Term Load Forecasting. In *Forecasting and Assessing Risk of Individual Electricity Peaks. Mathematics of Planet Earth*; Springer: Berlin/Heidelberg, Germany, 2020 [CrossRef]
3. Ma, J.; Ma, X. A review of forecasting algorithms and energy management strategies for microgrids. *Syst. Sci. Control. Eng. Tailor Fr.* **2018**, *6*, 237–248. [CrossRef]
4. Hossain, M.S.; Mahmood, H. Short-Term Photovoltaic Power Forecasting Using an LSTM Neural Network and Synthetic Weather Forecast. *IEEE Access* **2020**, *8*, 172524–172533. [CrossRef]

5. Chafi, Z.S.; Afrakhte, H. Short-Term Load Forecasting Using Neural Network and Particle Swarm Optimization (PSO) Algorithm. *Math. Probl. Eng.* **2021**, *2021*, 5598267. [CrossRef]
6. Arvanitidis, A.I.; Bargiotas, D.; Daskalopulu, A.; Laitsos, V.M.; Tsoukalas, L.H. Enhanced Short-Term Load Forecasting Using Artificial Neural Networks. *Energies* **2021**, *14*, 7788. [CrossRef]
7. Singh, S.; Hussain, S.; Bazaz, M.A. Short term load forecasting using artificial neural network. In Proceedings of the 2017 Fourth International Conference on Image Information Processing (ICIIP), Shimla, India, 21–23 December 2017; pp. 1–5. [CrossRef]
8. Buitrago, J.; Asfour, S. Short-term forecasting of electric loads using nonlinear autoregressive artificial neural networks with exogenous vector inputs. *Energies* **2017**, *10*, 40. [CrossRef]
9. Rafi, S.H.; Masood, N.A.; Deeba, S.R.; Hossain, E. A Short-Term Load Forecasting Method Using Integrated CNN and LSTM Network. *IEEE Access* **2021**, *9*, 32436–32448. [CrossRef]
10. Izzatillaev, J.; Yusupov, Z. Short-term Load Forecasting in Grid-connected Microgrid. In Proceedings of the 2019 7th International Istanbul Smart Grids and Cities Congress and Fair (ICSG), Istanbul, Turkey, 25–26 April 2019; pp. 71–75. [CrossRef]
11. Zhang, A.; Zhang, P.; Feng, Y. Short-term load forecasting for microgrids based on DA-SVM. *COMPEL-Int. J. Comput. Math. Electr. Electron. Eng.* **2019**, *38*, 68–80. [CrossRef]
12. Semero, Y.K.; Zhang, J.; Zheng, D. EMD–PSO–ANFIS-based hybrid approach for short-term load forecasting in microgrids. *IET Gener. Transm. Distrib.* **2020**, *14*, 470–475. [CrossRef]
13. Semero, Y.K.; Zhang, J.; Zheng, D.; Wei, D. An Accurate Very Short-Term Electric Load Forecasting Model with Binary Genetic Algorithm Based Feature Selection for Microgrid Applications. *Electr. Power Compon. Syst.* **2018**, *46*, 1570–1579. [CrossRef]
14. Cerne, G.; Dovzan, D.; Skrjanc, I. Short-Term Load Forecasting by Separating Daily Profiles and Using a Single Fuzzy Model Across the Entire Domain. *IEEE Trans. Ind. Electron.* **2018**, *65*, 7406–7415. [CrossRef]
15. Jimenez, J.; Donado, K.; Quintero, C.G. A Methodology for Short-Term Load Forecasting. *IEEE Lat. Am. Trans.* **2017**, *15*, 400–407. [CrossRef]
16. Guo, W.; Che, L.; Shahidehpour, M.; Wan, X. Machine-Learning based methods in short-term load forecasting. *Electr. J.* **2021**, *34*, 106884. [CrossRef]
17. Groß, A.; Lenders, A.; Schwenker, F.; Braun, D.A.; Fischer, D. Comparison of short-term electrical load forecasting methods for different building types. *Energy Inform.* **2021**, *4*, 1–16. [CrossRef]
18. Zhang, R.; Zhang, C.; Yu, M. A Similar Day Based Short Term Load Forecasting Method Using Wavelet Transform and LSTM. *IEEJ Trans. Electr. Electron. Eng.* **2022**, *17*, 506–513. [CrossRef]
19. Wang, R.; Chen, S.; Lu, J. Electric short-term load forecast integrated method based on time-segment and improved MDSC-BP. *Syst. Sci. Control. Eng.* **2021**, *9* (Suppl. S1), 80–86. [CrossRef]
20. Hafeez, G.; Javaid, N.; Riaz, M.; Ali, A.; Umar, K.; Iqbal, Z. Day Ahead Electric Load Forecasting by an Intelligent Hybrid Model Based on Deep Learning for Smart Grid. In *Advances in Intelligent Systems and Computing*; Springer: Berlin/Heidelberg, Germany, 2019; Volume 993. [CrossRef]
21. Kuster, C.; Rezgui, Y.; Mourshed, M. Electrical load forecasting models: A critical systematic review. In *Sustainable Cities and Society*; Elsevier: Amsterdam, The Netherlands, 2017; Volume 35, pp. 257–270. [CrossRef]
22. Zheng, X.; Ran, X.; Cai, M. Short-Term Load Forecasting of Power System based on Neural Network Intelligent Algorithm. *IEEE Access* **2020**. [CrossRef]
23. Moradzadeh, A.; Zakeri, S.; Shoaran, M.; Mohammadi-Ivatloo, B.; Mohammadi, F. Short-Term Load Forecasting of Microgrid via Hybrid Support Vector Regression and Long Short-Term Memory Algorithms. *Sustainability* **2020**, *12*, 7076. [CrossRef]
24. El Khantach, A.; Hamlich, M.; Belbounaguia, N.E. Short-term load forecasting using machine learning and periodicity decomposition. *AIMS Energy* **2019**, *7*, 382–394. [CrossRef]
25. Alquthami, T.; Zulfiqar, M.; Kamran, M.; Milyani, A.H.; Rasheed, M.B. A Performance Comparison of Machine Learning Algorithms for Load Forecasting in Smart Grid. *IEEE Access* **2022**, *10*, 48419–48433. [CrossRef]
26. Bashir, T.; Haoyong, C.; Tahir, M.F.; Liqiang, Z. Short term electricity load forecasting using hybrid prophet-LSTM model optimized by BPNN. *Energy Rep.* **2022**, *8*, 1678–1686. [CrossRef]
27. Ribeiro, A.M.N.C.; do Carmo, P.R.X.; Endo, P.T.; Rosati, P.; Lynn, T. Short-and Very Short-Term Firm-Level Load Forecasting for Warehouses: A Comparison of Machine Learning and Deep Learning Models. *Energies* **2022**, *15*, 750. [CrossRef]
28. Rao, S.N.V.B.; Padma, K. ANN based Day-Ahead Load Demand Forecasting for Energy Transactions at Urban Community Level with Interoperable Green Microgrid Cluster. *Int. J. Renew. Energy Res.* **2021**, *11*, 147–157. [CrossRef]
29. Rao, S.N.V.B.; Kumar, Y.V.P.; Pradeep, D.J.; Reddy, C.P.; Flah, A.; Kraiem, H.; Al-Asad, J.F. Power Quality Improvement in Renewable-Energy-Based Microgrid Clusters Using Fuzzy Space Vector PWM Controlled Inverter. *Sustainability* **2022**, *14*, 4663. [CrossRef]
30. Scott, D.; Simpson, T.; Dervilis, N.; Rogers, T.; Worden, K. Machine Learning for Energy Load Forecasting. *J. Phys. IOP Publ.* **2018**, *1106*, 012005. [CrossRef]
31. Hong, W.-C. Electric load forecasting by support vector model. *Appl. Math. Model.* **2009**, *33*, 2444–2454. [CrossRef]
32. Hu, Z.; Bao, Y.; Xiong, T. Electricity Load Forecasting Using Support Vector Regression with Memetic Algorithms. *Sci. World J.* **2013**, *2013*, 292575. [CrossRef] [PubMed]

33. Sandeep Rao, K.; Siva Praneeth, V.N.; Pavan Kumar, Y.V.; John Pradeep, D. Investigation on various training algorithms for robust ANN-PID controller design. *Int. J. Sci. Technol. Res.* **2020**, *9*, 5352–5360. Available online: https://www.ijstr.org/paper-references.php?ref=IJSTR-0220-30423 (accessed on 17 May 2022).
34. Kamath, H.G.; Srinivasan, J. Validation of global irradiance derived from INSAT-3D over India. *Int. J. Sol. Energy* **2020**, *202*, 45–54. [CrossRef]
35. Source: NASA/POWER CERES/MERRA2 Native Resolution Hourly Data. Available online: https://power.larc.nasa.gov (accessed on 17 May 2022).
36. Shohan, M.J.A.; Faruque, M.O.; Foo, S.Y. Forecasting of Electric Load Using a Hybrid LSTM-Neural Prophet Model. *Energies* **2022**, *15*, 2158. [CrossRef]

Article

Intelligent Control of Irrigation Systems Using Fuzzy Logic Controller

Arunesh Kumar Singh [1], Tabish Tariq [1], Mohammad F. Ahmer [2], Gulshan Sharma [3,*], Pitshou N. Bokoro [3] and Thokozani Shongwe [3]

1. Department of Electrical Engineering, Faculty of Engineering & Technology, Jamia Millia Islamia (A Central University), New Delhi 110025, India
2. Department of Electrical and Electronics Engineering, Mewat Engineering College, Nuh 122107, India
3. Department of Electrical Engineering Technology, University of Johannesburg, Johannesburg 2006, South Africa
* Correspondence: gulshans@uj.ac.za

Abstract: In this paper, we explain the design and implementation of an intelligent irrigation control system based on fuzzy logic for the automatic control of water pumps used in farms and greenhouses. This system enables its user to save water and electricity and prevent over-watering and under-watering of the crop by taking into account the climatic parameters and soil moisture. The irrigation system works without human intervention. The climate sensors are packaged using electronic circuits, and the whole is interfaced with an Arduino and a Simulink model. These sensors provide information that is used by the Simulink model to control the water pump speed; the speed of the water pump is controlled to increase or decrease the amount of water that needs to be pushed by the pump. The Simulink model contains the fuzzy control logic that manages the data read by the Arduino through sensors and sends the command to change the pump speed to the Arduino by considering all the sensor data. The need for human intervention is eliminated by using this system and a more successful crop is produced by supplying the right amount of water to the crop when it is needed. The water supply is stopped when a sufficient amount of moisture is present in the soil and it is started as soon as the soil moisture levels drops below certain levels, depending upon the environmental factors.

Keywords: intelligent control; irrigation system; fuzzy logic; automatic irrigation control system

1. Introduction

Agriculture plays an important role in the economy and is considered the backbone of the economic systems of emerging countries. Agriculture has been linked to the production of staple food crops for decades. To produce a successful crop, one must take into account the process of irrigation and the amount of water that is being used. The amount of water should only be that which is needed by the plants. Since water is one of the most precious resources we have, we should use it wisely. In this paper, we discuss the implementation of an intelligent control system based on fuzzy logic that, after considering the climatic conditions, decides how much water should be given to the crop, and a successful crop is produced if the right amount of water is supplied—not too much and not too little [1,2].

The fuzzy control system developed considers four input variables: soil moisture, solar irradiance, air temperature, and air humidity—as all these factors affect the evaporation rate of water from the soil [3–5]. By controlling the output variable of the fuzzy logic control system [1,3], i.e., the pump voltage of the water pump using a pulse-width modulation technique [6], we control the speed of the pump, which in turn results in the change in the rate of water supply. This makes the system different from previous work as more input variables are employed and a direct rule base is created based upon the relationship of each input variable with the output variable [1,7–12].

2. Fuzzy Logic Control System

A fuzzy logic control system is developed with the help of four blocks. The first is fuzzification, which converts a crisp input value into a fuzzy value by assigning a degree of membership to the input. Then, the second block is an inference engine that deduces the fuzzy result from fuzzified inputs on the basis of the if–then rules block. The if–then rules is the rule base that contains all the relevant combinations of inputs and outputs that are designed by the user to denote a mathematical relationship between them [1,3–5,13]. On the basis of membership functions, the fuzzified inputs and outputs are distributed into different sets. The controller provides a crisp output that is derived from the fuzzy output that the inference engine generated. This conversion of inference engine output from fuzzy to crisp is done by the defuzzification process. Figure 1 shows a general fuzzy-logic-based control system in the form of a block diagram.

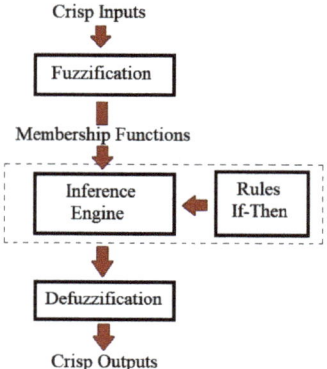

Figure 1. Block diagram of a fuzzy logic control system.

Figure 2 shows the fuzzy inference system developed in MATLAB. Our fuzzy inference system is designed using the MAMDANI approach of fuzzy inference. As a result, the AND operator is realized by calculating the minimum, whereas the OR operator is achieved by calculating the maximum [3,13–15]. Considering the four input variables, soil moisture, solar irradiance, air temperature, and air humidity, the output variable, pump voltage, is controlled based on the fuzzy rules defined in the MATLAB fuzzy rule base.

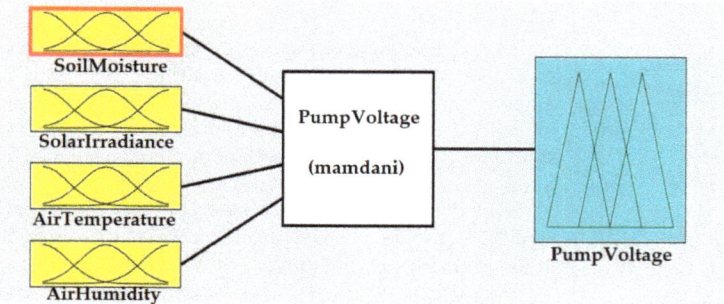

Figure 2. Fuzzy inference system.

We chose pump voltage as the output variable because the quantity of water required in the soil can be varied by varying the DC pump's terminal voltage since the pump speed is directly related to the pump input voltage. To prove that by changing the DC pump's terminal voltage, the rate of flow of water can be changed; we conducted an experiment in which the pump input voltage is changed manually, and at each voltage value the time needed to fill an empty mug is noted. The experimental data of pump voltage vs. pump water output are shown in Table 1.

Table 1. Pump voltage vs. time taken to fill 500 mL.

S. No.	Pump Voltage (Volts)	Time Taken to Push 500 mL (s)
1	2.0	106
2	2.5	78
3	3.2	54
4	4.0	41
5	5.0	33
6	6.0	29
7	7.0	24
8	8.0	22
9	9.0	20
10	10.0	18

2.1. Membership Functions

Each of the input variables is designed with three membership functions and the output variable with five membership functions [1,16]. All input and output variables are defined with the help of trapezoidal and triangular membership functions and linguistic variables [17,18]. The triangular membership function and the trapezoidal membership function were implemented solely for the sake of simplicity and for achieving good quality control. According to Lotfi A. Zadeh, the simplest methods work the best because they are intuitively clear and we can easily make use of our intuition and the mathematical formulas together, but if we use complex functions, we can only rely on the formulas as they are difficult to intuitively understand. This explanation is reasonable on the qualitative level and the quantitative explanation is provided in the research [3,19].

2.1.1. Soil Moisture

The soil moisture input variable is defined with the help of three linguistic variables, namely low, normal, and high, as shown in Figure 3. It ranges from 0 to 100 to denote the percentage of moisture content in the soil. The membership function parameters are given below:

- Low—trapezoidal membership function with params [−36 −4 20 35]
- Normal—triangular membership function with params [20 40 60]
- High—trapezoidal membership function with params [45 60 104 136]

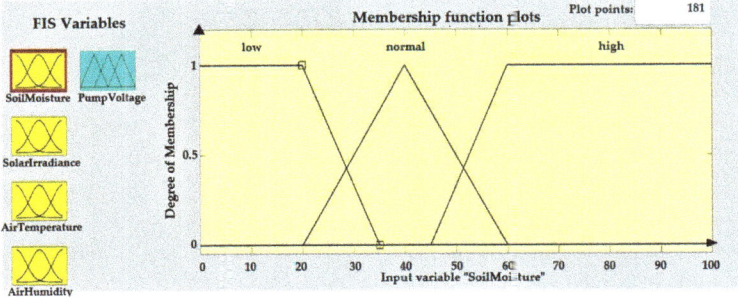

Figure 3. Soil moisture membership function plot.

2.1.2. Solar Irradiance

The solar irradiance input variable is defined with the help of three linguistic variables, namely dim, normal, and bright, as shown in Figure 4. It ranges from 0 to 1000 to denote

the amount of solar irradiance incident on the soil w.r.t. watts per square meter. The membership function parameters are given below:
- Dim—trapezoidal membership function with params [−360 −40 350 500]
- Normal—triangular membership function with params [350 500 675]
- Bright—trapezoidal membership function with params [500 675 1040 1360]

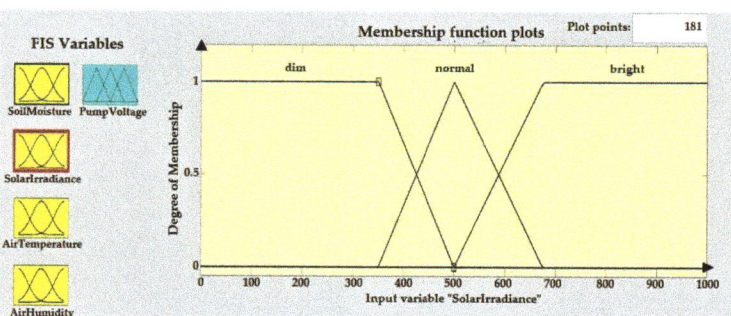

Figure 4. Solar irradiance membership function plot.

2.1.3. Air Temperature

The air temperature input variable is defined with the help of three linguistic variables, namely cold, normal, and hot, as shown in Figure 5. It ranges from 0 to 50 to denote the temperature of the air in degrees Celsius. The membership function parameters are given below:
- Cold—trapezoidal membership function with params [−18 −2 17.5 22.5]
- Normal—triangular membership function with params [17.5 22.5 27.5]
- Hot—trapezoidal membership function with params [22.5 27.5 52 68]

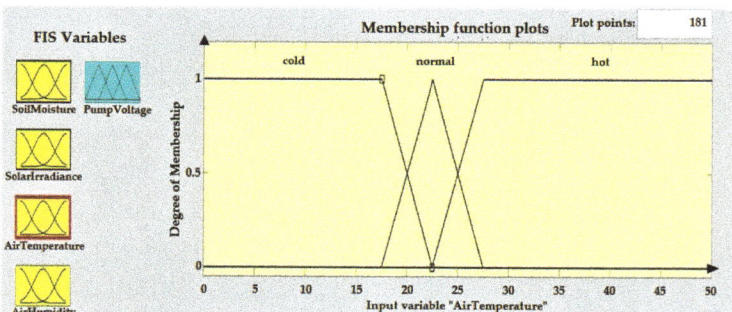

Figure 5. Air temperature membership function plot.

2.1.4. Air Humidity

The air humidity input variable is defined with the help of three linguistic variables, namely low, normal, and high, as shown in Figure 6. It ranges from 0 to 100 to denote the percentage of moisture content in the air. The membership function parameters are given below:
- Low—trapezoidal membership function with params [−36 −4 35 50]
- Normal—triangular membership function with params [35 50 70]
- High—trapezoidal membership function with params [52.5 70 104 136]

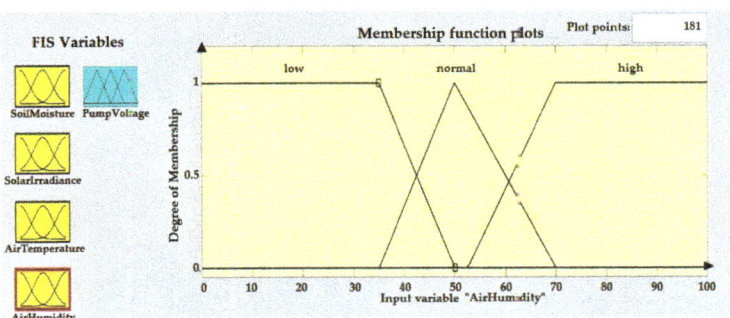

Figure 6. Air humidity membership function plot.

2.1.5. Pump Voltage

The pump voltage output variable is defined with the help of five linguistic variables, namely very low, low, normal, high, and very high, as shown in Figure 7. It ranges from 0 to 13 to denote the pump voltage in volts. The membership function parameters are given below:

- Very low—trapezoidal membership function with params [−2.43 −0.271 2 3.3]
- Low—triangular membership function with params [2 3.3 5]
- Normal—triangular membership function with params [3.3 5 7]
- High—triangular membership function with params [5 7 10]
- Very high—trapezoidal membership function with params [7 10 11.1 13.2]

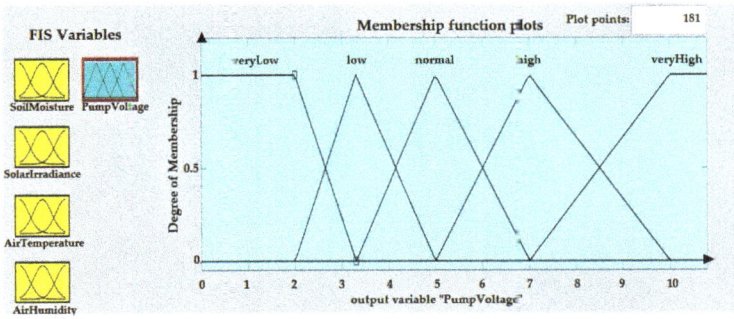

Figure 7. Pump voltage membership function plot.

2.2. Fuzzy Rules

Fuzzy rules are defined by taking into consideration how each of the input variables affects the amount of water that the plant needs. First, we check the soil moisture starting from high to low, then solar irradiance is checked starting from dim to bright, then air temperature from cold to hot, and finally air humidity from high to low, as in this sequence we can deduce that if soil moisture is high and solar irradiance is dim and air temperature is cold and air humidity is high then, the need to supply more water to the soil is minimal, and by permuting all the input variables in this order we can say that the water need is increasing, as this is the lowest case. The total count of fuzzy rules that need to be designed can be determined from the general formula for calculating the number of fuzzy rules, which is by multiplying the number of membership functions for each input variable, i.e., $3 \times 3 \times 3 \times 3 = 81$ [1,13].

We can verify that the rules defined by the user are correct by viewing the surface graphs automatically generated by the MATLAB software after inputting all the rules, as per the deduction that the amount of water needed by the plant increases directly with solar irradiance and air temperature, while it decreases with increases in soil moisture and air humidity [13].

We can see that the pump voltage increases with an increase in solar irradiance and air temperature, as shown in Figure 8, and the pump voltage decreases with an increase in soil moisture and air humidity, as shown in Figure 9.

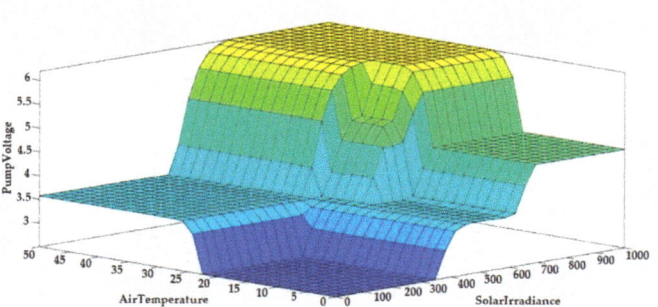

Figure 8. Surface graph of air temperature and solar irradiance vs. pump voltage.

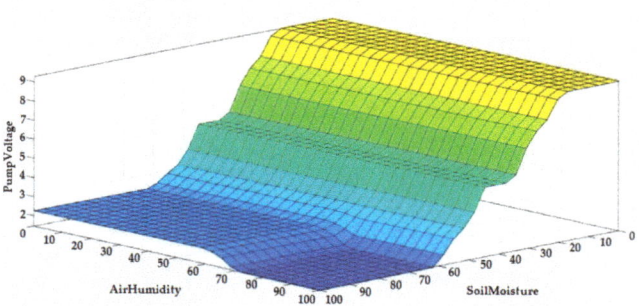

Figure 9. Surface graph of soil moisture and air humidity vs. pump voltage.

3. Verification of Fuzzy Controller

To test the fuzzy controller, we designed a prototype model in Simulink that can be used to test the fuzzy controller for a large set of input values for each input variable. Figure 10 shows the model used for testing the fuzzy logic controller in Simulink. For each input variable, a sine wave function block is used with different parameters to permute every possibility as defined in the fuzzy rule base. For soil moisture, a sine wave with 50 amplitude and 50 bias is used to get a range of 0–100 as soil moisture will always be in percentage. Solar irradiance is simulated as a sine wave of 500 amplitude and 500 bias to get a range of 0–1000 as solar irradiance can range from 0–1000 watts per meter square. Air temperature is simulated with the help of a sine wave with an amplitude equal to 25 and a bias of 25 to get a range of 0–50, which will be in degrees Celsius, and, finally, the air humidity is the same as soil moisture as it will also be in percent. Each of the waves is specified to engulf each other in all possible ways to get all the permutations of all input variable values. Figure 10 shows the testing model of the system designed in Simulink with the above specifications, Figure 11a shows the changing values of input variables throughout the simulation, and the pump voltage result is shown in Figure 11b.

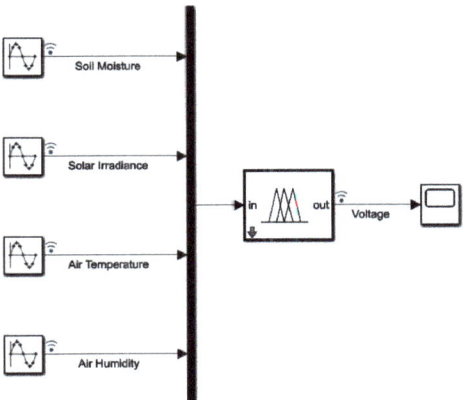

Figure 10. Simulink model for testing fuzzy controller.

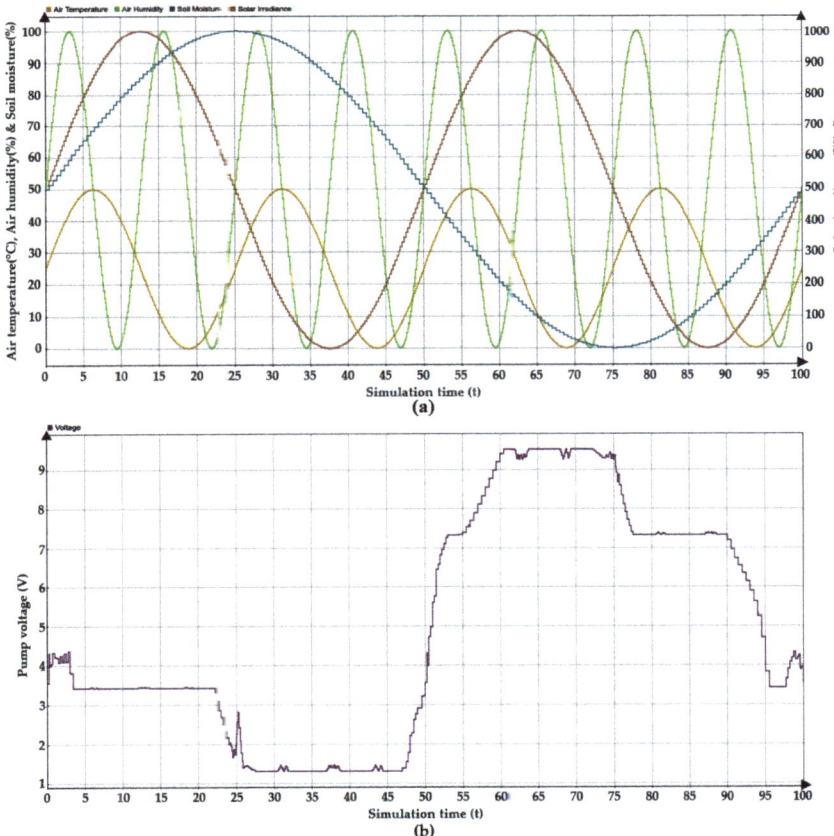

Figure 11. (**a**) Solar irradiance, air temperature, air humidity, and soil moisture input graph; (**b**) voltage output graph.

Test Results

According to Figure 11b for voltage output (simulation time is represented on the x-axis and the value of the respective variable w.r.t. to its unit of measurement is represented on the y-axis), the following observations were made:

- While moving from 20 to 30 on the x-axis, we can see that the voltage is decreasing and this happens when solar irradiance is decreasing while soil moisture and air humidity is increasing through the same value. Since soil moisture was almost at the peak and solar irradiance was decreasing, the need for more water was eliminated and pump voltage decreased to minimum.
- From 45–60 on the x-axis, we see that the voltage is increasing; this happens while the solar irradiance is also increasing and soil moisture is decreasing. This also suggests that more water is needed when soil moisture is low and solar irradiance is high.
- From 75–77, the voltage drops suddenly to around 6.7 volts; as solar irradiance decreases below 400 watts per meter square, soil moisture starts increasing above 0 and air humidity is also increasing. Since there is little sunlight and air humidity is high, the need for water is decreased.
- From 90–95, the voltage decreases as soil moisture increases.

From the above observations and Figure 11, we can clearly see the effect of each input variable on the pump voltage. The pump voltage graph mostly resembles the soil moisture graph but appears inverted, meaning that the soil moisture has the highest impact on the pump voltage, followed by solar irradiance, then air temperature, and lastly air humidity.

4. Analysis of Fuzzy Controller

The fuzzy logic controller was designed with respect to controlling the soil moisture content based on the instantaneous values of the pre-defined input variables that are soil moisture, solar irradiance, air temperature, and air humidity. The pump voltage was chosen as the output variable of the fuzzy controller because of the relationship between pump voltage and the flow rate of water, controlling which is the main purpose of our controller. Since the DC pump that was taken could be easily controlled by changing the terminal voltage and thus changing its speed. Using the observations from the experiment described above and in Table 1, the following mathematical equations can be written. Since an increase in input voltage results in less time needed to fill the mug completely, we can say that input voltage is inversely proportional to the time needed to fill the mug completely and can be written in mathematical form as:

$$\text{Input voltage} \propto 1/\text{time taken to fill 500 mL} \tag{1}$$

and the rate of flow of water can be defined by:

Rate of flow of water = change in volume/time taken to fill 500 mL = $(500 - 0)$/time taken to fill 500 mL,

The above equation can be rewritten as:

$$\text{Rate of flow of water} \propto 1/\text{time taken to fill 500 mL} \tag{2}$$

From Equations (1) and (2), we can say that the rate of flow of water is directly proportional to the input voltage of the DC water pump.

$$\text{Rate of flow of water} \propto \text{Input voltage} \tag{3}$$

Mathematically, the relationship between the input variables that are soil moisture, solar irradiance, air temperature, and air humidity and the pump voltage output variable can be defined by:

$$\text{Soil Moisture} \propto 1/\text{Pump Voltage} \tag{4}$$

$$\text{Solar Irradiance} \propto \text{Pump Voltage} \tag{5}$$

$$\text{Air Temperature} \propto \text{Pump Voltage} \qquad (6)$$

$$\text{Air Humidity} \propto 1/\text{Pump Voltage} \qquad (7)$$

These four equations can be proved by considering the effect of each input variable on the soil moisture content:

1. Equation (4) is a general equation considering that increasing soil moisture content indicates less need for water and, thus, the inverse relationship between the variable soil moisture and pump voltage;
2. When the value of the solar irradiance increases, it increases the rate of evaporation, and, thus, to maintain the moisture content of the soil, the rate of flow of water from the pump needs to be increased, and, thus, the pump voltage is increased;
3. Air temperature acts similarly as solar irradiance acts upon the moisture content of the soil and, thus, the direct relationship is formed;
4. Air humidity is defined by the inverse relationship because when the moisture content of the air is low the evaporation rate of water in the soil is increased as the dry air tends to absorb the moisture from the surface.

From Equations (4)–(7):

$$\text{Pump Voltage} \propto \text{Solar Irradiance} \times \text{Air Temperature})/(\text{Soil Moisture} \times \text{Air Humidity}) \qquad (8)$$

Equation (8) represents the mathematical model of the system.

4.1. System Response w.r.t Each Input Variable

The response of the fuzzy controller with respect to each input variable is analyzed in the same way the above results are found. A test model is designed for each input variable such as soil moisture, solar irradiance, air temperature, and air humidity. Each model is run three times by considering the values of the remaining variables as:

1. Minimum water needed;
2. Maximum water needed;
3. At standard operating conditions.

4.1.1. Soil Moisture

According to Equation (4), the effect of soil moisture on the pump voltage should be inversely proportional, and with an increase in the soil moisture value, the pump voltage should be decreased. This can be validated by the following graphs. Figure 12a,b show the graphs of soil moisture vs. time and voltage vs. time when all the other variables' need for water is at a minimum, respectively. When soil moisture is 0–20, we can see that the pump voltage is constant (around 7.3 V). As the soil moisture value increases from 21 to 60, the pump voltage is decreased simultaneously to somewhere around 1 V and then is maintained there as the soil moisture increases further to 100 and back to 60.

As soon as the soil moisture starts decreasing below 60, the pump voltage is increased gradually to 7.3 V. The pump's voltage is not increased above 7.3 V because the need for water is at minimum by the other variables such as solar irradiance and air temperature being set to minimum and the air humidity being set to maximum, signifying no effect of these variables on the evaporation rate of water from the soil. This is with respect to the fuzzy controller only and when the hardware model is connected to it the output will change significantly as the minimum voltage our system can work on is 3.3 V. So, as soon as the fuzzy controller outputs a value below 3.3, the actual value becomes 0 and the DC pump is shut down.

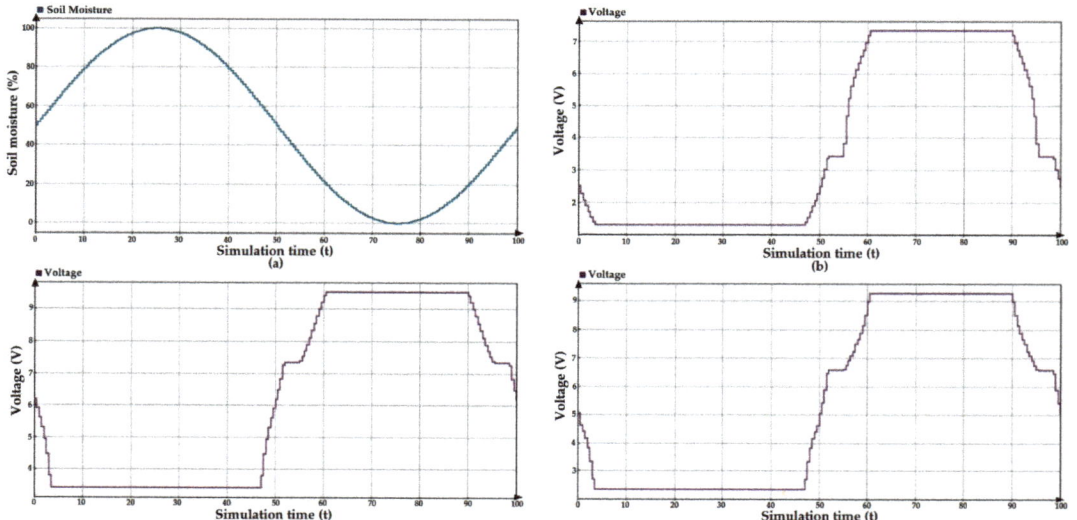

Figure 12. (**a**) Soil moisture vs. time; (**b**) voltage vs. time at minimum water need; (**c**) voltage vs. time at maximum water need; (**d**) voltage vs. time at standard operating conditions.

Similarly, considering that the need for water is at maximum w.r.t. to the other variables, the effect of soil moisture on the pump voltage is defined by the graph shown in Figure 12c, Considering the soil moisture value from Figure 12a. We can see that the range of pump voltage has shifted from 1–7.3 V to 3.2–9.5 V, but it is still in accordance with our Equation (7) as the voltage is increased when soil moisture decreases from 60–20 and vice versa. The range of pump voltage has changed slightly from 3.2–9.5 V to 2.2–9.2 V at standard operating conditions, according to Figure 12d considering soil moisture value from Figure 12a. To get the graph at standard operating conditions, we manually set solar irradiance to 600.00 W/m², air temperature to 25 °C, and air humidity to 50%. This slight decrease in the range is due to a decrease in the solar irradiance value from 1000.00 W/m² at max to 600.00 W/m², while the graph is almost identical.

4.1.2. Solar Irradiance

According to Equation (5), the effect of solar irradiance on the pump voltage should be directly proportional, and with an increase in the soil moisture value the pump voltage should also be increased [8]. This can be validated by the following graphs shown in Figure 13.

Figure 13a,b show the graphs of solar irradiance vs. time and voltage vs. time, respectively, when all the other variables' need for water is at a minimum. When solar irradiance is 0 to 400.00 W/m², we can see that the pump voltage is constant (around 1.35 V). As the solar irradiance value increases from 400.00 W/m² to 700.00 W/m², the pump voltage is increased simultaneously to somewhere around 3.4 V and then is maintained there as the solar irradiance increases further to 1000.00 W/m² and back to 700.00 W/m². As soon as the solar irradiance starts decreasing below 700.00 W/m², the pump voltage is decreased gradually to 1.35 V. The pump's voltage is not increased above 3.4 V because the need for water is at minimum by the other variables such as air temperature being set to minimum and the air humidity and soil moisture being set to maximum, signifying no effect of these variables on the evaporation rate of water from the soil.

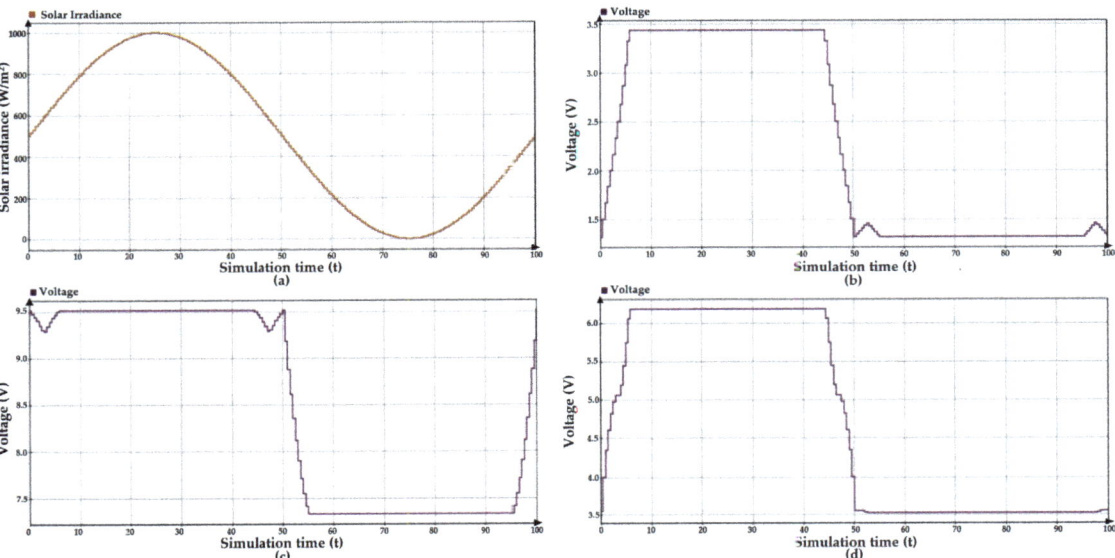

Figure 13. (a) Solar irradiance vs. time; (b) voltage vs. time at minimum water need; (c) voltage vs. time at maximum water need; (d) voltage vs. time at standard operating conditions.

Similarly, considering that the need for water is at maximum w.r.t. to the other variables, the effect of solar irradiance on the pump voltage is defined by the graph shown in Figure 13c, Considering solar irradiance value from Figure 13a. We can see that the range of the pump voltage has shifted from 1.35–3.4 V to 7.25–9.5 V, and it is still in accordance with our Equation (5), as the voltage is increased when solar irradiance increases from 300.00–500.00 W/m^2 and vice versa. The range of the pump voltage is changed slightly from 7.25–9.5 V to 3.5–6.25 V at standard operating conditions, according to Figure 13d, Considering the solar irradiance value from Figure 13a. To get the graph at standard operating conditions, we manually set soil moisture to 50%, air temperature to 25 °C, and air humidity to 50%. This slight decrease in the range is due to a decrease in solar irradiance value from 1000.00 W/m^2 at max to 600.00 W/m^2, while the graph is almost identical.

4.1.3. Air Temperature

According to Equation (6), the effect of air temperature on the pump voltage should be directly proportional, and with an increase in the air temperature value the pump voltage should also be increased [8]. This can be validated by the following graphs shown in Figure 14.

In Figure 14b, when the need for water is minimum because of other factors, then the effect of air temperature can be neglected as the value of pump voltage is below 3.3 V and the Arduino microcontroller will set the voltage to 0 V when below this level. Similarly, considering that the need for water is at maximum w.r.t. to the other variables, the effect of air temperature on the pump voltage is still negligible, as shown by the graph in Figure 14c.

From Figure 14d, we can see that the pump voltage changes according to Equation (6). The effect of air temperature is observed in standard conditions. To get the graph at the standard operating conditions we manually set solar irradiance to 600.00 W/m^2, soil moisture to 50 °C, and air humidity to 50%. The graph depicts a range of 3.6–6.1 V, while between the range, the pump voltage is seen to be increasing when the air temperature increases above 15 degrees Celsius and hits the maximum when the air temperature reaches

27 degrees Celsius. Since between the range, the pump voltage increases with an increase in air temperature, and Equation (6) is validated.

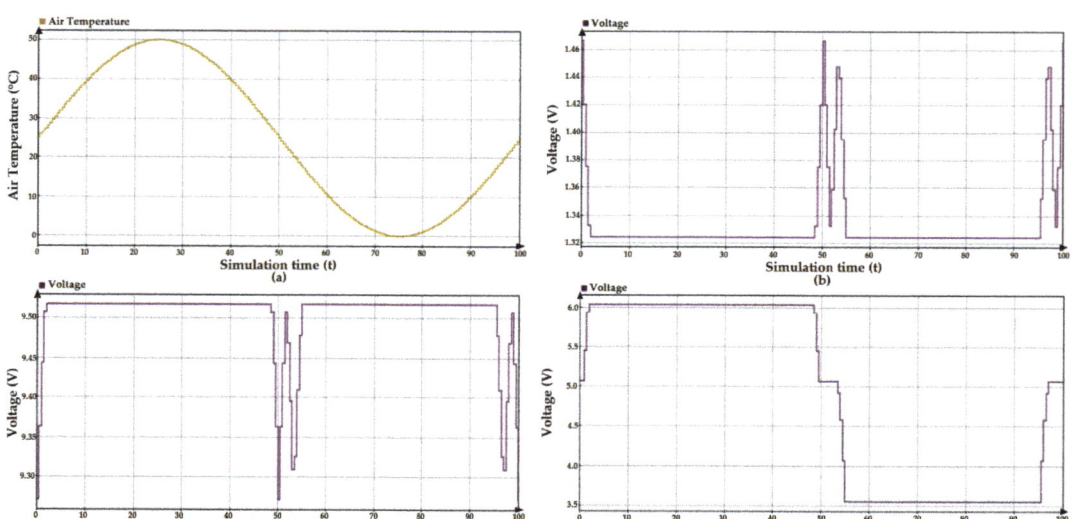

Figure 14. (**a**) Air temperature vs. time; (**b**) voltage vs. time at minimum water need; (**c**) voltage vs. time at maximum water need; (**d**) voltage vs. time at standard operating conditions.

4.1.4. Air Humidity

According to Equation (7), the effect of air humidity on the pump voltage should be inversely proportional, and with an increase in the air humidity value the pump voltage should be decreased [8]. This can be validated by the following graphs shown in Figure 15.

Figure 15a,b show the graphs of air humidity vs. time and voltage vs. time, respectively, when all the other variables' need for water is at a minimum. This graph can be neglected as the maximum voltage observed is less than 3.3 V. Similarly, considering that the need for water is at maximum w.r.t. to the other variables, the effect of air humidity on the pump voltage is defined by the graph shown in Figure 15c. This graph also does not provide enough information as the range observed is only 0.4 V and, based on our hardware resolution, it will be rounded off. From Figure 15d, we again observed that the graph lies between an overall range of 0.3 V and the actual output will be rounded off. This concludes that the effect of air humidity is negligible on the pump voltage.

The highest range observed from the above graphs was the effect of soil moisture on pump voltage and the second highest was solar irradiance. The effect of air temperature and air humidity is found to be negligible, while the graph of air temperature under standard conditions showed some response. We can conclude that the highest priority was given to soil moisture as it is also the control variable that we need to monitor, and then the second is solar irradiance as it is the largest factor that affects the amount of water needed by the soil, and the third and fourth are air temperature and air humidity, respectively. We can say that the system is highly sensitive to change in soil moisture and then moderately sensitive to change in solar irradiance, less sensitive to change in air temperature, and minutely sensitive to change in air humidity.

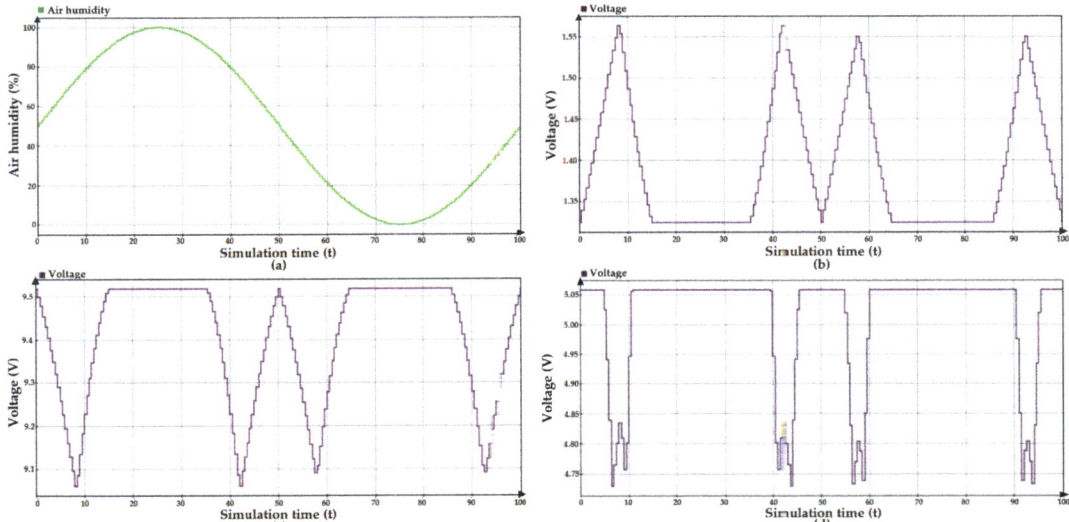

Figure 15. (a) Air humidity vs. time; (b) voltage vs. time at minimum water need; (c) voltage vs. time at maximum water need; (d) voltage vs. time at standard operating conditions.

5. Proposed Model

For data acquisition, the Arduino microcontroller is used with different types of sensors to collect the real-time readings of all the input variables. Arduino Nano is used for the base of our system. Serial communication is established between Arduino and Simulink to get all the input values and the output is then sent back to Arduino, which also controls the input voltage of the pump [2,20–24]. Figure 16 shows all the relevant components and their connectivity in the proposed model in the form of a block diagram.

Soil moisture is taken in by soil moisture sensor RC-A-4079, which comprises a probe with two electrodes and one eight-pin integrated circuit. The two electrodes on the probe act as a resistor with variable resistance (just like a potentiometer). The resistance of these two electrodes changes relative to the quantity of water in the soil. The sensor determines the soil moisture level by applying a voltage to the resistance and measuring the voltage drop from the source [2,23,25–27]. Solar irradiance is calculated by using solar panel characteristics as described in [28]; short-circuit current and solar panel temperature are substituted in the formula:

$$G_{Isc} = (G_{STC}/I_{SCstc})(I_{SC} - \mu_{Isc}(T_C - T_{Cstc})) \tag{9}$$

where

G_{Isc} = the solar irradiance calculated using the short-circuit current,
G_{STC} = the solar radiation in standard operating conditions = 1000 W/m^2,
I_{SCstc} = the short-circuit current in standard conditions (Appendix A),
I_{SC} = the short-circuit current read from the sensor,
μ_{Isc} = the temperature coefficient for the short-circuit current (Appendix A),
T_C = the measured temperature of the solar panel,
T_{Cstc} = the temperature of the solar panel at STC (298.15 K).

The short-circuit current of the solar panel is measured by short-circuiting the solar panel terminals using a 5 V DC relay and then taking the current value from an INA219 current sensor and the panel temperature from a thermistor attached to the solar panel [29–34]. Air temperature and air humidity are taken from a DHT11 sensor [23,26].

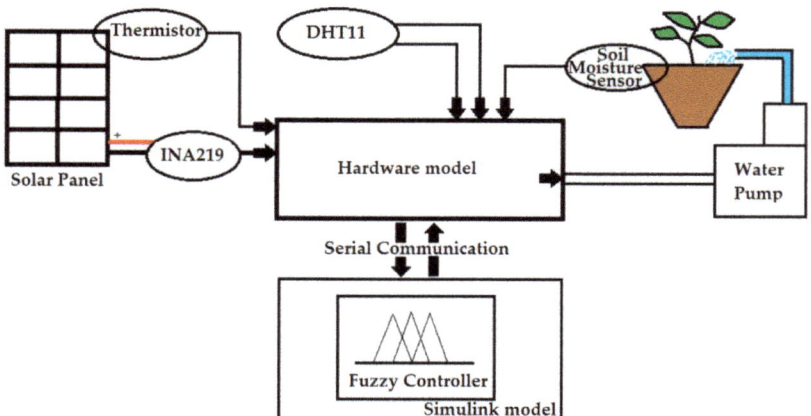

Figure 16. Block diagram of the control system.

All the values from the sensors are taken in by the Arduino and then sent to the Simulink model (block parameters can be found in Appendix A), as shown in Figure 17, with a serial receive block; solar irradiance is calculated by passing the panel temperature and short-circuit current through a subsystem block, which comprises the logic to find the solar irradiance value, as shown in Figure 18; input variables are passed to the fuzzy controller block and the output is then sent to the Arduino using the serial send block. The Arduino then changes the input voltage of the DC water pump using the PWM technique and motor driver [6,35]. After changing the voltage of the DC water pump, a delay is programmed into the Arduino to wait for 10 minutes and start again. All of these steps can be summarized by the flow chart defined in Figure 19. The flow chart does not have a stop block because the algorithm is designed to work recursively and keeps on repeating the process of gathering sensor data and changing the speed of the water pump based on them. The algorithm and the system were developed after extensive research to work efficiently based on all the input factors and sensor modules [7,10–12,24,27,36]. The final working model of the system is shown in Figure 20.

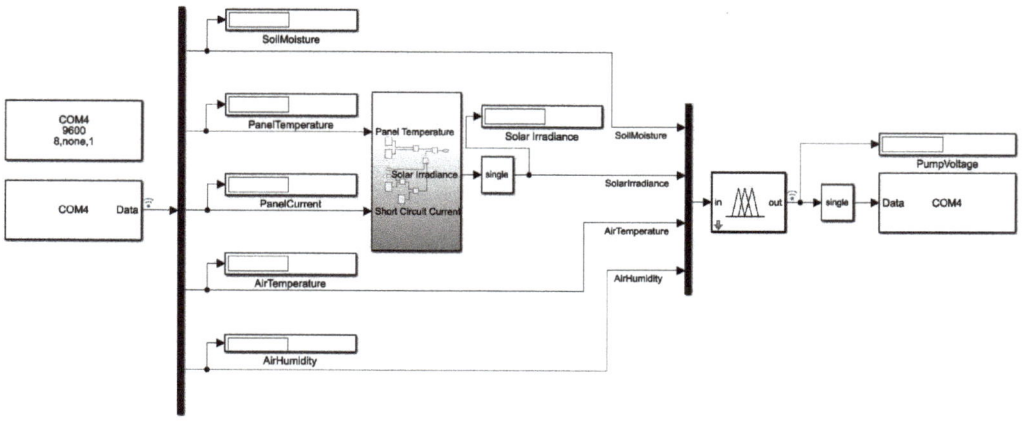

Figure 17. Simulink model.

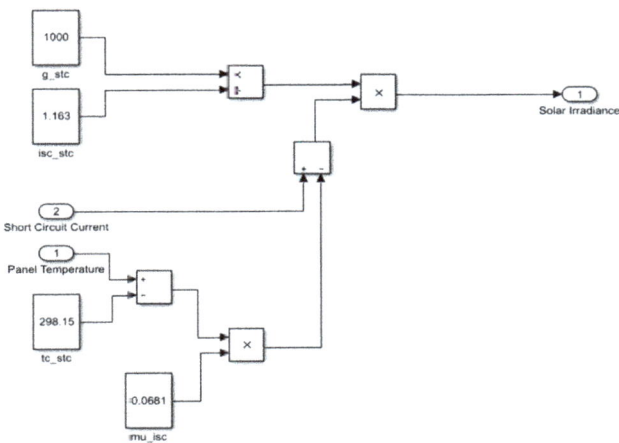

Figure 18. Subsystem to find solar irradiance value.

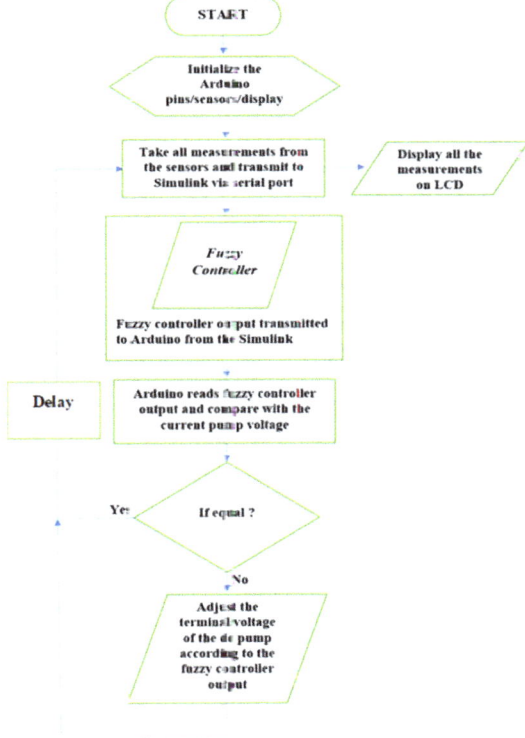

Figure 19. Flow chart of the control algorithm.

Figure 20. Working model of the system.

6. Results

The automatic irrigation control system is operated at different times during the day from sunrise to before sunset on different days to obtain different input values from various sensors to observe the change in output voltage and the pump's water flow rate. The results obtained are shown in Table 2.

Table 2. Results.

S.No.	Sensor Readings						Solar Irradiance (W/m^2)	Fuzzy ControllerOutput (Volts)	Actual Pump Voltage (Volts)
	Soil Moisture (%)	Panel Temperature (K)	Short Circuit Current (A)	Air Temperature (°C)	Air Humidity (%)				
1	0	294.18	0.500	22.70	57	662.61	9.36	9.4
2	85	294.70	0.496	23.00	56	628.72	2.73	0
3	22	297.29	0.326	29.74	37	330.76	7.24	7.2
4	34	293.76	0.247	30.20	57	469.03	5.65	5.7
5	89	297.29	0.897	26.00	49	821.53	3.44	3.5
6	100	294.80	0.504	22.79	58	629.98	2.70	0
7	55	295.22	0.498	22.10	59	600.49	4.12	4.1
8	84	293.13	0.273	21.10	60	528.21	1.85	0
9	4	293.45	0.314	21.79	57	545.68	8.93	9.0
10	90	297.71	0.753	26.79	50	673.42	3.40	3.4

According to Table 2, we observed that the input voltage of the water pump increases based on the environmental values measured from the various sensors. We can see that when soil moisture is measured to be 0% and 4%, the solar intensity ranges from 545.68 W/m^2 to 662.61 W/m^2, the air temperature is approximately equal, which is 22.7 degrees Celsius, and the air humidity is 57%. This results in the fuzzy logic controller giving the output of 9.36 V and 8.93 V, both of them being near 9 V, which is in the maximum range of our system, providing the most water to the plant. As the soil moisture value increases to 22% and there is a significant solar intensity of 330.76 W/m^2 and the air temperature increases to 29.74 degrees Celsius and the air humidity is decreased to 37%, then the fuzzy controller gives the output of 7.24 V, reducing the pump's voltage from before when there was no moisture in the soil.

We saw that the minimum voltage observed during the day was 1.85 V when soil moisture was around 84% with a solar intensity of around 528.21 W/m^2 and the air temperature being 21.1 degrees Celsius and air humidity equaling 60%, Our system cannot modulate the voltage to below 3.3 V, so it turns the pump off at this range indicating 0 V. At soil moisture of 84% and 85%, we can see the solar irradiance values at 528.21 W/m^2 and 628.72 W/m^2, respectively. We can ignore the changes in air temperature and humidity as they are much less. We saw that the fuzzy controller output changed from 1.85 V to 2.73 V because of the increase in the value of solar irradiance. Now, when the soil moisture is further increased to 90%, the value of solar irradiance is 673.42 W/m^2, while comparing this with the values of other variables at a soil moisture level of 85%, we saw a further increase in the solar irradiance value, a 3.79-degree increase in air temperature, and a 6% decrease in air humidity—all variables indicating more need for water. Thus, the fuzzy controller output can be seen increasing to 3.4 V and the actual pump is switched ON. The circuit diagram, the Arduino program and other supplementary materials can be downloaded from github (link to github repository is mentioned in the supplementary materials section below).

7. Conclusions

The automatic control system was developed using the fuzzy logic controller, which is able to handle different values and to run and control the DC water pump's input voltage based on environmental factors like soil moisture, solar irradiance, air temperature, and air humidity. By controlling the water flow rate, thereby controlling the soil moisture and taking it as an input parameter, a closed-loop control system that is getting continuous feedback is formed. This, in turn, saves water when sufficient moisture is present in the soil and it is not required; also, the plant/crop is not overwatered and is provided with only the sufficient amount of water that is required for optimal growth. The sufficient amount required by the plant/crop is always dependent on the type of plant/crop and the type of soil, and thus the pump voltage membership function values needs to be tweaked for different types of plant/crop and soil. By automating the irrigation process, we also achieved the prevention of under-watering of the plant/crop since the system would automatically provide water to the plant/crop when it senses the need depending upon the soil moisture and the various environmental factors incorporated in the system through different sensors. In addition, our system uses solar panel characteristics to measure the solar irradiance instead of using a pyranometer, which is expensive, or using commonly available radiation sensors, which are not suitable to capture the radiation for the whole spectrum. This solar irradiance measurement not only provides us with a better precision while measuring solar irradiance but also reduces the cost of the system.

Supplementary Materials The supplementary materials, such as the Arduino program (MAIN.ino); the circuit diagram (Schematic_main_2022-01-08); the fuzzy inference system file for MATLAB (Fuzzy_Controller.fis); the Simulink model (MASTER.slx). This can be downloaded at the following link: https://github.com/TabTQ/Intelligent-Control-of-Irrigation-System-using-Fuzzy-Logic-Controller.git (accessed on 16 August 2022).

Author Contributions: All authors planned the study, contributed to the idea and field of information; conceptualization, T.T. and A.K.S.; investigation, A.K.S.; methodology, T.T. and A.K.S.; project administration, A.K.S., M.F.A., G.S., P.N.B. and T.S.; software, T.T.; analysis, A.K.S., T.T., M.F.A. and G.S.; resources, T.T.; conclusion, A.K.S. and T.T.; writing—original draft preparation, T.T.; writing—review and editing, A.K.S., T.T., M.F.A. and G.S.; validation, A.K.S. and G.S.; visualization, T.T.; supervision, A.K.S. and P.N.B. All authors have read and agreed to the published version of the manuscript.

Funding: This research received no external funding.

Data Availability Statement: Not applicable here.

Conflicts of Interest: The authors declare no conflict of interest.

Appendix A

Solar panel specifications	
I_{SCstc}	1.163 A
μ_{Isc}	0.0681 AK^{-1}
Simulink serial configuration block parameters	
Baud rate	9600 bps
Data bits	8
Parity	None
Stop bits	1
Byte order	little-endian
Flow control	None
Timeout	10
Simulink serial receive block parameters	
Terminator	CR/LF
Data size	[1 5]
Data type	Single
Block sample time	0.01
Simulink serial send block parameters	
Terminator	CR/LF

References

1. Khatri, V. Application of Fuzzy logic in water irrigation system. *Int. Res. J. Eng. Technol.* **2018**, *5*, 3372.
2. Akter, S.; Mahanta, P.; Mim, M.H.; Hasan, M.R.; Ahmed, R.U.; Billah, M.M. Developing a smart irrigation system using arduino. *Int. J. Res. Stud. Sci. Eng. Technol.* **2018**, *6*, 31–39.
3. Espitia, H.; Soriano, J.; Machón, I.; López, H. Design methodology for the implementation of fuzzy inference systems based on boolean relations. *Electronics* **2019**, *8*, 1243. [CrossRef]
4. Wang, L.X. *A Course in Fuzzy Systems and Control*; Prentice-Hall, Inc.: Hoboken, NJ, USA, 1996; Available online: https://dl.acm.org/doi/abs/10.5555/248374 (accessed on 22 August 2022).
5. Li, J.; Gong, Z. SISO Intuitionistic Fuzzy Systems: IF-*t*-Norm, IF-*R*-Implication, and Universal Approximators. *IEEE Access* **2019**, *7*, 70265–70278. [CrossRef]
6. Petru, L.; Mazen, G. PWM control of a DC motor used to drive a conveyor belt. *Procedia Eng.* **2015**, *100*, 299–304. [CrossRef]
7. Debo-Saiye, Y.; Okeke, H.S.; Mbamaluikem, P.O. Implementation of an Arduino-Based Smart Drip Irrigation System. *Int. J. Trend Sci. Res. Dev.* **2020**, *5*, 1130–1133.
8. Sheikh, S.S.; Javed, A.; Anas, M.; Ahmed, F. Solar based smart irrigation system using PID controller. *IOP Conf. Ser. Mater. Sci. Eng.* **2018**, *414*, 012040. [CrossRef]
9. Rajeshkumar, N.; Vaishnavi, B.; Saraniya, K.; Surabhi, C. Smart crop field monitoring and automation irrigation system using IoT. *Int. Res. J. Eng. Technol.* **2019**, *6*, 7976–7979.
10. Vij, A.; Vijendra, S.; Jain, A.; Bajaj, S.; Bassi, A.; Sharma, A. IoT and machine learning approaches for automation of farm irrigation system. *Procedia Comput. Sci.* **2020**, *167*, 1250–1257. [CrossRef]
11. Priyanka, M.; Gajendra, R.B.; Amravati, P. Smart Irrigation System Using Machine Learning. *Zeich. J.* **2021**, *7*, 18–25.
12. Imam, M.A.; Francis, M.S. PLC Based Automated Irrigation System. *Int. J. Res. Appl. Sci. Eng. Technol.* **2021**, *9*, 1519–1525. [CrossRef]
13. Iqbal, K.; Khan, M.A.; Abbas, S.; Hasan, Z.; Fatima, A. Intelligent transportation system (ITS) for smart-cities using Mamdani fuzzy inference system. *Int. J. Adv. Comput. Sci. Appl.* **2018**, *9*. [CrossRef]
14. Alpay, Ö.; Erdem, E. The Control of Greenhouses Based on Fuzzy Logic Using Wireless Sensor Networks. *Int. J. Comput. Intell. Syst.* **2019**, *12*, 190–203. [CrossRef]

15. Rahim, R. Comparative analysis of membership function on Mamdani fuzzy inference system for decision making. *J. Phys. Conf. Ser.* **2017**, *930*, 012029. [CrossRef]
16. Izzuddin, T.A.; Johari, M.A.; Rashid, M.Z.A.; Jali, M.H. Smart Irrigation using Fuzzy Logic Method. *ARPN J. Eng. Appl. Sci.* **2018**, *13*, 517–522. Available online: http://www.arpnjournals.org/jeas/research_papers/rp_2018/jeas_0118_6698.pdf (accessed on 14 March 2022).
17. Siddique, N.; Adeli, H. *Computational Intelligence: Synergies of Fuzzy Logic, Neural Networks and Evolutionary Computing*; John Wiley & Sons: Hoboken, NJ, USA, 2013; ISBN 1118534816, 9781118534816.
18. Cueva-Fernandez, G.; Espada, J.P.; García-Díaz, V.; Gonzalez-Crespo, R. Fuzzy decision method to improve the information exchange in a vehicle sensor tracking system. *Appl. Soft Comput.* **2015**, *35*, 708–716. [CrossRef]
19. Kreinovich, V.; Kosheleva, O.; Shahbazova, S.N. Why Triangular and Trapezoid Membership Functions: A Simple Explanation. In *Studies in Fuzziness and Soft Computing*; Shahbazova S., Sugeno, M., Kacprzyk, J., Eds.. Recent Developments in Fuzzy Logic and Fuzzy Sets; Springer: Cham, Switzerland, 2020; Volume 391. [CrossRef]
20. Hilali, A.; Alami, H.; Rahali, A. Control Based On the Temperature and Moisture, Using the Fuzzy Logic. *Int. J. Eng. Res. Appl.* **2017**, *7*, 60–64. [CrossRef]
21. Trigui, M.; Barrington, S.; Gauthier, L. Gauthier. Structures and environment: A strategy for greenhouse climate control, part I: Model development. *J. Agric. Eng. Res.* **2001**, *78*, 407–413. [CrossRef]
22. Rahali, A.; Guerbaoui, M.; Ed-Dahhak, A.; El Afou, Y.; Tannouche, A.; Lachhab, A.; Bouchikhi, B. Development of a data acquisition and greenhouse control system based on GSM. *Int. J. Eng. Sci. Technol.* **1970**, *3*, 297–306. [CrossRef]
23. Naik, P.; Kumbi, A.; Katti, K.; Telkar, N. Automation of irrigation system using IoT. *Int. J. Eng. Manuf. Sci.* **2018**, *8*, 77–88. Available online: https://www.ripublication.com/ijems_spl/ijemsv8n1_08.pdf (accessed on 14 March 2022).
24. Derib, D. Cooperative Automatic Irrigation System using Arduino. *Int. J. Sci. Res.* **2019**, *6*, 1781–1787.
25. Farahani, H.; Wagiran, R.; Hamidon, M.N. Humidity Sensors Principle, Mechanism, and Fabrication Technologies: A Comprehensive Review. *Sensors* **2014**, *14*, 7881–7939. [CrossRef] [PubMed]
26. Pawar, S.B.; Rajput, P.; Shaikh, A. Smart irrigation system using IOT and raspberry pi. *Int. Res. J. Eng. Technol.* **2018**, *5*, 1163–1166.
27. Mane, S.S.; Mane, M.S.; Kadam, U.S.; Patil, S.T. Design and Development of Cost Effective Real Time Soil Moisture based Automatic Irrigation System with GSM. *IRJET* **2019**, *6*, 1744–1751.
28. Vigni, V.L.; Manna, D.L.; Sanseverino, E.R.; Di Dio, V.; Romano, P.; Di Buono, P.; Pinto, M.; Miceli, R.; Giaconia, C. Proof of Concept of an Irradiance Estimation System for Reconfigurable Photovoltaic Arrays. *Energies* **2015**, *8*, 6641–6657. [CrossRef]
29. Cruz-Colon, J.; Martinez-Mitjans, L.; Ortiz-Rivera, E.I. Design of a Low-Cost Irradiance Meter using a Photovoltaic Panel. In Proceedings of the IEEE Photovoltaic Specialists Conference, Austin, TX, USA, 3–8 June 2012. [CrossRef]
30. Fuentes, M.; Vivar, M.; Burgos, J.; Aguilera, J.; Vacas, J. Design of an accurate, low-cost autonomous data logger for PV system monitoring using Arduino™ that complies with IEC standards. *Sol. Energy Mater. Sol. Cells* **2014**, *130*, 529–543. [CrossRef]
31. Ortiz-Rivera, E.I.; Peng, F.Z. Analytical Model for a Photovoltaic Module using the Electrical Characteristics provided by the Manufacturer Data Sheet. In Proceedings of the 2005 IEEE 36th Power Electronics Specialists Conference, Recife, Brazil, 12–16 June 2005; pp. 2087–2091. [CrossRef]
32. Jazayeri, M.; Uysal, S.; Jazayeri, K. A simple MATLAB/Simulink simulation for PV modules based on one-diode model. In Proceedings of the 2013 High Capacity Optical Networks and Emerging/Enabling Technologies, Magosa, Cyprus, 11–13 December 2013; pp. 44–50. [CrossRef]
33. Celik, A.N.; Acikgoz, N. Modelling and experimental verification of the operating current of mono-crystalline photovoltaic modules using four-and five-parameter models. *Appl. Energy* **2007**, *84*, 1–15. [CrossRef]
34. Skoplaki, E.; Palyvos, J.A. On the temperature dependence of photovoltaic module electrical performance: A review of efficiency/power correlations. *Sol. Energy* **2009**, *83*, 614–624. [CrossRef]
35. Uyanik, I.; Catalbas, B. A low-cost feedback control systems laboratory setup via Arduino–Simulink interface. *Comput. Appl. Eng. Educ.* **2018**, *26*, 718–726. [CrossRef]
36. Sadiq, M.T.; Hossain, M.M.; Rahman, K.F.; Sayem, A.S. Automated irrigation system: Controlling irrigation through wireless sensor network. *Int. J. Electr. Electron. Eng.* **2019**, *7*, 33–37. [CrossRef]

Article

Real-Time Validation of a Novel IAOA Technique-Based Offset Hysteresis Band Current Controller for Grid-Tied Photovoltaic System

Bhabasis Mohapatra [1], Binod Kumar Sahu [1,*], Swagat Pati [1], Mohit Bajaj [2,*], Vojtech Blazek [3,*], Lukas Prokop [3], Stanislav Misak [3] and Mosleh Alharthi [4]

1. Department of Electrical Engineering, ITER, Siksha 'O' Anusandhan (Deemed to be University), Odisha 751030, India
2. Department of Electrical Engineering, Graphic Era (Deemed to be University), Dehradun 248002, India
3. ENET Centre, VSB—Technical University of Ostrava, 708 00 Ostrava, Czech Republic
4. Department of Electrical Engineering, College of Engineering, Taif University, P.O. Box 11099, Taif 21944, Saudi Arabia
* Correspondence: binoditer@gmail.com (B.K.S.); thebestbajaj@gmail.com (M.B.); vojtech.blazek@vsb.cz (V.B.)

Abstract: Renewable energy sources have power quality and stability issues despite having vast benefits when integrated with the utility grid. High currents and voltages are introduced during the disconnection or injection from or into the power system. Due to excessive inverter switching frequencies, distorted voltage waveforms and high distortions in the output current may be observed. Hence, advancing intelligent and robust optimization techniques along with advanced controllers is the need of the hour. Therefore, this article presents an improved arithmetic optimization algorithm and an offset hysteresis band current controller. Conventional hysteresis band current controllers (CHCCs) offer substantial advantages such as fast dynamic response, over-current, and robustness in response to impedance variations, but they suffer from variable switching frequency. The offset hysteresis band current controller utilizes the zero-crossing time of the current error for calculating the lower/upper hysteresis bands after the measurement of half of the error current period. The duty cycle and hysteresis bands are considered as design variables and are optimally designed by minimizing the current error and the switching frequency. It is observed that the proposed controller yields a minimum average switching frequency of 2.33 kHz and minimum average switching losses of 9.07 W in comparison to other suggested controllers. Results are validated using MATLAB/Simulink environment followed by real-time simulator OPAL-RT 4510.

Keywords: arithmetic optimization algorithm (AOA); conventional hysteresis band current controller (CHCC); improved arithmetic optimization algorithm (IAOA); offset hysteresis band current controller (OFHCC); particle swarm optimization (PSO)

1. Introduction

Global warming and climate issues are considerably increased due to the unfeasible energy consumption of fossil fuel resources. By 2050, the global temperature is expected to increase around 2^0 Celsius because of the emissions by non-renewable energy sources [1,2]. Economical and renewable energy sources (RESs) must be explored to reduce the worse environmental impacts through effective actions. Various research works have been carried out on RES methods in the last few decades to improve overall efficiency. According to the International Renewable Energy Agency (IRENA) report, the cost of energy from renewables has gradually decreased in the previous few years [3]. Many countries are installing RESs in their power systems due to reduced power energy costs. The increase in urbanization and growth of world industry has made power generation by RESs widespread.

The solar energy available for trapping changes throughout the day. The weather parameters play a crucial part in the reliability of solar energy trapping [4]. The reliability of renewable energy technologies, reduction in carbon emissions, reduction in costs of energy generation, and competitive nature of the market are the different reasons to adopt the RESs in large proportions. The yearly consumption of natural gas and oil used in power production will double by 2050 if world consumption increases linearly [5–7]. In 2017, approximately 77% of new installations were based on the extraction of electric energy from wind and solar energies [8]. In the period between 2010 and 2017, the cost of energy from PVs was reduced by three-quarters due to technological advancement [9]. The cost of wind energy generation has decreased by about half due to the reduction in wind turbine prices.

The advancement of PV technology throughout the globe is represented in two stages: improvement in research and development and growth of PV projects [10]. The contribution of research papers has increased considerably relating to the optimization techniques in PV systems in the last few years. PV installation has improved due to the evolution of scientific research articles. Thus, optimization techniques have a crucial role in the effectiveness and reliability of photovoltaic systems.

Wind speed, solar radiation, and ambient temperature are the weather factors strongly related to the PV optimization technique. Various mathematical models of PV technologies are illustrated in [11]. The PSO technique is the most reliable and efficient algorithm for trapping maximum PV performance and higher power outputs [12]. In terms of speed and ability, the PSO algorithm reached positive impressions in the optimization method [13]. The key points of optimization of photovoltaic systems are varying convergence rates, optimal scheduling operations, computational complexity, and accurate performance. Compared to the conventional technique, the intelligent technique has proven more accurate and robust because of precise convergence speed and calculation and the exploitation and exploration to reach the global optimal solutions [14].

The deployment of different optimization techniques in photovoltaic systems and wind turbines has increased the production capacity of RESs to meet the rise in energy demand in the world market. In 2020, the total power-delivering capacity increased by 9% compared to 2016 [15]. Since 2017, the renewable energy power production cost has significantly decreased [16]. However, the deployment and development of RES technologies need additional investment and policies that must be examined thoroughly [17]. In addition, more awareness must be created about the quality and efficiency of using renewable energy. Renewable energy is estimated as 70% of the whole world's power generation capacity as per the Global Report of 2019. A huge investment is being made worldwide for research and development to improve the efficiency of PV systems. The intermittent nature of renewable energy sources is the main drawback, but renewable energy proves more reliable in operational parameters [18].

Metaheuristic optimization algorithms (MOAs) have been very popular in engineering applications. The reasons for this increasing demand are (i) avoidance of local optima, (ii) simple and effective hardware implementation, (iii) derivation-free mechanisms, (iv) flexible and simple structure and concepts, etc. The optimization problems are solved by nature-inspired MOAs simulating physical or biological phenomena [19]. For the achievement of the global optimum solution, various nature-inspired algorithms such as particle swarm optimization (PSO) [20], grey wolf optimizer (GWO) [21], gravitational search algorithm (GSA), teaching–learning-based optimization (TLBO) [22], and artificial bee colony (ABC) algorithm [23] are applied in different areas of research. A mathematics-based model known as the arithmetic optimization algorithm (AOA) technique [24] was recently proposed to solve optimization problems. In some cases, the original AOA technique has some drawbacks such as local optima and premature coverage. The main goal of this paper is to find the optimal values of the hysteresis bands and duty cycle. A hybrid algorithm technique, i.e., an improved arithmetic optimization algorithm (IAOA) technique, is proposed by combining the arithmetic optimization algorithm with particle swarm optimization in this article. The drawbacks such as being trapped in fast convergence and local optima of the traditional AOA technique are addressed by IAOA. Thus, IAOA may be used to improve the performance of AOA. In this article, some popular benchmark functions are considered to prove the efficacy of the proposed IAOA technique over the AOA technique.

There are different current control techniques proposed for grid-tied inverter systems. Still, the hysteresis band current controller is easy to implement and has faster current controllability compared to other current control methods. The interval between the two consecutive switching actions varies within the power frequency cycle in most PWM applications. Therefore, the switching frequency varies in time with conditions and operation points. The increasing switching frequency causes an increase in switching losses, EMF-related problems, and audible noise. An extra offset hysteresis band is added over the existing two-level hysteresis band to develop an offset hysteresis band current controller, which reduces the switching frequency to a lower value. The stress on inverter switching is reduced, as this strategy uses the zero switching condition of the inverter. Conventional hysteresis band current controllers do not consider the inverter zero switching condition [25]. The merits of the OFHCC over the conventional HCC are shown in Figure 1. The performance characteristics of grid-tied inverters basically depend on the controller strategy [26]. The methods proposed in [27] are centered on the current control strategies which include linear and nonlinear controllers. The linear controllers include repetitive current (RC), proportional-integral (PI), and proportional-resonant (PR) controllers. On the other hand, the nonlinear techniques include hysteresis current controllers, predictive controllers, and dead-beat (DB) controllers [28].

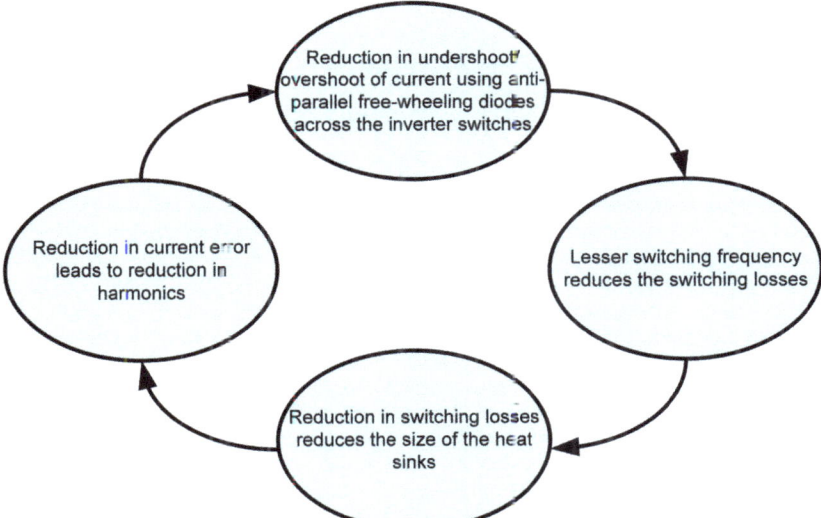

Figure 1. Advantages of offset hysteresis band current controller (OFHCC).

The contributions and key highlights of the paper are as follows:
1. Design of a novel IAOA optimization technique for a microgrid-connected PV system.
2. Application of conventional and offset hysteresis band current controller in a PV-based microgrid.
3. Realization of enhanced performance with an IAOA-based offset hysteresis band current controller.
4. Establishment of the effectiveness of the proposed control algorithm in mitigating harmonics from the grid current.
5. Comparative analysis of novel metaheuristic algorithm-based conventional and offset hysteresis band current controllers with MATLAB/Simulink and OPAL-RT simulator with linear and nonlinear loads.

The architecture of the article is as follows: Section 2 describes the system modeling i.e., PV module, single-phase PWM-VSI, proposed methodology, and reference current technique. An analysis of different hysteresis current controllers is presented in Section 3. Common benchmark functions are depicted in Section 4. A detailed analysis of the algorithms along with respective pseudocodes is presented in Section 5. Finally, Section 6 demonstrates the features and validity of the proposed methodology through simulation and experimental results.

2. System Modelling

2.1. PV Module Model

Various I-V characteristics are obtained for photovoltaic panels under fluctuating solar irradiance and temperature. PV cells are generally modeled using the single and double-diode models [29,30]. Due to its wide applicability and simplicity, the single diode is used in various PV cell modeling applications [31]. The development of simulation strategies requires precise details of power generation and behavior. The five unknown parameters I_{PH}, α, R_P, R_S, and I_S can be illustrated as I = f(V, I) as shown in Equation (1). PV cell modeling follows the circuit-based depiction of the PV module [32].

The quality of current flow in the PV panel is as follows:

$$I_{PV} = I_{PH} - I_S \left[e^{q\left(\frac{V_{PV} + I_{PV}R_S}{\alpha KT}\right)} - 1 \right] - \left(\frac{V_{PV} + I_{PV} R_S}{R_P} \right) \quad (1)$$

where I_{PV} = output current of PV module (a), I_{PH} = irradiance produced current (a), I_S = reverse saturation current (a), V_{PV} = output of PV module (V), R_S = series resistance (Ω), R_P = parallel resistance (Ω), q = charge of electron, K = Boltzmann constant, T = operating temperature of PV module (K), and α = diode ideality parameter.

2.2. Modeling of Single-Phase PWM-VSI

Distributed power generation based on PV energy systems and wind mostly uses grid-integrated VSIs as a fragment of conversion systems. The performance evaluation of power electronic devices can be performed by analyzing the total harmonic distortion (THD), switching losses, transient response, and energy efficiency. In addition, power electronic devices are used to convert DC into AC form [33–38].

Wavelet analysis is utilized to calculate the high distortions in the output voltage [39]. THD estimates the static load variations of the output voltage. The desired quality of VSI output voltage is challenging to obtain in the present scenario. For dynamic load variations, advanced controllers such as CHCCs and OFHCCs reduce the output voltage distortions [40].

$$V_{DC} = L_f \frac{di_o}{dt} + V_g \quad (2)$$

$$i_o - i_{ref} = i_e \quad (3)$$

$$V_{DC} = L_f \frac{d(i_{ref} + i_e)}{dt} + V_g \quad (4)$$

Taking into account the dynamic conditions of the system,

$$V_{DC} - V_g = L_f \frac{di_e}{dt} \quad (5)$$

$$\frac{di_e}{dt} = \frac{V_{DC} - V_g}{L_f} \quad (6)$$

where i_o is the actual current of the inverter, i_e is the error current, and i_{ref} is the reference current.

2.3. Proposed Methodology

The proposed methodology is the integration of IAOA-OFHCC and IAOA-CHCC with a grid PV system. There are basically seven components: (1) PV array, (2) boost converter with optimized duty cycle, (3) algorithms for optimization of hysteresis bands and duty cycle, (4) conventional hysteresis band current controller, (5) offset hysteresis band current controller, (6) utility grid, and (7) reference current technique. Figure 2 presents the schematic diagram of the grid-tied PV system. The voltage of PV is increased to a higher level through the optimized duty cycle of the boost converter. The boost converter's output is fed to a single-phase inverter where DC power is converted into AC power, making it suitable with reduced current rippling for feeding to the utility grid. CHCC and OFHCC control the single-phase inverter's switching. The error current is passed through both the controllers between the optimized bands. The proposed schemes are compared with the PSO-CHCC, FBI-CHCC, AOA-CHCC, PSO-OFHCC, FBI-OFHCC, and AOA-OFHCC.

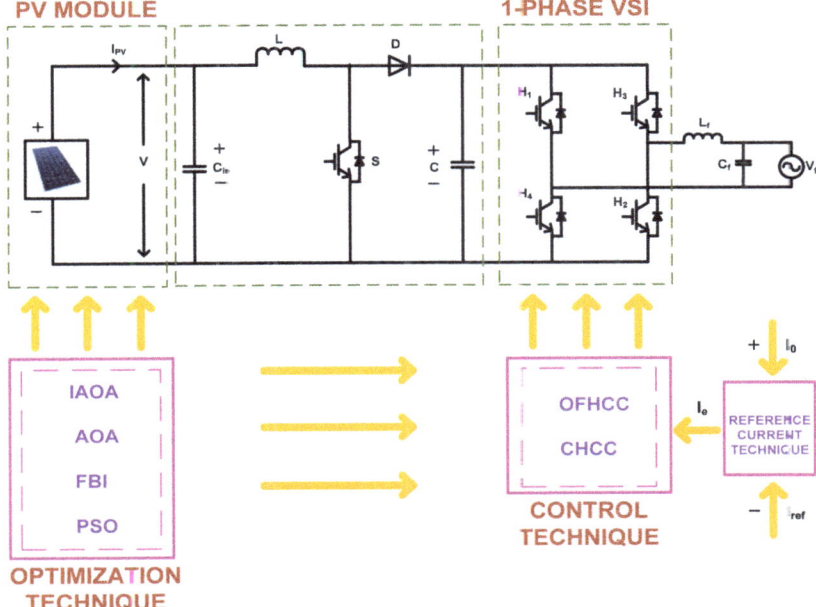

Figure 2. System block diagram.

2.4. Reference Current Technique

The reference current method in [41] is implemented in this paper. After the measurement of the grid voltage, the manipulation is carried out to calculate the reference current. The scaling factor (α) is regularly updated with the load variation. The grid-connected reference current technique is shown in Figure 3.

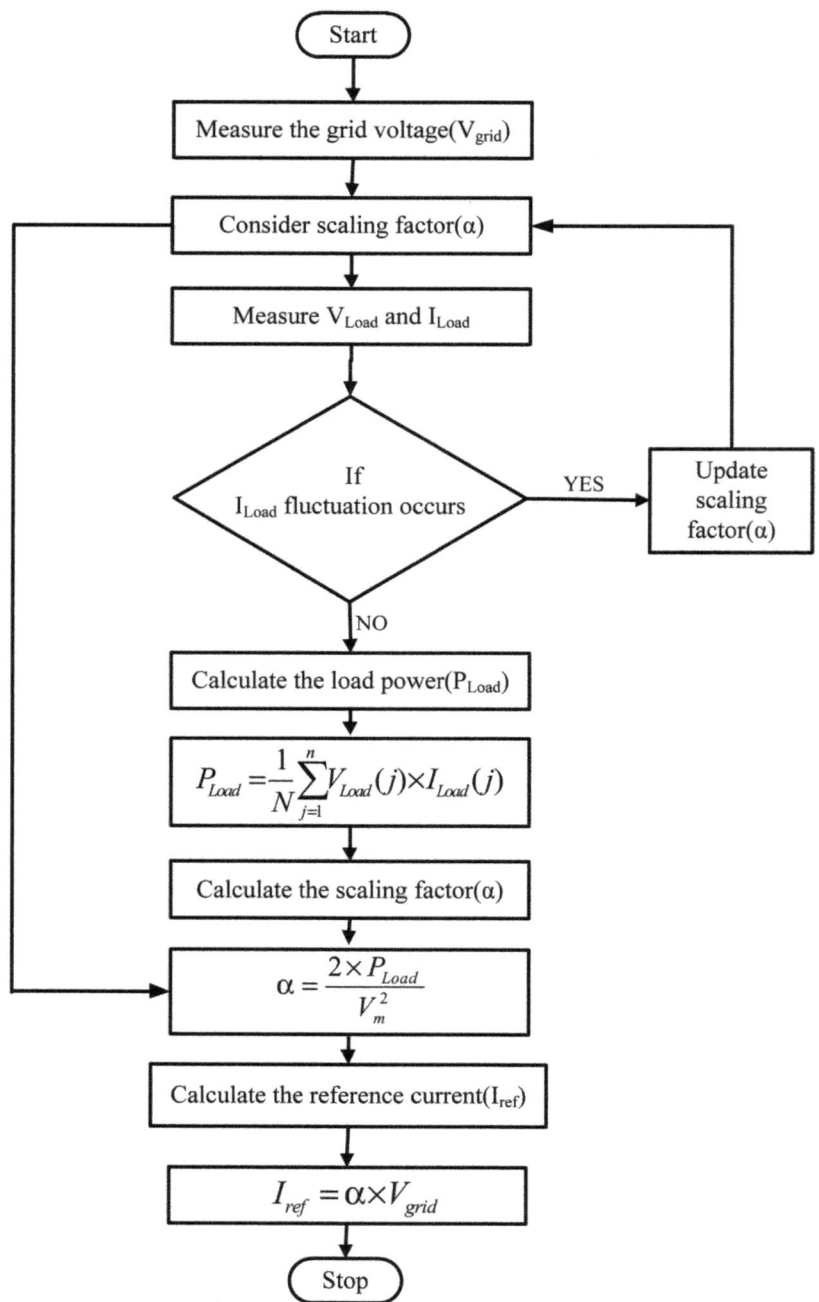

Figure 3. Reference current technique algorithm.

3. Analysis of Advanced Controllers

3.1. Conventional Hysteresis Band Current Controller (CHCC)

The conventional hysteresis current controller is very well regarded due to its unconditioned stability and excellent transient response for grid-connected inverters [42–46]. The actual output of the inverter is compared with the current reference to generate the current error. The current error generated by the actual and reference current difference is restricted within the optimized hysteresis bands. When the current error reaches the lower band, the inverter switches S_2 and S_3 are turned on, and inverter switches S_1 and S_4 are turned on when the error touches the upper band. The switching signals of the IGBTs are generated by restricting the error current within the fixed hysteresis bands. The bandwidth of the hysteresis current controller is calculated in [42]. With decreased hysteresis bandwidth, the error is minimized, but the average switching frequency will increase, resulting in increased average switching losses in the system [47]. The operation of the conventional hysteresis band current controller is explained in Figure 4.

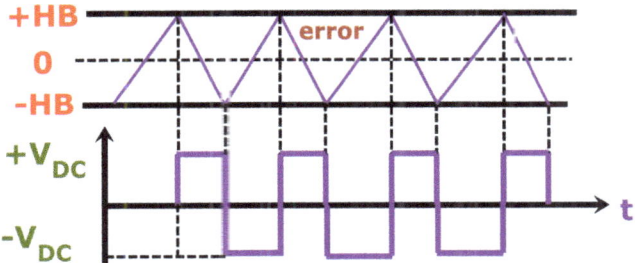

Figure 4. Analysis of conventional hysteresis band current controller [48].

3.2. Offset Hysteresis Band Current Controller (OFHCC)

The conventional hysteresis band current controller does not utilize the zero output condition of the inverter, which leads to a high average switching frequency deviation. The offset hysteresis band current controller is implemented by using an extra hysteresis band and considering the inverter zero output condition which results in a reduction in current error, average switching frequency, and average switching losses. An optimized hysteresis band is proposed which overlaps the existing upper and lower hysteresis bands. When the error current reaches the inner hysteresis band, the inverter output is set to zero condition. Similarly, reversal of the error current occurs when the error current passes an outer hysteresis band, making the inverter output either positively or negatively active. The error current will continue through the inner band to the next outer band, and the error current will reverse. The operating strategy of OFHCC is depicted in Figure 5.

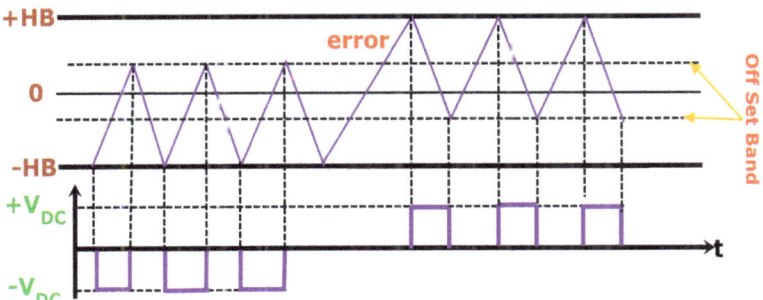

Figure 5. Analysis of offset hysteresis band current controller [48].

A new inverter switching process introduces the output current with either positive or negative DC offset error depending on the active output voltage. The error current is restricted between the upper–outer and lower–inner hysteresis bands for a negative inverter output, and the error current is restricted between the lower–outer and upper–inner hysteresis bands for a positive inverter output, as shown in Figures 6 and 7, respectively [49].

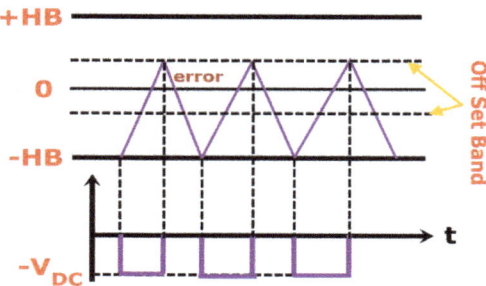

Figure 6. Analysis of offset hysteresis band current controller with negative inverter output [48].

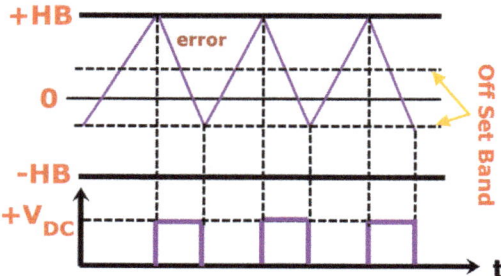

Figure 7. Analysis of offset hysteresis band current controller with positive inverter output [48].

In an offset hysteresis band current controller, the average switching frequency is reduced by a factor of 4.4, and thus the switching losses are reduced by a factor of 4.4 compared to those of a conventional hysteresis band current controller. Thus, the performance of the OFHCC is improved.

The switching cycle of the offset hysteresis band current controller is as follows: $0 \to t_1 \to \frac{T}{2}$.

For Cycle $0 \to t_1$,

$$\Delta I = -1.1 HB, \quad V_{DC} = 0, \quad \Delta t = t_1$$

and thus,

$$t_1 = T_{ON} = \frac{1.1 L_f HB}{V_g} \tag{7}$$

For Cycle $t_1 \to \frac{T}{2}$,

$$\Delta I = 1.1 HB, \quad V_{DC} = +V_{DC}, \quad \Delta t = \frac{T}{2} - t_1$$

and thus,

$$\frac{T}{2} - t_1 = T_{OFF} = \frac{1.1 L_f HB}{V_{DC} - V_g} \tag{8}$$

The summation of Equations (7) and (8) gives the average switching frequency of the OFHCC:

$$T = \frac{2.2\, V_{DC}\, L_f\, HB}{V_g(V_{DC} - V_g)} \quad (9)$$

$$f_{s,av}^{OFHCC} = \frac{1}{T} = \frac{V_g(V_{DC} - V_g)}{4.4 V_{DC} L_f\, HB} \quad (10)$$

The switching strategy is illustrated in Figure 8.

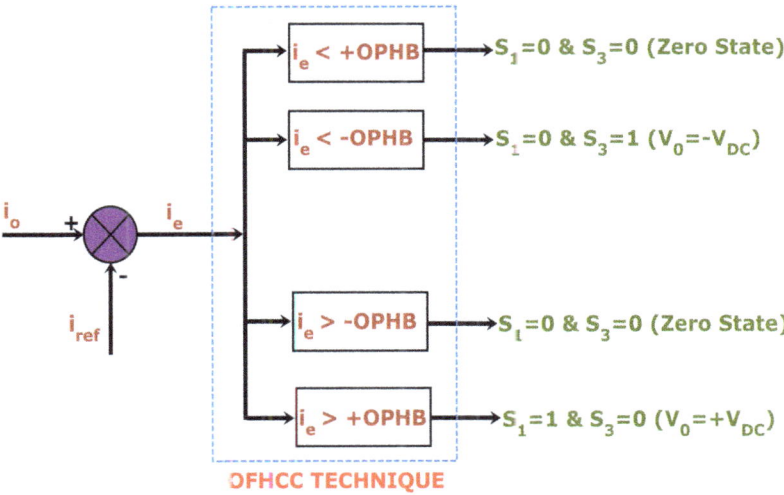

Figure 8. Switching strategy of offset hysteresis band current controller [48].

The average switching frequency is time-varying and is a function of the inductor (L_f) and hysteresis band (HB), which can be observed from Equation (10). The selection of the inductor (L_f) value is made in such a way that it compromises between the ripple current and the average switching frequency. The reasonable harmonics and average switching frequency are achieved by choosing an optimum value for the hysteresis band.

4. Common Benchmark Functions Used in the Study

This article analyzes six benchmark functions, namely Beale, Powell, Matyas, Griewank, Eggholder, and Shubert, to show the superiority of the IAOA over the traditional AOA [50]. For each function, the expression, the range of search space, and the dimension are presented in Table 1. IAOA and AOA were coded in MATLAB and run 30 times by taking the maximum number of population and number of iterations as 100. The performance parameters, such as mean, maximum, minimum, and standard deviation, are presented in Table 2 to prove the supremacy of the IAOA technique. Table 2 shows that mean, maximum, minimum, and standard deviation values are less for IAOA than for AOA. The number of iterations needed for convergence to the global optimum value is lower for the IAOA technique, but due to the inclusion of an additional update phase, the computation time is slightly longer in the case of the IAOA algorithm. Figure 9 depicts the convergence plots of different benchmark functions.

Table 1. Description of benchmark functions used in the study.

Function	Function's Expression	Dimension	Range
Beale (F1)	$f(x) = (1.5 - x_1 + x_1x_2)^2 + (2.25 - x_1 + x_1x_2^2)^2 + (2.625 - x_1 + x_1x_2^3)^2$	2	$[-4.5, 4.5]$
Powell (F2)	$f(x) = \sum_{i=1}^{d/4}\left[(x_{4i-3} + 10x_{4i-2})^2 + 5(x_{4i-1} + x_{4i})^2 + (x_{4i-2} - 2x_{4i-1})^4 + 10(x_{4i-3} - x_{4i})^4\right]$	10	$[-4, 5]$
Matyas (F3)	$f(x) = 0.26(x_1^2 + x_2^2) - 0.48x_1x_2$	2	$[-10, 10]$
Griewank (F4)	$f(x) = \sum_{i=1}^{d}\frac{x_i^2}{4000} - \prod_{i=1}^{d}\cos\left(\frac{x_i}{\sqrt{i}}\right) + 1$	30	$[-600, 600]$
Eggholder (F5)	$f(x) = -(x_2 + 47)\sin\left(\sqrt{\left|x_2 + \frac{x_1}{2} + 47\right|}\right) - x_1\sin\left(\sqrt{|x_1 - (x_2 + 47)|}\right)$	2	$[-512, 512]$
Shubert (F6)	$f(x) = \left(\sum_{i=1}^{5} i\cos((i+1)x_1 + i)\right)\left(\sum_{i=1}^{5} i\cos((i+1)x_2 + i)\right)$	2	$[-5.12, 5.12]$

Table 2. Performance analysis for AOA and IAOA algorithms.

Algorithm	Function	Optimum Value	Minimum	Maximum	Mean	Standard Deviation	Computational Time (s)
IAOA	F1	0	3.5828×10^{-16}	1.9598×10^{-13}	4.2215×10^{-14}	4.9633×10^{-14}	0.0834
AOA			1.0785×10^{-15}	7.7380×10^{-13}	8.6741×10^{-14}	1.7924×10^{-13}	0.0595
IAOA	F2	0	0	2.6215×10^{-20}	8.7385×10^{-22}	4.7863×10^{-21}	0.0766
AOA			0	6.5352×10^{-18}	2.2641×10^{-19}	1.1925×10^{-18}	0.0563
IAOA	F3	0	0	1.1962×10^{-63}	3.9877×10^{-65}	2.1840×10^{-64}	0.0542
AOA			0	5.6773×10^{-63}	1.8930×10^{-64}	1.0365×10^{-63}	0.0423
IAOA	F4	0	0	0	0	0	0.6315
AOA			0	0	0	0	0.5263
IAOA	F5	−959.640	−959.4607	−959.4607	−959.4607	1.0283×10^{-12}	0.0457
AOA			−959.4607	−959.4607	−959.4607	2.0283×10^{-12}	0.0133
IAOA	F6	−186.731	−186.7309	−186.7309	−186.7301	1.4597×10^{-9}	0.0958
AOA			−186.7309	−186.7305	−186.7309	8.2884×10^{-4}	0.0757

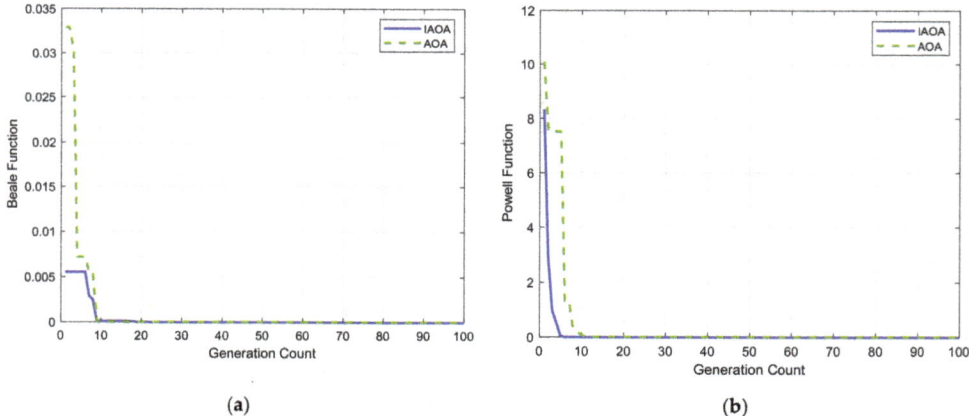

Figure 9. Cont.

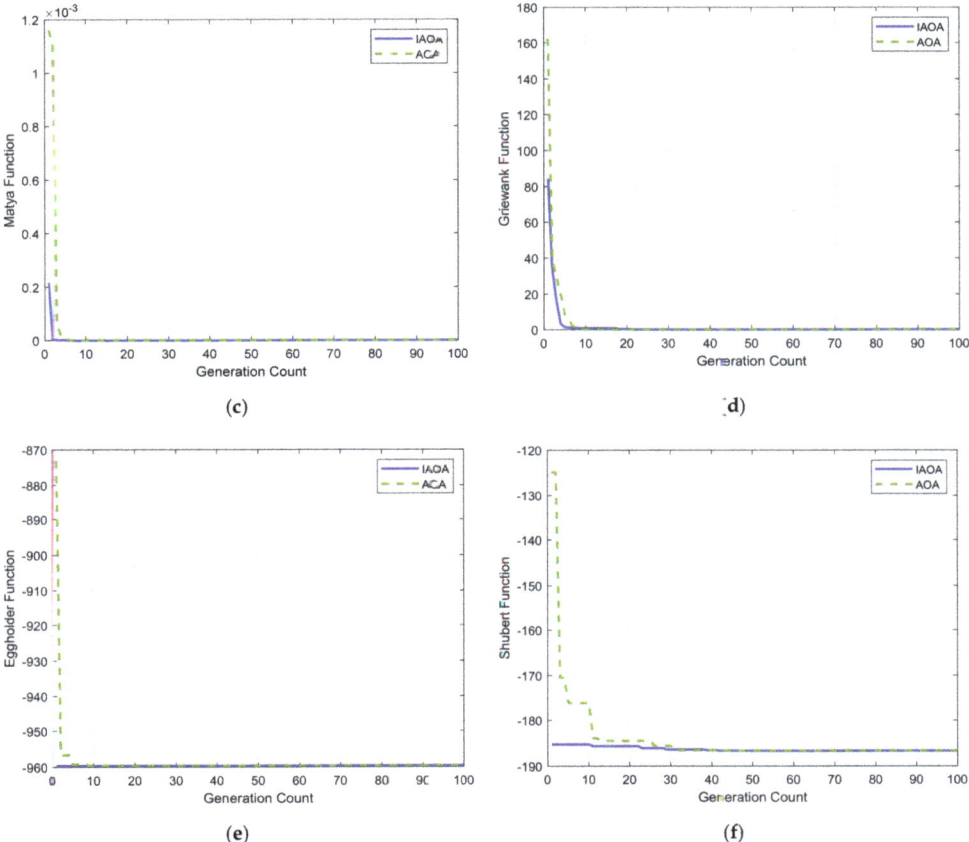

Figure 9. Convergence plots of benchmark functions. (**a**) Convergence characteristics of Beale function. (**b**) Convergence characteristics of Powell function. (**c**) Convergence characteristics of Matyas function. (**d**) Convergence characteristics of Griewank function. (**e**) Convergence characteristics of Eggholder function. (**f**) Convergence characteristics of Shubert function.

5. Analysis of Algorithms

The performance of any metaheuristic algorithm mainly depends on the balance between two phases: exploitation and exploration. In the exploration phase, the algorithm finds new areas of solution time, and the exploitation phase extracts valuable information related to neighborhood regions of the search space [51]. Initially, in this work, PSO, FBI, and AOA techniques are used to determine the optimal values of the design parameters of the grid-tied PV system. However, some of the optimization algorithms fail to converge to global minima as they tend to become stuck in local minima. Algorithms are modified or hybridized and tested on many benchmark functions [52]. In this work, an IAOA optimization technique, where the PSO algorithm is implemented to find the values of some parameters of AOA optimally, is presented and tested on six popular benchmark functions. Finally, the IAOA technique is implemented to design the same parameters to obtain better results.

5.1. Forensic-Based Investigation (FBI) Algorithm

Chou and Nguyen proposed a metaheuristic optimization technique known as the forensic-based investigation algorithm (FBI) [53]. Location, suspected investigation, and stalking are the important ideas of the FBI. Opening the case, analysis of discoveries, examination of directions, actions, and trials are the five steps involved in a large-scale forensic investigation process [54]. In the beginning, the facts about the crime are gathered by the police team, and then this evidence helps the police team to begin the investigation. The inquiry team analyzes the crime position, probable suspects, target, and data about the crime. The team interprets the knowledge and correlates it with the impressions gained during the investigation to assess the probable suspects [55].

The steps involved in the FBI algorithm are as follows:

Step A1: Interpretation of findings:

$$X'_{A_{ij}} = X_{A_{ij}} + \frac{rand * \left(\sum_{1}^{a_1} X_{A_{aj}}\right)}{a_1} \tag{11}$$

$$X'_{A_{ij}} = X_{A_{ij}} + rand * \frac{X_{A_{ij}} - \left(X_{A_{kj}} + X_{A_{hj}}\right)}{2} \tag{12}$$

Step A2: Direction of inquiry:

$$\text{Prob}\left(X_{A_{ij}}\right) = \left(P_{A_i} - P_{min}\right)/(P_{max} - P_{min}) \tag{13}$$

$$X'_{A_{ij}} = X_{min} + \sum_{1}^{a_2} \alpha * X_{A_{bj}} \tag{14}$$

$$X'_{A_{ij}} = X_{min} + X_{A_{dj}} + rand * \left(X_{A_{kj}} - X_{A_{hj}}\right) \tag{15}$$

Step B1: Action taken after receiving the reports:

$$X'_{B_{ij}} = rand1 * X_{B_{ij}} + rand2 * \left(X_{min} - X_{B_{ij}}\right) \tag{16}$$

Step B2: Extension of process of action:

$$X'_{B_{ij}} = X_{B_{rj}} + rand3 * \left(X_{B_{rj}} - X_{B_{ij}}\right) + rand4 * \left(X_{min} - X_{B_{rj}}\right) \tag{17}$$

$$X'_{B_{ij}} = X_{B_{ij}} + rand3 * \left(X_{B_{ij}} - X_{B_{rj}}\right) + rand4 * \left(X_{min} - X_{B_{ij}}\right) \tag{18}$$

where,

α = effectiveness coefficient, i.e., $[-1, 1]$;

$j = 1, 2, \ldots, D$, and D is the number of dimensions;

a_1 and a_2 are the numbers of individuals that affect the movement of $X_{A_{ij}}$ assumed to be 2 and 3;

d, k, h, and i are four suspected locations; {d, k, h, i} ε {1, 2, ..., NP}; d, k, and h are chosen randomly; and NP is the number of suspected locations;

$X_{A_{ij}}$ = suspected location;

$X'_{A_{ij}}$ = new suspected location;

P_{max} = lowest possibility value corresponding to the worst objective value;

P_{min} = highest possibility position corresponding to the best objective value;

P_{A_i} = possibility that the suspect is at location X_{A_i};

X_{min} = highest possibility position corresponding to the best solution;

rand is a random number in the range $[-1, 1]$;

rand1, rand2, rand3, and rand4 are random numbers in the range $[0, 1]$.

5.2. Particle Swarm Optimization (PSO) Algorithm

Eberhart and Kennedy proposed a stochastic optimization algorithm based on swarming in 1995. The social behavior of animals such as birds, fishes, and insects is simulated in the PSO algorithm. Each member in the swarm changes its search pattern and confirms a cooperative pattern to find food under the gained experiences of other members and its own experience. PSO mainly operates on two basic ideas: one is based on artificial life, which provides the artificial structures with life features, and the other is the swarm mode, in which the swarm searches for the prey in a large section in the solution space of the optimized objective functions [56].

PSO involves the following steps:

1. Initialization: Within the specific search range, the initial population and initial size velocity [NP × D] are generated. Here, 'D' is the dimension of the problem and 'NP' is the number of the population.
2. Velocity update: Equation (19) is utilized to update the velocity in this step.

$$V_{new} = w \times v_{old} + C_1 \times rand_1 \times (p_{best} - x) - C_2 \times rand_2 \times (g_{best} - x) \quad (19)$$

where 'C_1' and 'C_2' are acceleration constants generally taken as 2.05; p_{best} is the local best, i.e., the best solution so far achieved by a particle; g_{best} is the global best i.e., the best solution in the population; $rand_1$ and $rand_2$ are random numbers within the range [0, 1]; and 'w' is called the inertia weight, which is decreased from 0.9 to 0.4 with iterations.

3. Position update: The newly generated velocity is combined with the initial population to update the initial population.

$$x_{new} = x_{old} + v_{new} \quad (20)$$

5.3. Arithmetic Optimization Algorithm (AOA)

Abualigah et al. developed a new mathematical optimization technique known as the arithmetic optimization algorithm (AOA) based on addition, subtraction, multiplication, and division. Two essential phases for achieving global optimum solutions are the exploitation and exploration stages.

The exploitation phase achieves a nearer optimal solution as it provides low dispersion in search space utilizing the addition and subtraction operators.

In the exploration stage, the search space is explored in various regions and trends to achieve a better optimal solution using the multiplication and division operators [24].

The various stages in the AOA technique are as follows:

1. Initialization: The initial population size [NP × D] is developed randomly within the predefined search space. Equation (21) evaluates the math optimizer acceleration (MOA).

$$MOA = min_a + iter \times \left(\frac{max_a - min_a}{iter_{max}} \right) \quad (21)$$

where 'iter' and '$iter_{max}$' are the iteration count and the maximum number of iterations; 'min_a' and 'max_a' are the minimum and maximum values of the acceleration function taken as 0.2 and 0.9, respectively.

2. Update phase: Using Equation (22), the math optimizer probability (MOP) is generated.

$$MOP = 1 - \frac{(iter)^{\frac{1}{\alpha}}}{(iter_{max})^{\frac{1}{\alpha}}} \quad (22)$$

where the solution is updated by generating three random numbers, r1, r2, and r3, and 'α' is taken as 5.

if r1 < MOA
   if r2 > 0.5

$$x_{new} = \frac{g_{best}}{MOP + \varepsilon} \times ((U_1 - L_1) \times \mu + L_1) \quad (23)$$

   else

$$x_{new} = g_{best} \times MOP \times ((U_1 - L_1) \times \mu + L_1) \quad (24)$$

   end
else
   if r3 < 0.5

$$x_{new} = g_{best} - MOP \times ((U_1 - L_1) \times \mu + L_1) \quad (25)$$

else
$$x_{new} = g_{best} + MOP \times ((U_l - L_l) \times \mu + L_l) \qquad (26)$$
end

where 'U_l' and 'L_l' are the upper and lower limits of the variables to be designed and 'μ' is taken as 0.5.

5.4. Improved Arithmetic Optimization Algorithm (IAOA)

The traditional AOA technique attracts attention due to the exploration of search spaces. Conversely, in the non-optimal solutions, the traditional AOA technique suffers from premature stagnation. The poor exploration capability of the traditional AOA technique in the early stages causes the quick loss of population diversity. The exploration and exploitation stages are clearly shown in the previous section. In order to overcome the inadequacy of the traditional AOA, an improved variant of traditional AOA known as the improved arithmetic optimization algorithm (IAOA) is proposed and then employed for the optimization of the hysteresis band and duty cycle for the PV microgrid system. The robustness and efficiency of the improved arithmetic optimization algorithm technique are investigated through six standard benchmark functions.

In the suggested IAOA technique, two important parameters 'α and μ' of the AOA technique are optimally designed with the help of the PSO algorithm. The upper and lower limits of 'α and μ' are 5.0 and 0.5, respectively. Figure 10 depicts the intuitive and detailed process of the IAOA technique. The IAOA pseudocode is elaborated as follows:

1. Generate the initial population for design variables and the constants 'α and μ' of the AOA technique.
2. Evaluate the objective function and identify the best-performing solution (g_{best}).
3. Update the solution with the AOA technique using Equations (21)–(26).
4. Update the values of 'α and μ' with the PSO algorithm using Equations (19) and (20).
5. Repeat the previous two steps until the stopping criterion is met.

The switching frequency and current error are multiplied with suitable weighing factors and combined as a single objective function that is to be optimized with different optimization techniques. The objective function expression is given as follows:

$$f = w_1 \times e_i + w_2 \times f_s \qquad (27)$$

The average switching losses are calculated in [57] as shown in Equation (28):

$$P_{sl} = f_s * mean(E_{on} + E_{off}) \qquad (28)$$

where f_s is the switching frequency; e_i is the current error; and w_1 and w_2 are the weighting factors taken as 0.85 and 0.15. respectively.

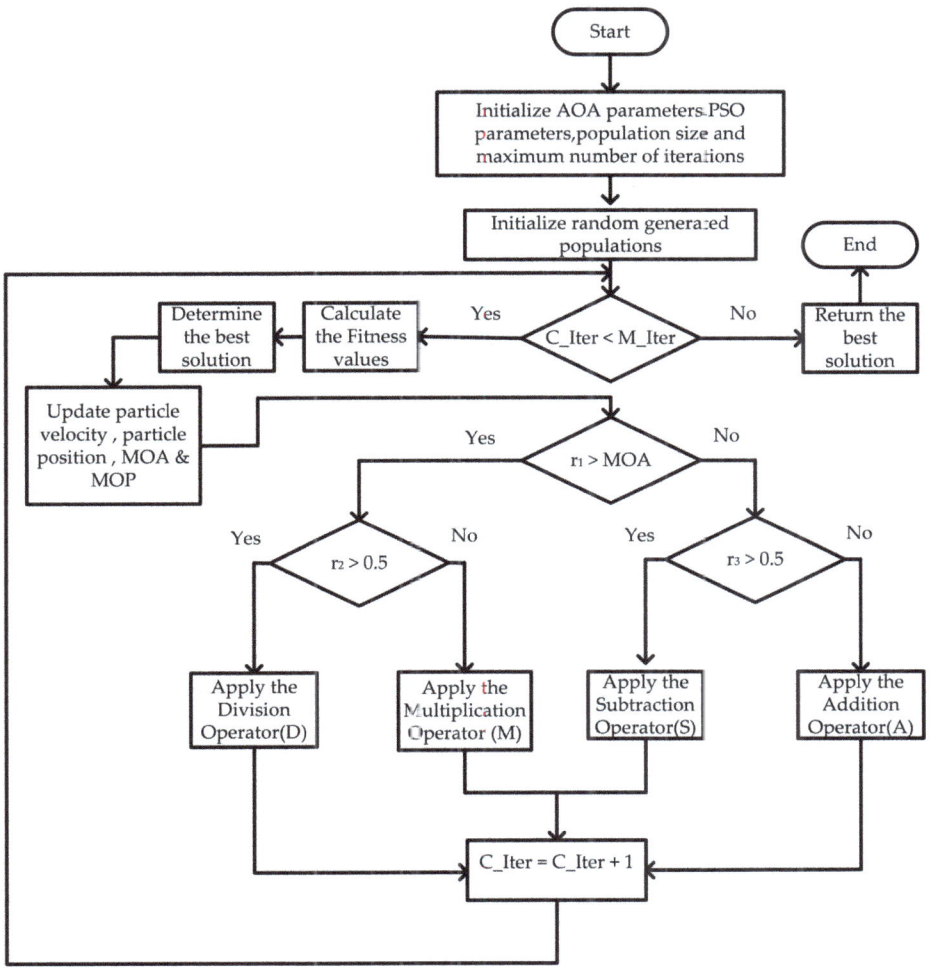

Figure 10. Flowchart of IAOA.

6. Results and Discussion

The lower and upper bands of the CHCC and OFHCC and the duty cycle of the converter were optimally tuned using the PSO, FBI, AOA, and IAOA techniques to improve the dynamic performance of the grid-tied PV system. Various parameters of the grid-tied system are presented in Table 3. Comparative performance analysis of different controllers was performed by taking various performance indicators such as average switching frequency (Avg_{SF}), average switching losses (Avg_{SFL}), maximum switching frequency (Max_{SF}), minimum switching frequency (Min_{SF}), zero-crossing switching frequency (ZSF), and %THD. Optimal hysteresis bands, duty cycles, and various performance indicators are indicated in Table 4. The optimized values of hysteresis bands and duty cycles obtained from the MATLAB/Simulink environment were input into the OPAL-RT 4510 real-time simulator for experimental verification of power quality profiles. The obtained results are compared with different hysteresis band current controller techniques for rooftop microgrid systems at a constant hysteresis band of 0.5, as reported in [58].

Table 3. System parameters.

Parameter	Numerical Value
Grid frequency	50 Hz
Line inductance	15 mH
Irradiance	500 W/m^2
E_{on}	2.2 mJ
E_{off}	1.7 mJ
Cell temperature	25 °C
Load variation	1000 W to 2000 W

Table 4. Performance analysis of various controllers.

Controller	D	HB$_1$	HB$_2$	HB$_3$	HB$_4$	Max$_{SF}$ (in kHz)	Min$_{SF}$ (in kHz)	Avg$_{SF}$ (in kHz)	Avg$_{SFL}$ (in W)	ZSF (in kHz)	%THD
PSO-CHCC	0.215	0.375	−0.572	-	-	9.25	8.75	7.85	30.60	8.50	0.49
FBI-CHCC	0.319	0.462	−0.868	-	-	8.75	6.25	7.58	29.56	8.00	0.54
AOA-CHCC	0.169	0.620	−0.587	-	-	7.25	3.25	5.50	21.48	6.50	0.57
IAOA-CHCC	0.245	0.809	−0.932	-	-	6.25	3.75	5.30	20.67	5.75	0.74
PSO-OFHCC	0.279	0.953	−0.874	0.874	−0.953	6	3.50	3.87	15.09	4.25	0.73
FBI-OFHCC	0.112	0.913	−0.827	0.827	−0.913	5.5	2.50	2.89	11.26	3.25	0.68
AOA-OFHCC	0.141	0.817	−0.645	0.645	−0.817	6	2.75	2.68	10.46	3.75	0.82
IAOA-OFHCC	**0.117**	**0.964**	**−0.532**	**0.532**	**−0.964**	**6.25**	**2.75**	**2.33**	**9.07**	**3.50**	**1.45**
SBHCC [58]	-	0.5	−0.5	-	-	20	-	-	-	-	4.42
DBHCC−1 [58]	-	0.5	−0.5	-	-	10	-	-	-	-	4.33
DBHCC−2 [58]	-	0.5	−0.5	-	-	20	-	-	-	-	2.65
MDBHCC [58]	-	0.5	−0.5	-	-	5.5	-	-	-	-	4.33
VBHCC [58]	-	0.5	−0.5	-	-	15	-	-	-	-	4.17

The microgrid-connected dynamic simulation model was developed using real-time simulation OPAL-RT 4510 with Xilinx Kintex-7 FPGA software. The real-time simulation was realized in RT-LAB using the Simulink models on multi-core CPU computers. RT-LAB builds the parallel tasks from the original models, and each task is assigned to one CPU to enhance the overall simulation time. In addition, the RT-LAB toolbox and power system solver were used to improve the accuracy and simulation time of the grid-connected system. A snapshot of the proposed schemes, along with the other techniques in OPAL-RT 4510, is shown in Figure 11.

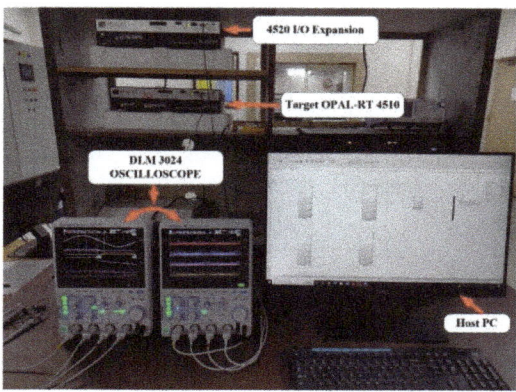

Figure 11. Experimental set-up using OPAL-RT 4510.

The reference current, actual current, and switching pulse simulations for one cycle of IAOA-CHCC and IAOA-OFHCC are shown in Figure 12. The experimental results for corresponding profiles are shown in Figures 13 and 14 for PSO-CHCC, FBI-CHCC, AOA-CHCC, IAOA-CHCC, PSO-OFHCC, FBI-OFHCC, AOA-OFHCC, and IAOA-OFHCC. The switching frequencies are calculated from the inverter switching pulses. The proposed IAOA-OFHCC technique due to the optimum utilization of inverter switches has reduced switching frequency and the lowest switching losses, resulting in high efficiency compared to other techniques.

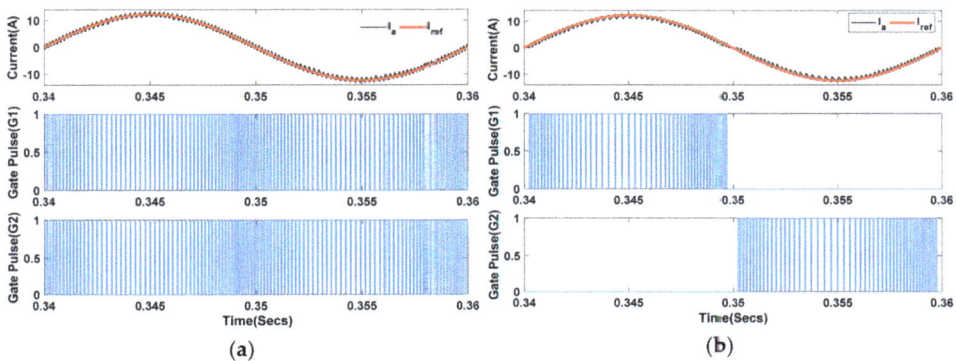

Figure 12. Simulation results showing I_{ref}, I_a, and switching pulses for (**a**) IAOA-CHCC and (**b**) IAOA-OFHCC.

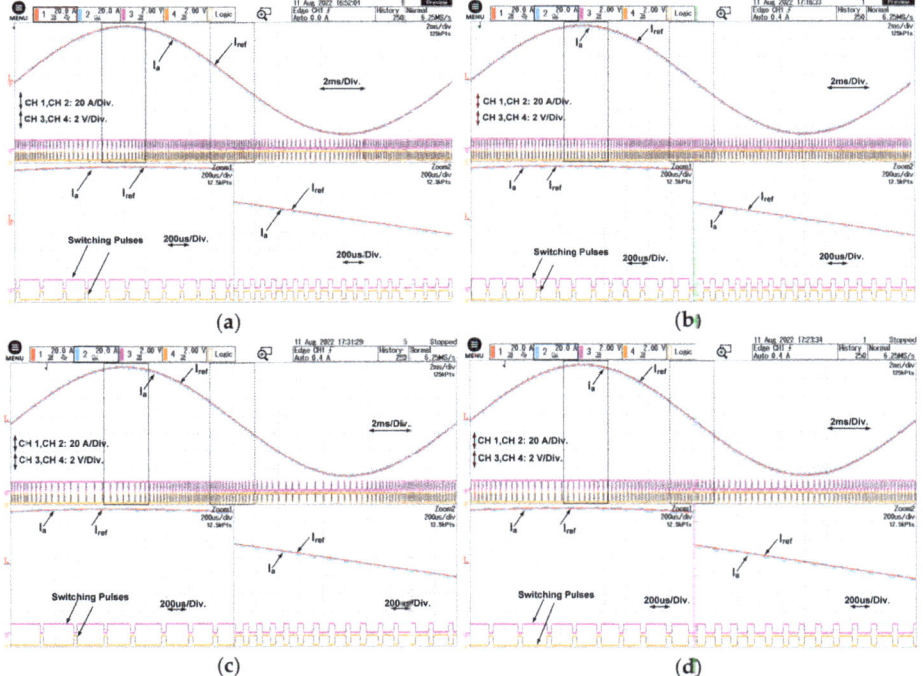

Figure 13. Experimental results showing I_{ref}, I_a, and switching pulses for (**a**) PSO-CHCC, (**b**) FBI-CHCC, (**c**) AOA-CHCC, and (**d**) IAOA-CHCC.

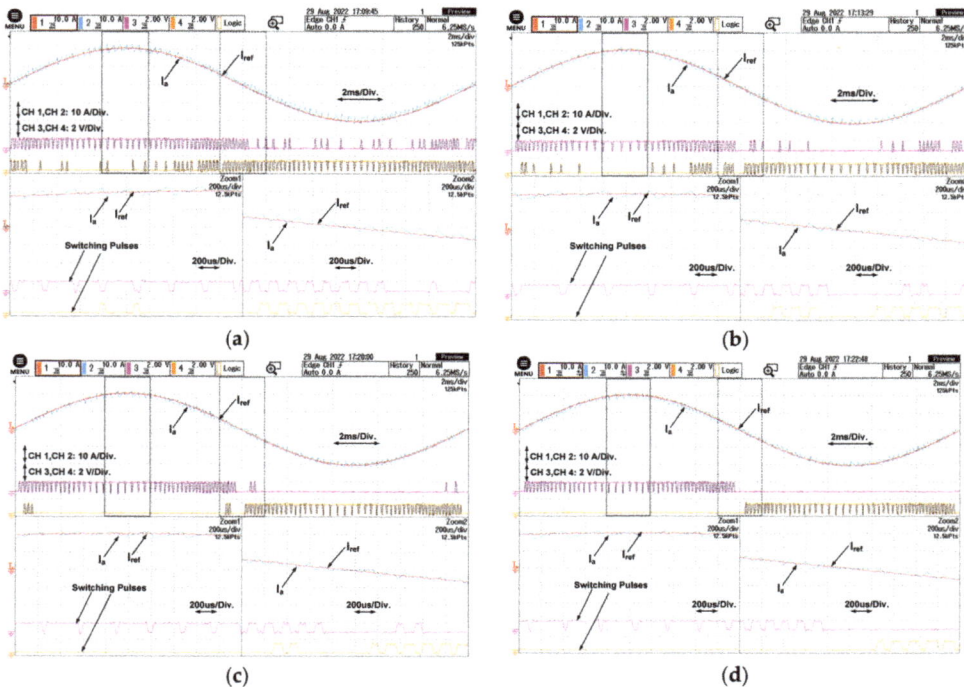

Figure 14. Experimental results showing I_{ref}, I_a, and switching pulses for (**a**) PSO-OFHCC, (**b**) FBI-OFHCC, (**c**) AOA-OFHCC, and (**d**) IAOA-OFHCC.

The simulation and experimental results for the dynamic state performance of I_a and I_{ref} under load change from 1000 W to 2000 W at 4.95 s to 5.05 s are shown in Figures 15 and 16, respectively. The shape of the current waveform is not distorted during the transient and maintains a sinusoidal shape, indicating that both the actual current and reference current are in-phase. The proposed algorithm with controllers has a fast dynamic response with variation in load. In addition, the proposed scheme has the lowest average switching frequency, which makes the IGBT device operate under a safe operating range.

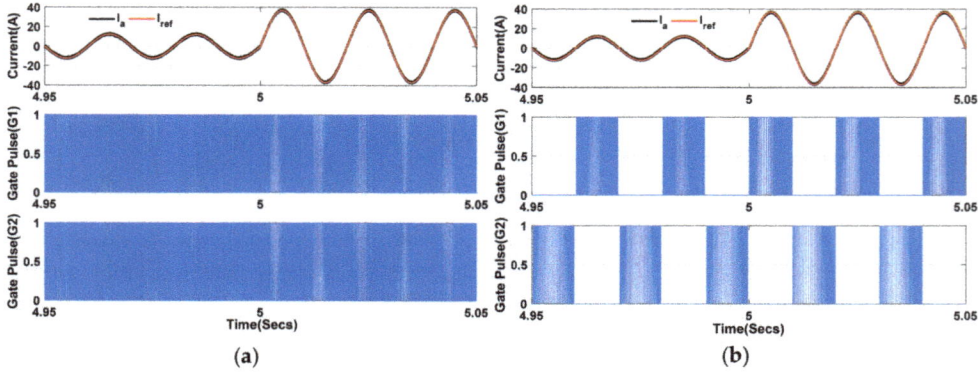

Figure 15. Simulation results showing I_{ref}, I_a, and switching pulses under load change for (**a**) IAOA-CHCC and (**b**) IAOA-OFHCC.

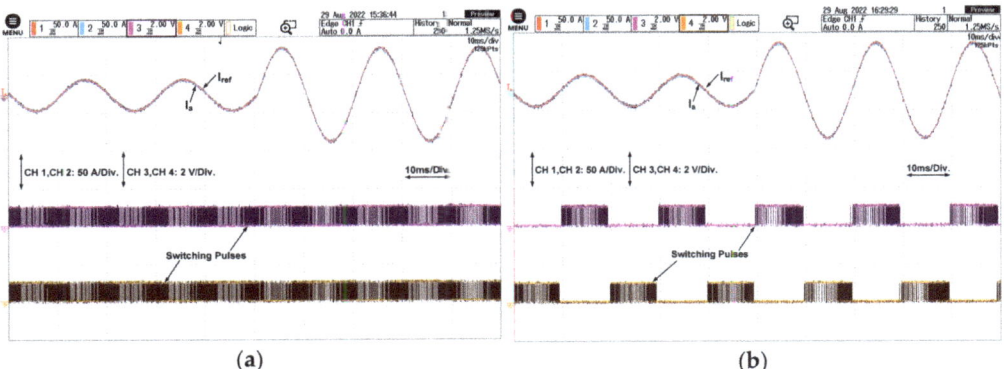

Figure 16. Experimental results showing I_{ref}, I_a, and switching pulses under load change for (**a**) IAOA-CHCC and (**b**) IAOA-OFHCC.

The error current of the proposed algorithm with offset hysteresis band and other algorithms and controllers are shown in Figures 17–20 with simulation and experimental results. It can be observed that the error current remains within the hysteresis bands in both cases. The inverter current spectrum for the grid-tied system is shown in Figures 21 and 22 for CHCC and OFHCC with respective algorithms for the PV microgrid. It can be observed that PSO-CHCC has the lowest %THD of 0.49 but has the average switching frequency and average switching losses of 7.85 kHz and 30.60 W, respectively, so the proposed IAOA-OFHCC technique dominates all other control algorithms with minimum average switching frequency and minimum average switching losses of 2.33 kHz and 9.07 W, respectively.

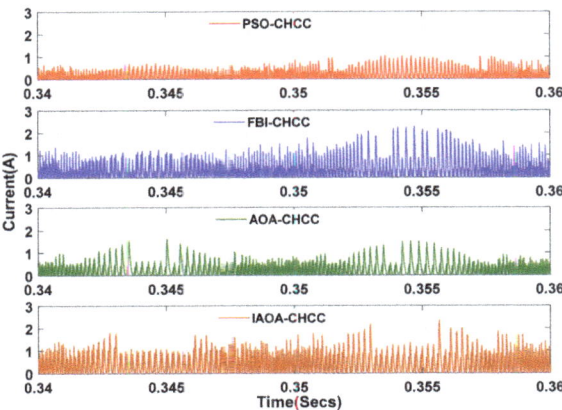

Figure 17. Simulation results showing current error for PSO-CHCC, FBI-CHCC, AOA-CHCC, and IAOA-CHCC.

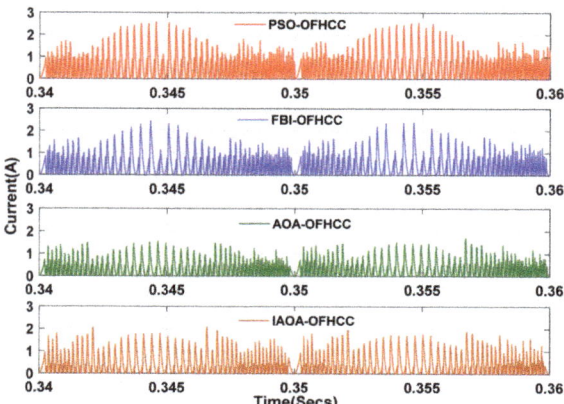

Figure 18. Simulation results showing current error for PSO-OFHCC, FBI-OFHCC, AOA-OFHCC, and IAOA-OFHCC.

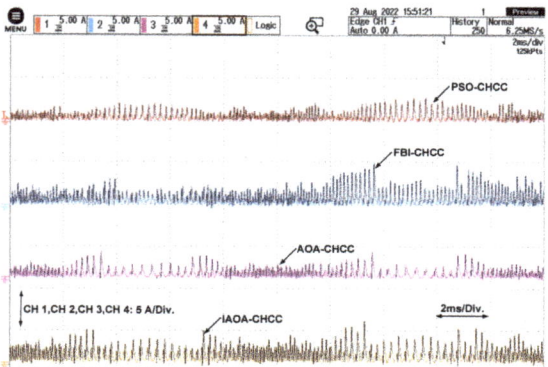

Figure 19. Experimental results showing current error for PSO-CHCC, FBI-CHCC, AOA-CHCC, and IAOA-CHCC.

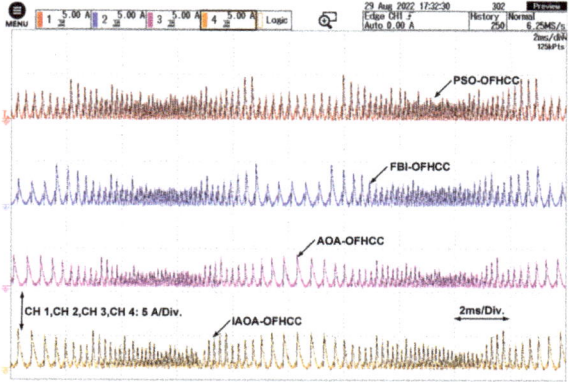

Figure 20. Experimental results showing current error for PSO-OFHCC, FBI-OFHCC, AOA-OFHCC, and IAOA-OFHCC.

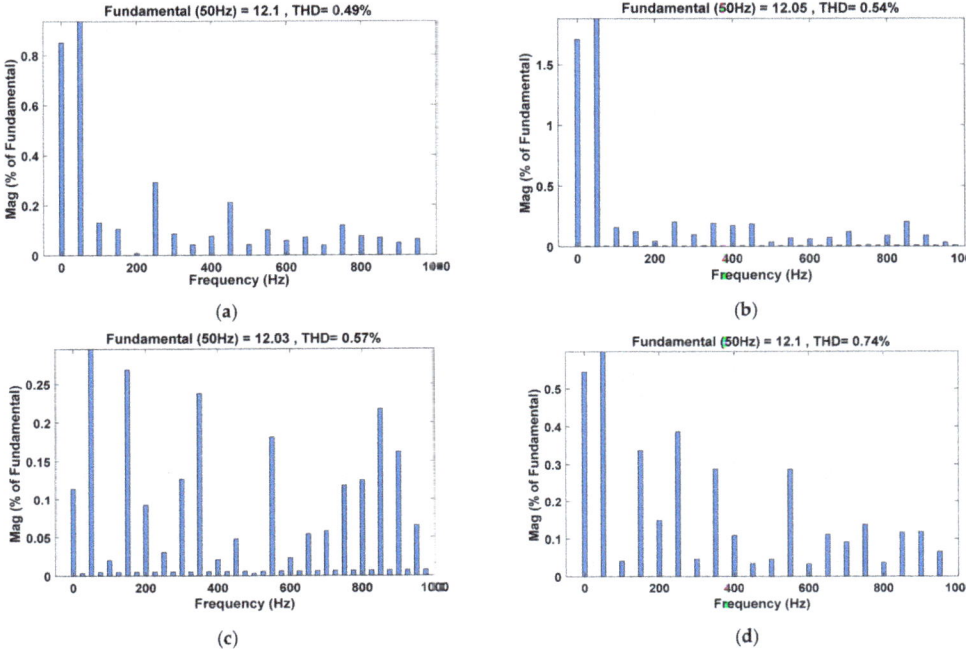

Figure 21. Harmonic spectra for (**a**) PSO-CHCC, (**b**) FBI-CHCC, (**c**) AOA-CHCC, and (**d**) IAOA-CHCC.

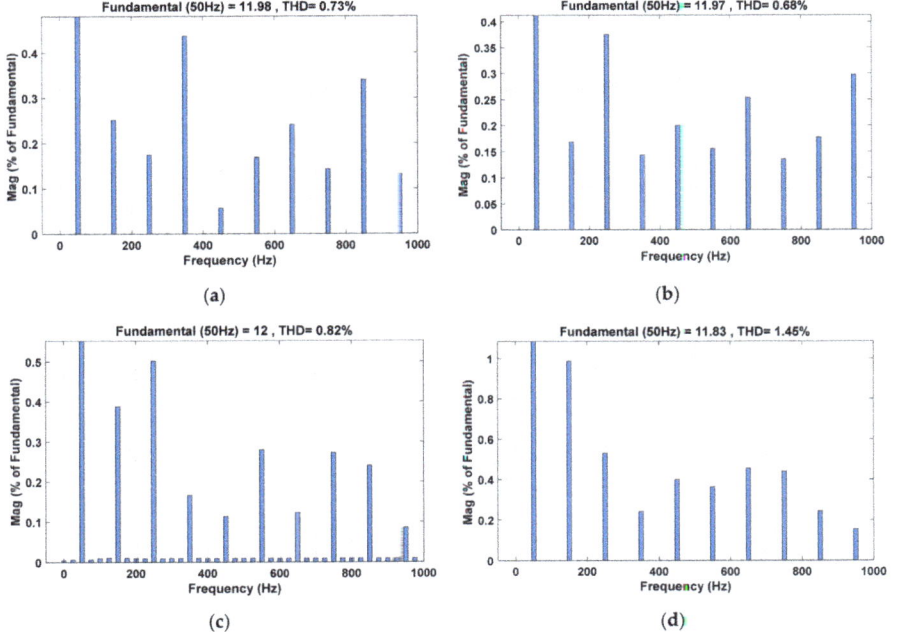

Figure 22. Harmonic spectra for (**a**) PSO-OFHCC, (**b**) FBI-OFHCC, (**c**) AOA-OFHCC, and (**d**) IAOA-OFHCC.

Figure 23 represents the performance indices of different controllers. The proposed IAOA-OFHCC technique addresses the problems of the high variations in average switching frequency and average switching losses. The proposed method's low average switching losses are indicative of a more efficient system. For step change in load feeding to the grid, all the techniques dynamic and steady-state characteristics were experimentally verified using the digital simulator OPAL-RT 4510. As a result, power can be delivered to the microgrid at higher efficiency, mitigating the power quality problems.

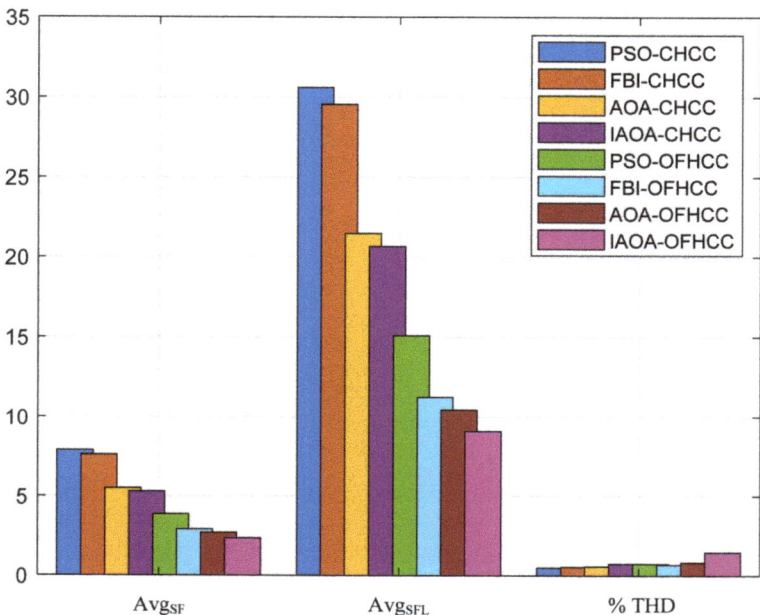

Figure 23. Bar plot of various performance indices for different controllers.

Behavior of the Proposed Control Algorithm under Partial Shading Condition

During the day, it is crucial to extract the maximum quantity of power without a change in irradiance level. However, due to the partial shading effect, the PV output power is reduced, the cost increases, and thus the efficiency decreases. Many conventional techniques fail to extract the maximum power point due to the formation of multiple hot spots in PV strings. In order to handle this drawback, the proposed IAOA-based OFHCC control algorithm has been studied. Initially, it is assumed that the PV module receives an insolation of 800 W/m^2. The system is then subjected to two different percentages of partial shading, i.e., 30% and 50%. The load power is considered to be constant throughout the partial shading operation. The variations in converter input voltage (Vpv), converter output voltage (Vo), and converter output current (Io) due to partial shading are shown in Figure 24. It can be clearly seen from Figure 24 that as the shading increases, it causes a reduction in the PV array output voltage. The boost converter efficiently maintains the DC voltage at 450 V, as shown in Figure 24.

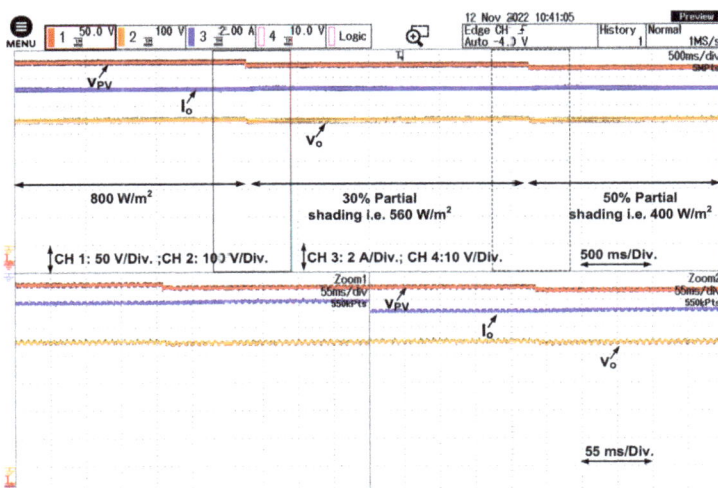

Figure 24. Experimental waveforms under partial shading of V_{PV}, V_O, and I_O curves.

The grid voltage (V_{Grid}), inverter current (I_{Inv}), load current (I_{Load}), and grid current (I_{Grid}) responses are shown in Figure 25. From Figure 25, it can be clearly noticed that the inverter currents, grid currents, and load currents remain unaltered even after partial shading. Due to the robustness of the proposed control algorithm, the boost converter is capable of maintaining a constant DC voltage. The boost converter maintains a fairly constant voltage at the DC side, which nullifies the effect of partial shading on the AC side of the proposed system.

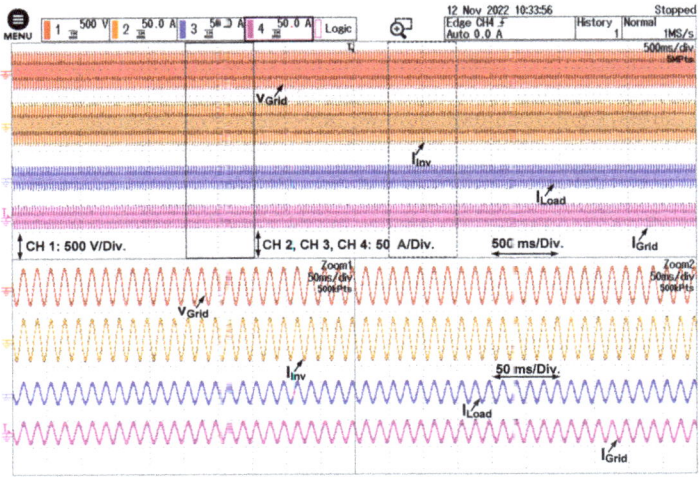

Figure 25. Experimental waveforms under partial shading of V_{Grid}, I_{Inv}, I_{Load}, and I_{Grid} curves.

7. Conclusions

In this article, experimental validation of IAOA-OFHCC and IAOA-CHCC inverter control has been provided to achieve better power quality at reduced switching frequency so that sinusoidal current is injected into the grid. Compared to the switching frequency obtained by PSO, FBI, and AOA, the novel IAOA technique with an OFHCC controller has reduced switching frequency which justifies the acceptance of grid-connected IAOA-

OFHCC design. The offset hysteresis band current controller, in addition to preserving the benefits of the conventional hysteresis band current controller, also delivers extra benefits such as fast dynamic response, reduced average switching frequency and losses, and higher accuracy and makes the microgrid robust. According to the mathematical presentation, IAOA has a straightforward implementation for addressing new optimization difficulties in microgrids. It does not need numerous constraints, only requiring a stopping criterion, a population size, and the standard parameters for optimizing hysteresis bands and duty cycles. The current fed from the inverter is sinusoidal with low total harmonic distortion (THD) following the IEEE 519 standard. The single-phase inverter provides a reduced ripple output using the proposed hybrid methodology. The performance of the novel method was assessed under load variation and proved robust in reference current tracking.

Author Contributions: Conceptualization, B.M., B.K.S. and S.P.; methodology, B.M., B.K.S., M.A. and S.P.; software, S.P., M.A. and M.B.; validation, S.P. and M.B.; formal analysis, B.M.; investigation, V.B. and S.M.; resources, V.B. and S.M.; data curation, M.B. and M.A.; writing—original draft preparation, V.B. and S.M.; writing—review and editing, V.B. and S.M.; visualization, M.B.; supervision, B.K.S. and S.P.; project administration, M.B.; funding acquisition, M.B., V.B., S.M. and L.P. All authors have read and agreed to the published version of the manuscript.

Funding: This paper was supported by the following projects: the doctoral grant competition VSB—the Technical University of Ostrava, reg. No. CZ.02.2.69/0.0/0.0/19 073/0016945 within the Operational Programme Research, Development, and Education, under project DGS/TEAM/2020-017 "Smart Control System for Energy Flow Optimization and Management in a Microgrid with V2H/V2G Technology"; FV40411 Optimization of process intelligence of parking system for Smart City, project TN01000007, National Centre for Energy; and Taif University Researchers Supporting Project TURSP 2020/122, Taif University, Taif, Saudi Arabia.

Data Availability Statement: Not applicable.

Acknowledgments: The authors appreciate the doctoral grant competition VSB—the Technical University of Ostrava, reg. No. CZ.02.2.69/0.0/0.0/19 073/0016945 within the Operational Programme Research, Development, and Education, under project DGS/TEAM/2020-017 "Smart Control System for Energy Flow Optimization and Management in a Microgrid with V2H/V2G Technology"; FV40411 Optimization of process intelligence of parking system for Smart City, project TN01000007, National Centre for Energy; and Taif University Researchers Supporting Project TURSP 2020/122, Taif University, Taif, Saudi Arabia for supporting this work.

Conflicts of Interest: The authors declare no conflict of interest.

References

1. Bastida, L.; Cohen, J.J.; Kollmann, A.; Moya, A.; Reichl, J. Exploring the role of ICT on household behavioural energy efficiency to mitigate global warming. *Renew. Sustain. Energy Rev.* **2019**, *103*, 455–462. [CrossRef]
2. Al-Shetwi, A.Q.; Hannan, M.A.; Jern, K.P.; Mansur, M.; Mahlia, T.M. Grid-connected renewable energy sources: Review of the recent integration requirements and control methods. *J. Clean. Prod.* **2020**, *253*, 119831. [CrossRef]
3. Irena, I. *Renewable Power Generation Costs in 2017*; Report; International Renewable Energy Agency: Abu Dhabi, United Arab Emirates, 2018.
4. Bhandari, B.; Lee, K.T.; Lee, G.Y.; Cho, Y.M.; Ahn, S.H. Optimization of hybrid renewable energy power systems: A review. *Int. J. Precis. Eng. Manuf.-Green Technol.* **2015**, *2*, 99–112. [CrossRef]
5. Lund, H.; Mathiesen, B.V. Energy system analysis of 100% renewable energy systems—The case of Denmark in years 2030 and 2050. *Energy* **2009**, *34*, 524–531. [CrossRef]
6. Harrouz, A.; Abbes, M.; Colak, I.; Kayisli, K. Smart grid and renewable energy in Algeria. In Proceedings of the IEEE 6th International Conference on Renewable Energy Research and Applications (ICRERA), San Diego, CA, USA, 5–8 November 2017; pp. 1166–1171.
7. Qazi, A.; Hussain, F.; Rahim, N.A.; Hardaker, G.; Alghazzawi, D.; Shaban, K.; Haruna, K. Towards sustainable energy: A systematic review of renewable energy sources, technologies, and public opinions. *IEEE Access* **2019**, *7*, 63837–63851. [CrossRef]
8. Al Maamary, H.M.; Kazem, H.A.; Chaichan, M.T. Renewable energy and GCC States energy challenges in the 21st century: A review. *Int. J. Comput. Appl. Sci. IJOCAAS* **2017**, *2*, 11–18.
9. Hannan, M.A.; Ghani, Z.A.; Hoque, M.M.; Hossain Lipu, M.S. A fuzzy-rule-based PV inverter controller to enhance the quality of solar power supply: Experimental test and validation. *Electronics* **2019**, *8*, 1335. [CrossRef]
10. Gul, M.; Kotak, Y.; Muneer, T. Review on recent trend of solar photovoltaic technology. *Energy Explor. Exploit.* **2016**, *34*, 485–526. [CrossRef]

11. Sathishkumar, R.; Malathi, V.; Premka, V. Optimization and design of PV-wind hybrid system for DC micro grid using NSGA II. *Circuits Syst.* **2016**, *7*, 1106. [CrossRef]
12. Sawant, P.T.; Bhattar, C.L. Optimization of PV system using particle swarm algorithm under dynamic weather conditions. In Proceedings of the IEEE 6th International Conference on Advanced Computing (IACC), Bhimavaram, India, 27–28 February 2016; pp. 208–213.
13. Ram, J.P.; Babu, T.S.; Rajasekar, N. A comprehensive review on solar PV maximum power point tracking techniques. *Renew. Sustain. Energy Rev.* **2017**, *67*, 826–847. [CrossRef]
14. Zakaria, A.; Ismail, F.B.; Lipu, M.H.; Hannan, M.A. Uncertainty models for stochastic optimization in renewable energy applications. *Renew. Energy* **2020**, *145*, 1543–1571. [CrossRef]
15. Hannan, M.A.; Tan, S.Y.; Al-Shetwi, A.Q.; Jern, K.P.; Begum, R.A. Optimized controller for renewable energy sources integration into micro-grid: Functions, constraints and suggestions. *J. Clean. Prod.* **2020**, *256*, 120419. [CrossRef]
16. Ilas, A.; Ralon, P.; Rodriguez, A.; Taylor, M. *Renewable Power Generation Costs*; International Renewable Energy Agency: Abu Dhabi, United Arab Emirates, 2017.
17. Gooding, P.A.; Makram, E.; Hadidi, R. Probability analysis of distributed generation for island scenarios utilizing Carolinas data. *Electr. Power Syst. Res.* **2014**, *107*, 125–132. [CrossRef]
18. Blondeau, J.; Mertens, J. Impact of intermittent renewable energy production on specific CO_2 and NOx emissions from large scale gas-fired combined cycles. *J. Clean. Prod.* **2019**, *221*, 261–270. [CrossRef]
19. Khishe, M.; Mosavi, M.R. Chimp optimization algorithm. *Expert Syst. Appl.* **2020**, *149*, 113338. [CrossRef]
20. Eberhat, R.; Kennedy, J. A new optimizer using particle swarm theory. In Proceedings of the Sixth International Symposium on Micro Machine and Human Science, Piscataway, NJ, USA, 4–6 October 1995; pp. 39–43.
21. Mirjalili, S.; Mirjalili, S.M.; Lewis, A. Grey Wolf Optimizer. *Adv. Eng. Softw.* **2014**, *69*, 46–61. [CrossRef]
22. Rao, R.V.; Savsani, V.J.; Vakharia, D.P. Teaching–learning-based optimization: A novel method for constrained mechanical design optimization problems. *Comput.-Aided Des.* **2011**, *43*, 303–315. [CrossRef]
23. Xiang, W.L.; An, M.Q. An efficient and robust artificial bee colony algorithm for numerical optimization. *Comput. Oper. Res.* **2013**, *40*, 1256–1265. [CrossRef]
24. Abualigah, L.; Diabat, A.; Mirjalili, S.; Abd Elaziz, M.; Gandomi, A.H. The arithmetic optimization algorithm. *Comput. Methods Appl. Mech. Eng.* **2021**, *376*, 113609. [CrossRef]
25. Varaprasad, O.V.; Sarma, D.S. An improved three level Hysteresis Current Controller for single phase shunt active power filter. In Proceedings of the IEEE 6th India International Conference on Power Electronics, Kurukshetra, India, 8–10 December 2014; pp. 1–5.
26. Zeb, K.; Uddin, W.; Adil Khan, M.; Ali, Z.; Umair Al., M.; Christofides, N.; Kim, H.J. A comprehensive review on inverter topologies and control strategies for grid connected photovoltaic system. *Renew. Sustain. Energy Rev.* **2018**, *94*, 1120–1141. [CrossRef]
27. Parvez, M.; Elias, M.F.M.; Rahim, N.A.; Osman, N. Current control techniques for three-phase grid interconnection of renewable power generation systems: A review. *Sol. Energy* **2016**, *135*, 29–42. [CrossRef]
28. Hassan, Z.; Amir, A.; Selvaraj, J.; Rahim, N.A. A review on current injection techniques for low-voltage ride-through and grid fault conditions in grid-connected photovoltaic system. *Sol. Energy* **2020**, *207*, 851–873. [CrossRef]
29. Tian, H.; Mancilla David, F.; Ellis, K.; Muljadi, E.; Jenkins, P.A. Cell-to-module-to-array detailed model for photovoltaic panels. *Sol. Energy* **2012**, *86*, 2695–2706. [CrossRef]
30. Vengatesh, R.P.; Rajan, S.E. Investigation of cloudless solar radiation with PV module employing Matlab-Simulink. In Proceedings of the International Conference on Emerging Trends in Electrical and Computer Technology, Nagercoil, India, 23–24 March 2011; pp. 141–147.
31. Zhang, Y.; Gao, S.; Gu, T. Prediction of IV characteristics for a PV panel by combining single diode model and explicit analytical model. *Sol. Energy* **2017**, *144*, 349–355. [CrossRef]
32. Jalil, M.F.; Khatoon, S.; Nasiruddin, I.; Bansal, R.C. Review of PV array modelling, configuration and MPPT techniques. *Int. J. Model. Simul.* **2022**, *42*, 533–550. [CrossRef]
33. Koutroulis, E.; Blaabjerg, F. Methodology for the optimal design of transformer less grid-connected PV inverters. *IET Power Electron.* **2012**, *5*, 1491–1499. [CrossRef]
34. Reddak, M. An improved control strategy using RSC of the wind turbine based on DFIG for grid harmonic currents mitigation. *Int. J. Renew. Energy Res. (IJRER)* **2018**, *8*, 266–273.
35. Sharma, S.; Singh, B. Control of permanent magnet synchronous generator-based stand-alone wind energy conversion system. *IET Power Electron.* **2012**, *5*, 1519–1526. [CrossRef]
36. Kean, Y.W.; Ramasamy, A.; Sukumar, S.; Marsadek, M. Adaptive controllers for enhancement of stand-alone hybrid system performance. *Int. J. Power Electron. Drive Syst.* **2018**, *9*, 979. [CrossRef]
37. Panda, A.; Pathak, M.K.; Srivastava, S.P. A single phase photovoltaic inverter control for grid connected system. *Sadhana* **2016**, *41*, 15–30. [CrossRef]
38. Reddak, M.; Mesbahi, A.; Nouaiti, A.; Berdai, A. Design and implementation of nonlinear integral back stepping control strategy for single-phase grid-connected VSI. *Int. J. Power Electron. Drive Syst.* **2019**, *10*, 19.

39. Dyga, Ł.; Rymarski, Z.; Bernacki, K. The wavelet-aided methods for evaluating the output signal that is designated for uninterruptible power supply systems. *Przegląd Elektrotechniczny* **2020**, *96*, 50–54. [CrossRef]
40. Rymarski, Z.; Bernacki, K. Different Features of Control Systems for Single-Phase Voltage Source Inverters. *Energies* **2020**, *13*, 4100. [CrossRef]
41. Torrey, D.; Al-Zamel, A. Single-phase active power filters for multiple nonlinear loads. *IEEE Trans. Power Electron.* **1995**, *10*, 263–272. [CrossRef]
42. Abd Rahim, N.; Selvaraj, J. Implementation of hysteresis current control for single-phase grid connected inverter. In Proceedings of the International Conference on Power Electronics and Drive Systems, Bangkok, Thailand, 27–30 November 2007; pp. 1097–1101.
43. Dahono, P.A. New hysteresis current controller for single-phase full-bridge inverters. *IET Power Electron.* **2009**, *2*, 585–594. [CrossRef]
44. Yao, Z.; Xiao, L. Control of single-phase grid-connected inverters with nonlinear loads. *IEEE Trans. Ind. Electron.* **2011**, *60*, 1384–1389. [CrossRef]
45. Elsaharty, M.A.; Hamad, M.S.; Ashour, H.A. Digital hysteresis current control for grid-connected converters with LCL filter. In Proceedings of the 37th Annual Conference of the IEEE Industrial Electronics Society, Melbourne, Australia, 7–10 November 2011; pp. 4685–4690.
46. Ichikawa, R.; Funato, H.; Nemoto, K. Experimental verification of single phase utility interface inverter based on digital hysteresis current controller. In Proceedings of the International Conference on Electrical Machines and Systems, Beijing, China, 20–23 August 2011; pp. 1–6.
47. Chatterjee, A.; Mohanty, K.B. Current control strategies for single phase grid integrated inverters for photovoltaic applications-a review. *Renew. Sustain. Energy Rev.* **2018**, *92*, 554–569. [CrossRef]
48. Jena, S.; Mohapatra, B.; Panigrahi, C.K.; Mohanty, S.K. Power quality improvement of 1-φ grid integrated pulse width modulated voltage source inverter using hysteresis Current Controller with offset band. In Proceedings of the 3rd International Conference on Advanced Computing and Communication Systems (ICACCS), Coimbatore, India, 22–23 January 2016; pp. 1–7. [CrossRef]
49. Karuppanan, P.; Ram, S.K.; Mahapatra, K. Three level hysteresis current controller based active power filter for harmonic compensation. In Proceedings of the International Conference on Emerging Trends in Electrical and Computer Technology, Nagercoil, India, 23–24 March 2011; pp. 407–412.
50. Yang, X.S. *Nature-Inspired Optimization Algorithms*; Academic Press: Cambridge, MA, USA, 2020.
51. Gupta, S.; Deep, K.; Mirjalili, S. An efficient equilibrium optimizer with mutation strategy for numerical optimization. *Appl. Soft Comput.* **2020**, *96*, 106542. [CrossRef]
52. Chauhan, S.; Vashishtha, G. Mutation-based arithmetic optimization algorithm for global optimization. In Proceedings of the International Conference on Intelligent Technologies (CONIT), Hubli, India, 25–27 June 2021; pp. 1–6.
53. Chou, J.S.; Nguyen, N.M. FBI inspired meta-optimization. *Appl. Soft Comput.* **2020**, *93*, 106339. [CrossRef]
54. Salet, R. Framing in criminal investigation: How police officers (re) construct a crime. *Police J.* **2017**, *90*, 128–142. [CrossRef]
55. Fathy, A.; Rezk, H.; Alanazi, T.M. Recent approach of forensic-based investigation algorithm for optimizing fractional order PID-based MPPT with proton exchange membrane fuel cell. *IEEE Access* **2021**, *9*, 18974–18992. [CrossRef]
56. Wang, D.; Tan, D.; Liu, L. Particle swarm optimization algorithm: An overview. *Soft Comput.* **2018**, *22*, 387–408. [CrossRef]
57. Hasari, S.A.; Salemnia, A.; Hamzeh, M. Applicable method for average switching loss calculation in power electronic converters. *J. Power Electron.* **2017**, *17*, 1097–1108.
58. Singh, J.K.; Behera, R.K. Hysteresis current controllers for grid connected inverter: Review and experimental implementation. In Proceedings of the IEEE International Conference on Power Electronics, Drives and Energy Systems (PEDES), Chennai, India, 18–21 December 2018; pp. 1–6.

Review

Operational Issues of Contemporary Distribution Systems: A Review on Recent and Emerging Concerns

Kabulo Loji [1,2], Sachin Sharma [3], Nomhle Loji [1], Gulshan Sharma [2,*] and Pitshou N. Bokoro [2]

1. Department of Electrical Power Engineering, Durban University of Technology, Durban 4001, South Africa
2. Department of Electrical Engineering Technology, University of Johannesburg, Johannesburg 2006, South Africa
3. Department of Electrical Engineering, Graphic Era Deemed to be University, Dehradun 248002, India
* Correspondence: gulshans@uj.ac.za

Abstract: Distribution systems in traditional power systems (PS) constituted of passive elements and the distribution issues were then limited to voltage and thermal constraints, harmonics, overloading and unbalanced loading, reactive power compensation issues, faults and transients, loss minimization and frequency stability problems, to name a few. Contemporary distribution systems are becoming active distributed networks (ADNs) that integrate a substantially increasing amount of distributed energy resources (DERs). DERs include distributed generation (DG) sources, energy storage resources and demand side management (DSM) options. Despite their evidenced great benefits, the large-scale deployment and integration of DERs remain a challenge as they subsequently lead to the network operational and efficiency issues, hampering PS network reliability and stability. This paper carries out a comprehensive literature survey based on the last decade's research on operational challenges reported and focusing on dispatchable and non-dispatchable DGs grid integration, on various demand response (DR) mechanisms and, on battery energy storage system (BESS) charging and discharging challenges, with the aim to pave the way to developing suitable optimization techniques that will solve the coordination of multiple renewable sources, storage systems and DRs to minimize distribution systems' operational issues and thus improve stability and reliability. This paper's findings assist the researchers in the field to conduct further research and to help PS planners and operators decide on appropriate relevant technologies that address challenges inherent to DG grid integration.

Keywords: demand response strategies; demand side management; distributed energy resources; battery energy storage systems; distribution generation; operational challenges; optimization techniques

1. Introduction

1.1. Traditional vs. Contemporary PS Networks

From the first built PS network, more than 100 years have seen a huge development of the electricity generation and supply systems. Points of generation of electric power were indeed situated several kilometres away from points of consumption as shown in Figure 1, since for economic reasons and a secure supply of electrical power, long distance bulk power transfer was essential [1]. Until the 1990s, the electric power industry was inclined to have a vertical integration approach to generation and transmission, justified mostly by economic reasons as mentioned above, rather than the improvement of the overall efficiency/reliability of the system [2].

The quasi-increasing amount of diverse electrical nature's loads over time resulted in the change of the grid topology, prompting grid complexity growth as illustrated in Figure 2, requiring much more attention on PS operational issues than ever. Current transformations have been driven for the last two decades by the increasing integration of renewable energy sources (RES), particularly solar and wind sources, known for their intermittency and unpredictability, into national grids.

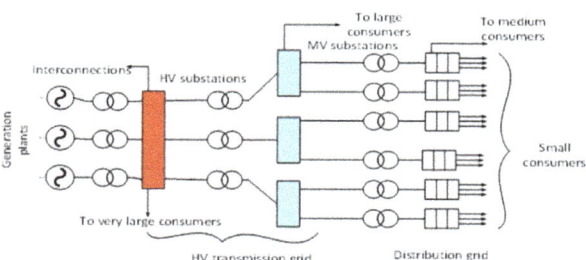

Figure 1. Traditional PS network topology.

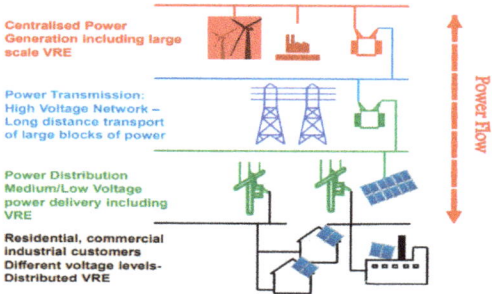

Figure 2. Contemporary network topology [3].

These transformations, commonly referred to as Smart Grid (SG), provide a combination of electrical power infrastructure with modern distributed computing facilities and communication networks [4]. Interest in RES grid integration has indeed developed because of the exponentially increasing demand for power delivery, a more secure energy future and energy policies adopted by governments in an effort to reduce CO_2 and greenhouse gas emissions [5–7]. Integration of renewable DGs, the most popular of them being solar and wind, into PS has positive and negative impacts on both power utilities and customers. Indeed, RES grid integration has evidenced substantial technical, environmental and economic benefits but at the same time, their increasing penetration leads to technical issues such as reverse energy flow from the customer end back into the transmission system [4], with negligible or reduced reactive power contribution [8,9]. One essential criterion for PS stability is to continuously balance power generation and consumption. Since for various reasons, the demand is volatile, generation must be flexible to accommodate the demand at any time [10]. The need to curtail consumers' peak hours and fill the gap caused by the mismatch between the amount of power generated and consumed at a specific time has become a very challenging task [11] for PS network researchers and operators.

1.2. Wind and PV Solar Trends and Contribution to Global Energy

The wind and solar-based RES footprint has been growing rapidly as shown in Table 1. Data extracted from the U.S. Energy Information Administration (EIA) indicate that from 2011 the global energy production increased by 28.59 % over a period of 10 years [12]. From a 2.36 % contribution to the global energy production in 2011, the combined solar PV and wind energy production has reached 10.41 % contribution to the world electricity generation in 2021.

Table 1. Solar PV and wind electricity generation growth from 2011 to 2021.

Energy Source/Activity	2011	2016	2021
Total Generation [billion kWh]	21,226	23,971	27,295
Solar [billion kWh]	66	341	1035
Wind [billion kWh]	435	957	1808

According to the International Renewable Energy Agency (IRENA), it is projected that in the next three decades, about 65% of the world energy will be produced from RES as shown in Figure 3 [13]. Consequently, the complexity of distribution issues will continue to grow with the increase of the number of DERs and micro grids (MG), especially because of the fact that most solar is connected to the DS rather than to the transmission system. PS networks will continue to encounter numerous variabilities, affecting fundamentally the planning and the operation of the electric distribution system, both technically and economically, prompting on one side the upgrade of the aging electricity infrastructure and on the other side, a subsequent transformation of the PS into Smart Grid (SG) process otherwise referred to as Grid modernisation. Uluski et al. in [14] project that the next generation distribution management shall need to incorporate more intelligence and advanced functionality to support these changes in the operation, monitoring and control of the distribution grid.

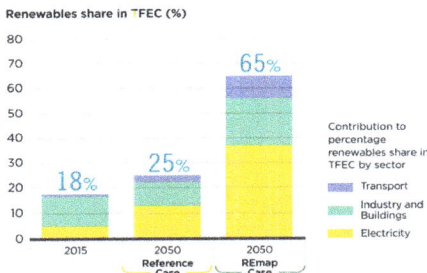

Figure 3. Renewable energy shares in the total final energy consumption [13].

1.3. DS Grid-Integration Challenges

DS Grid-integration challenges can be categorised into technical and non-technical. Technical challenges can be subdivided into operational and non-operational challenges. Operational challenges may be described as those pertaining to the hindrance of accomplishing the PS's main functions of secured and efficient generation, transmission and delivery of quality and reliable power. These may include actions or/and decisions taken timeously by PS operators to assure and maintain satisfactory PS operation. Non-operational challenges may be termed as those related to particularly the planning and the design activities that are initiated to minimize operational challenges. Non-technical challenges are essentially socio-economic and environmental challenges. Only operational challenges will be considered in this review paper. Figure 4 provides a comprehensive and up to date categorization of DG challenges.

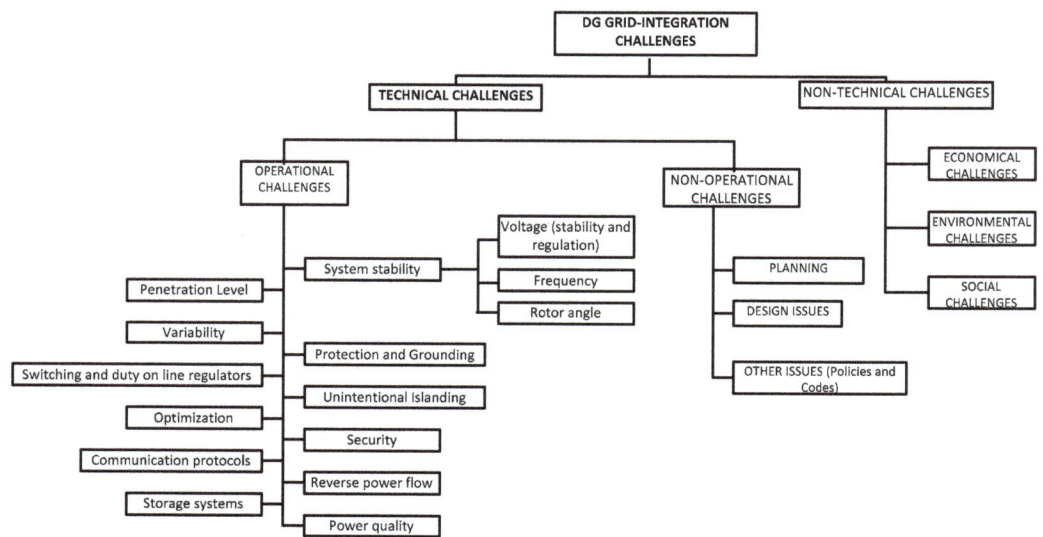

Figure 4. Categorization of DS grid integration challenges.

In order to approach those emerging challenges and address them efficiently, the authors in Ref. [15] suggest that both evolutionary and revolutionary technological changes would be required together with grid-integration technologies and techniques as well as substantial financial resources and strategies. The severity impact of the challenges resides mainly in:

- The difficulty to predict the system behaviour due to the fact that the optimal distribution network solutions must include the types of DG technologies, with their associated intricacies such as climatic conditions dependency and power generation compatibility.
- Higher penetration and inappropriate accommodation of DGs can jeopardise the system's protection and coordination and ultimately lead to system instability and excessive network losses [16,17].
- Energy storage capacity and location [18].
- Load demand scenarios and DR strategies for maximum possible utility and consumers' benefits.

2. Research Methodology and Organization of the Paper

2.1. Paper Methodology

Research data were collected using various search libraries with the major ones including IEEE Xplore, MDPI, Science Direct, Springer Link and Wiley Online Library. Figure 5 presents the summary of the data collection distribution from various sources used with the most relevant information coming from the IEEE Xplore. The category "others" included Elsevier, the National Renewable Energy Laboratory (NREL), Research gate and Google Scholar, particularly by searching up on pre-selected authors from the major libraries mentioned above. Six keywords were used to search for the data, namely: "Distribution systems issues", "Distributed Energy resources: review", "Review on operational issues on DS", "Renewable energy optimization techniques and objectives: review" and "Smart Grids". The search was conducted and finalized using a three-level filtering process for each search library as follows:

- Step 1: Relevance: using the above-mentioned keywords, run the search and obtain journal and conference papers that match the searching keywords. A total of 168 articles deemed to be relevant were collected at this stage.
- Step 2: Year of publication: from previous filter level results, only research papers of the last decade were selected, and 41 papers were excluded from step 1 selection. However, due to their strong relevance and valuable contribution to the topic in discussion, 17 papers out of this selection, published between 1994 and 2011 were rescued from the exclusion [1,2,6,19–28]. Figure 6 provides the collected data breakdown's distribution based on the year of publication.
- Step 3: Titles and abstracts: the last step was concerned with filtering using paper titles and abstracts.

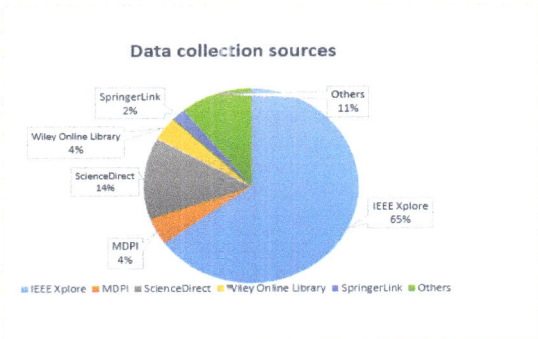

Figure 5. Distribution of collected data with reference to sources used.

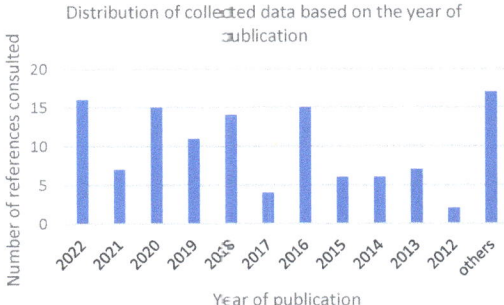

Figure 6. Distribution of collected data based on the year of publication.

2.2. Contributions of the Paper

The prime drive of this paper is to discuss and summarize various reported concerns on DS operational issues due to renewable energy grid integration and available remedies thus far. The main contributions of this article are:

- The paper reviews and presents DG grid integration challenges with regards to techno-economic aspects. The challenges addressed include intermittency and the no-dispatchability of RES, network power quality, stability and reliability, electricity market penetration and (de)regulation.
- Existing solutions and strategies are aggregated, packaged and presented in ready-to-use formats that are simple to refer to. The discussed solutions include DR strategies, charging and discharging techniques of battery energy storage, optimization techniques used for DERs in smart grids, coordination of multiple renewable sources, storage systems and DRs to minimize distribution systems' operational issues.

- The findings of this research paper will assist fellow PS renewable energy scholars and researchers to undertake further investigations and development in the field.

The rest of the paper is structured as follows: Section 3 highlights technical aspects of RE grid integration challenges with emphasis on DS grid integration operational challenges. Section 4 discusses the current status of solution strategies to overcome these challenges while identifying all possible gaps in the implementation of the strategies. Section 5 presents and discusses the conclusions as well as some future research directions that are corollary to the discussions.

3. Highlights of RES Grid-Integration Challenges

3.1. Review of Past Reviews on DS Issues

For more than a decade, DGs grid integration challenges have been the subjects of research surveys from several authors with numerous operational issues substantially reviewed. As earlier as 2009, Basso in Ref. [29] amongst others, documented and evaluated through a National Renewable Energy Laboratory (NREL) report, system impacts of DG penetration into transmission and distribution systems with a focus on renewable distributed resources technologies. The objectives of the report were to identify: (1) critical impact areas on transmission and distribution systems, (2) the best practices studies and challenge mitigation techniques related to the resolution of the system impacts, as well as (3) the then challenges and needs for further development to improve DG grid penetration. In [29], the author suggests that system impacts be categorised under the following headings: voltage regulation, power quality, voltage and system frequency stability, protection coordination, grounding, unintentional DG islanding, special issues related to DGs on secondary distribution network systems and special issues related to RES. Adding to the above categories, Prakash and Darbari in Ref. [30] spotted the development of secured and trusted system as a critical issue and identified the following security critical issues: methodologies to assess the security level of any system and monitoring of the system security including the development of security matrices, implementation of novel techniques for secure data communications, application of middle ware in DS security and applications of web services in security purposes. With solar and wind energy suppliers ramping up their energy capacity, Palmintier et al. [15] identified and reported further emerging challenges of concern, namely reverse power flow, increased duty on line regulators leading to equipment wear and tear, variability due weather uncertainty and capacitor switching. In the last past five years, research and the need for further development to improve DG grid penetration have been focussing on system efficiency, optimal planning and optimal integration. Researchers' attention is being drawn particularly on the following concerns: optimization techniques under various scenarios to enable higher penetration capacity, DSM and DR [31], energy storage systems to improve reliability, communication protocols, and cyber security. There is a general consensus that RES grid integration is an ongoing field for investigation and to respond to the anticipated RER integration challenges highlighted above, PS researchers propose advanced technologies and solution methodologies that will be discussed later in the paper.

The literature review conducted in this paper considered a number of relevant previous review papers that cover specific areas of DS challenges associated with DG grid-integration such as: reviews on DG penetration issues [15,32–38], flexibility issues in DS [10,39,40], protection issues [16,17,41,42], voltage stability and voltage regulation [43–46], uncertainty analysis and assessment [47,48], DR programs and DSM [31,49–53], unintentional islanding [54], cyber security issues [4,30], islanding [54], and vehicle grid system integration and applications [55–57]. Table 2 provides a comprehensive summary of the review papers samples used for the literature review of this paper.

Table 2. Sample review papers considered in the literature review.

Review Aspect	Paper Title	Ref.
Penetration issues	On the Path to Sun Shot: Emerging issues and Challenges in Integrating Solar with the Distribution System	[15]
	Integration of renewable distributed generators into the distribution system: a review	[32]
	Integrating Variable Renewable Energy: Challenges and Solutions	[33]
	Distributed generation: A review of factors that can contribute most to achieve a scenario of DG units embedded in the new distribution networks	[34]
	On the Path to Sun Shot: Emerging issues and Challenges in Integrating High Levels of Solar into the Electrical Generation and Transmission System	[35]
	A critical review of the integration of renewable energy sources with various technologies,	[36]
	Photovoltaic penetration issues and impacts in distribution network—A review	[37]
	Grid Integration Challenges and Solution Strategies for Solar PV Systems: A Review	[38]
Flexibility issues in DS	Research and Practice of Flexibility in Distribution Systems: A Review	[10]
	A review of demand side flexibility potential in Northern Europe	[40]
	Aggregation of Demand-Side Flexibilities: A Comparative Study of Approximation Algorithms	[39]
Wind and hybrid-systems operational issues	Solar–wind hybrid renewable energy system: A review	[58]
	Hybrid renewable energy systems for off-grid electric power: Review of substantial issue	[59]
	Wind Resources and Future Energy Security: Environmental, Social, and Economic Issues,	[60]
Protection issues	Renewable Energy Integration Challenge on Power System Protection and its Mitigation for Reliable Operation	[16]
	Renewable distributed generation: The hidden challenges—A review from protection perspective	[17]
	A comprehensive review on issues, investigations, control and protection trends, technical challenges and future directions for Microgrid technology	[41]
	A review of protection systems for distribution networks embedded with renewable generation	[42]
Voltage stability and voltage regulation	Voltage Stability Analysis with High Distributed Generation (DG) Penetration,	[43]
	A comprehensive review of the voltage stability indices	[44]
	Impact of distributed generation on protection and voltage regulation of distribution systems: A review	[45]
	Grid-connected photovoltaic system in Malaysia: A review on voltage issues,	[46]
DR programs and DSM strategies	Survey on Demand Response Programs in Smart Grids: Pricing Methods and Optimization Algorithms	[31]
	Residential peak electricity demand response—Highlights of some behavioural issues	[49]
	Particle Swarm Optimization in Residential Demand-Side Management: A Review on Scheduling and Control Algorithms for Demand Response Provision	[50]
	Residential Sector Demand Side Management: A Review	[51]
	A Survey of Efficient Demand-Side Management Techniques for the Residential Appliance Scheduling Problem in Smart Homes	[52]
	A review on price-driven residential demand response,	[53]

Table 2. Cont.

Review Aspect	Paper Title	Ref.
Vehicles grid system integration and applications	Comprehensive review & impact analysis of integrating projected electric vehicle charging load to the existing low voltage distribution system	[56]
	A comprehensive analysis of Vehicle to Grid (V2G) systems and scholarly literature on the application of such systems,	[55]
	A review on the state-of-the-art technologies of electric vehicle, its impacts and prospects	[57]
Unintentional Islanding	A review on islanding operation and control for distribution network connected with small hydro power plant	[54]

3.2. Impacts of Operational Challenges

The power output of most dominant DG resources is dependent on weather conditions, making these resources characterized by a variable generation property [38,61] that constitutes its own challenge. In traditional grids, operational uncertainties usually result from the demand side only. Distributed energy sources (DES) grid integration introduces new challenges, as the operational uncertainties emanate from both the demand and the generation sides [38] and have consequently significant impact on optimal planning of DGs [62]. Beside the technical considerations, Liu et. al. in Ref. [63] warn that these uncertainties can influence electricity users' economic benefits. Shafiullah et al. [38] note that accurate prediction of PV power for instance, has become an essential task for safe and stable PS operation and the prediction can focus on energy output or rate of change. The prediction types depend on the tools and information available from the meteorological stations. Ref. [38] also present the prediction models that were developed by [64,65]. Recent reported models for the prediction of power output are based on machine learning techniques as presented in Refs. [66–68].

The following challenges have been highlighted and dealt with by several researchers worldwide:

- Design and sizing of the system [5,15,32,33,47,69–74];
- Power balancing and voltage stability [7,43,69,71,75–78];
- Optimal energy management [11,79–94];
- Optimal DG allocation and penetration level [8,9,16,34,38,69,71,90,95–98];
- System cost minimization [22,82,99];
- Energy storage: operation strategies, coordination, optimization;
- Optimal coordination of various DERs [62,80,92,100–103];
- Localized overloading due to electric vehicle chargers [55,56,104].

Table 3 provides some references addressing design and integration, power quality and voltage stability, protection coordination, optimal distributed generation allocation, level of penetration as well as energy storage issues.

Table 3. Sample of some references and issues that they are addressing.

Ref.	Design and Integration of the System	Power Quality and Voltage Stability	Protection Coordination	Optimal DG Allocation	Penetration	Energy Storage
[5]	✓				✓	
[43]	✓	✓			✓	
[69]	✓				✓	
[98]	✓				✓	
[33]	✓					
[7]		✓		✓		
[8]				✓	✓	
[16]		✓	✓	✓		

Table 3. Cont.

Ref.	Design and Integration of the System	Power Quality and Voltage Stability	Protection Coordination	Optimal DG Allocation	Penetration	Energy Storage
[38]				✓		
[45]		✓	✓	✓		✓
[71]		✓		✓		✓
[105]		✓		✓		✓
[106]	✓					✓

4. Solutions Strategies for DS Grid Integration

To overcome the above challenges, researchers are exploring solutions that will provide satisfactory results to the power system network as a whole, as well as to procure benefits to both the utilities and the customers. The solutions that have been provided are summarized below.

4.1. Optimal Integration and Planning of Renewable Distributed Generation

DG optimal integration can improve network performance [103]. The optimal integration of DGs can be achieved through several strategies, the most popular one being through use of mathematical optimization models. Ehsan and Yang [62] have provided a good account of analytical techniques that are used for optimal integration and planning of renewable DG in the power distribution network. The strategies can, in a particular context and environment, invariably be used to address most of the challenges that have been mentioned in the previous section.

- The following, researched and presented by Georgilakis et al. in Ref. [107], are the mathematical formulations components for optimization approaches: a general problem statement, problem objectives whether single or multi, number of DGs and type of DG technology and a number of constraints to be considered.
- This is in the agreement that indeed, as mentioned by [32,94,95], the performance benefits depend mainly on the optimal sizing and location of the DG units, the DS configuration and the types of DG technologies used for conversion of energy. In Ref. [76], Esmaili was one of the earliest researchers to propose a multi-objective framework for placing and sizing DG units with the combination of the number of DGs, voltage stability margin and minimization of power loss into one objective function.
- In Ref. [108], the authors reviewed probabilistic optimization techniques (POT) in Smart Power Systems and noted that in order to account for uncertainties in optimization processes, stochastic optimization is essential. From their review, probabilistic optimization techniques were classified into stochastic optimization (SO), robust optimization (RO), distributionally robust optimization (DRO) and chance constraints optimization (CCO), each of which having their own advantages and drawbacks over the others, with the common drawback to all being their high computational requirements. Riaz et. al. [108] further proposed that the most advanced and less costly technique is the robust optimization in which a deterministic, set-based uncertainty model is used instead of a stochastic one. The authors suggest that POTs must be used in combination in order to deal with new challenges to achieve prolific outcomes.
- The authors in [92,101–103,109–112] have worked on various aspects related to DG grid integration optimization. The solutions proposed include the following benefits: more energy savings, improvement of voltage profile, reduced purchased power from the DGs, increased sold power to the distributed grid, decreased non-supply load, reduced overall cost of smart grid and mitigation of fault severity.
- Fast dispatch is one of the techniques that helps manage the variability of renewable generation because it reduces the need for regulating resources, improves efficiency and provides access to a broader set of resources to balance the system [33].

With regards to planning, to handle the high complexity of the investment planning problem for instance, Ref. [102] used a bi-level optimization framework that maximizes the net present value (NPV). Level-1 determines the optimal sizing of BESS in the presence of high PV penetration with the aim of minimizing the net present cost (NPC). The optimal BESS power dispatch in coordination with the DR aggregator is obtained in level-2, aiming to minimize NPC for voltage deviation penalty and PS losses with the scheduling of BESS and DR only [102]. In Ref. [61], a method to determine the optimal location, power and energy capacity of storage by creating an independent objective function for the voltage profile and power losses was proposed. The authors used the symbiotic organism search algorithm (SOSA) to solve the optimization problem with the following objectives: improvement of voltage profile, loss reduction, network reliability and minimization of storage costs including investment, operation and maintenance costs. SOSA has the advantage over other conventional algorithm (PSO and GA) of having specific adjusting parameters allowing for the conversion rate increase.

4.2. DER Coordination

Sharma et al. in [80] investigated the coordination of multiple DERs to address the techno-economic aspects of distribution network operation. The study aimed to find optimal dispatches of BESS in coordination with DR for wind generation and shunt capacitor with the target of minimizing distribution power loss. In [103], the authors developed a bi-level optimization framework for impact analysis of DR on PV and BESS accommodation in DS. The study was motivated by the undergoing intensive research on responsive loads driven by dynamic pricing that have shown benefits for utilities and consumers by shifting the demand peak to off-peak periods by utilizing renewable energy.

Achieving optimal integration of DGs is a complex problem involving many components, variables and constraints, network status, load dynamics and faults events, protection schemes, weather conditions and consumers' behaviour. Optimal integration requires the minimization as much as possible of operational issues. This is largely achievable through the coordination of multiple DERs. The following types of coordination have been under research with progressive results to achieve efficient, reliable and economical use of grid-integrated renewable energy resources:

- Coordination of DGs, BESS and DR for multi optimization of distribution networks [80,102];
- Energy scheduling with BESS cost [87];
- Energy management with electricity price;
- Accommodation of PV, DR and BESS [103];
- Solar PV with BESS under uncertain environment [112];
- Investment planning of DG resources with DR [102];
- DR analysis for optimal allocation of wind and solar [90];
- Optimal sizing of PV/wind and hybrid considering DSM [113];
- DR trends: users, network services, markets, and DERs [114];
- DR and intermittent RERs [115];
- Price-driven DR [53];
- Household appliances and DR [116];
- Joint allocation and operational management of DG and BESS in presence of DR [92];
- Pricing schemes, optimization objectives and solution methodologies of DSM [11];
- DR: Pricing, optimization and appliance scheduling [117];
- DSM model and optimization;
- Optimal planning and investment benefits of shared BESS;
- DGs, power losses and voltage stability.

4.3. Energy Storage Systems and Complementary Technologies

Kucur et al. [18] examined worldwide energy storage applications, their best location, applied technologies, total energy and power capacity and quality. Pumped storage are the most common type of grid-scale energy storage, but lead acid and lithium ion batteries are the most prominent for solar PV systems [18]. Although they have relatively high capital costs as indicated amongst other drawbacks by Liu et al. in Ref. [72], energy storage systems are essential technologies as they provide support to overcome the challenge of balancing supply and demand [18] and to cope with the intermittent renewable generation as well as to reduce the user's electricity purchase cost [63].

Installing BESS at any location with any random and non-optimum size can lead to high costs [61]. Using a storage device in the operation indices depends more on the installation location than the storage capacity. The authors in Ref. [118] have assessed the simultaneous impact of BESS, controllable load and network reconfiguration on contemporary distribution networks under uncertainties. The multi-constraint complex optimization problem was solved using an improved water evaporation optimization algorithm and the authors found that the coordination strategy reduced network loss while improving the voltage profile of the systems. The impact of multiple BESS strategies on energy loss of ADNs was investigated by Sharma et al. in [110]. With regards to the function of regulating the voltage on the utility side, Gamage et al. in Ref. [105] proposed an approach to integrate BESS to curb grid voltage violations.

- Ref. [63] proposed an approach of optimal planning of shared energy storage based on cost–benefit analysis to minimize the electricity procurement of retailers. They found that ES can effectively reduce the cost of retailers and high matching degree can be used as the selection criterion to obtain greater benefits from the shared ES [63].
- Ref. [72] proposed a comprehensive optimal allocation model of BESS considering operation strategy with the optimal capacity problem solved by cost–benefit analysis taking into account the reliability improvement benefits of BESS.
- The authors in Ref. [72] proposed system reliability improvements with BESS in planning operation strategies. The optimal BESS capacity and sizing problem was solved by cost–benefit analysis. The authors concluded that from an economic point of view, the distributed mode is preferable to centralized modes and the benefits of BESS can be improved by increasing the peak–valley difference of electricity price within a certain range.
- Ref. [72] was one of the earlier studies that proposed a comprehensive optimal allocation model of BESS that considered reliability benefits.

Table 4 presents a useful summary of important contribution, challenges, methodologies used and potential solutions to DG.

Table 4. Summary of major contributions, challenges, methodologies and potential solutions.

Ref.	Challenges or Issues	Solution Methodology	Research Objectives	Constraints/Objective Function	Paper Contribution
[90]	■ Optimal accommodation in coordination with DR ■ Impact on planning of wind-based and solar-based DGs in DS	MISOCP	■ Energy savings ■ Improvement of voltage profile	■ Energy losses ■ Minimum voltage ■ Average voltage deviation ■ DG penetration level ■ Peak demand	■ Integrating DR with planning of DGs leads to more energy savings and improvement of voltage profile
[16]	■ Secure protection for DS network ■ Protection blinding issue	Adaptive Over Current Protection (AOCP)		■ High DER penetration ■ Grid connected and islanded modes	■ Provided an alternative protection scheme working regardless connection to grid or islanding condition,

Table 4. Cont.

Ref.	Challenges or Issues	Solution Methodology	Research Objectives	Constraints/Objective Function	Paper Contribution
[119]	<ul><li>PV boosting development</li><li>LSPV modelling and simulation techniques</li><li>LSPV integration impacts on grid static and dynamic characteristics</li><li>Key techniques for improving LSPV transmission and consumption</li></ul>		<ul><li>RE integration</li><li>Large-scale PV development in China</li></ul>		<ul><li>Review of large scale PV integration (LSPV)</li><li>Recommendations for further research with regards to modelling and simulations, system integration and power generation delivery and co consumption</li></ul>
[115]	<ul><li>Reduce overall cost of smart grid and maximize reliability</li></ul>	<ul><li>PSO</li></ul>	<ul><li>Optimal size of units for the smart grid</li></ul>	<ul><li>Cooling and heating management</li><li>Impact on smart grid cost</li></ul>	<ul><li>Power consumption of heating and cooling systems resulted in decreasing the size of DGs</li><li>Reduced the purchase power from the DGs</li><li>Increased the sold power to distributed grid</li><li>Decreased non-supply loads</li><li>Reduced the overall cost of smart grids</li></ul>
[74]	<ul><li>Mitigation of fault severity brought by DG penetration</li><li>Causes protection devices not to operate properly</li></ul>	<ul><li>Fault at various locations</li><li>Balanced three-phase faults are used</li></ul>	<ul><li>Protection planning and coordination without and in the presence of DGs</li></ul>	<ul><li>Voltage constraint</li><li>Thermal limits</li></ul>	<ul><li>Addressed challenges associated with the operation of DS in both normal and contingency operation states</li></ul>
[86]	<ul><li>Optimal use of DERs</li></ul>	<ul><li>MPSO</li></ul>	<ul><li>Minimization of operating cost of a microgrid</li></ul>	<ul><li>Optimization problem of a community micro-grid</li></ul>	<ul><li>Problem with optimization of a community micro-grid</li><li>However, solutions had significant deviations due to prediction errors</li></ul>
[87]	<ul><li>Overall Minigrid cost reduction</li></ul>	<ul><li>PSO (for model optimization) and Rainflow (for battery degradation cost)</li></ul>	<ul><li>Electricity cost management through efficient control of BESS</li></ul>	<ul><li>Battery degradation cost</li><li>Dynamic electricity price</li></ul>	<ul><li>Proposed a day-ahead energy management for a community micro grid with consideration of battery degradation costs</li><li>40% cost reduction compared to the baseline approach</li></ul>

Table 4. Cont.

Ref.	Challenges or Issues	Solution Methodology	Research Objectives	Constraints/Objective Function	Paper Contribution
[92]	<ul><li>Energy losses</li><li>Voltage deviation (stability)</li></ul>	<ul><li>Mixed integer second order conic programming (MISOCP)</li></ul>	<ul><li>To propose framework for joint allocation and operational management of wind DG and BESS</li><li>Optimally allocate wind-based DG and BESS</li></ul>	<ul><li>Power flow</li><li>Wind-based DG constraints</li><li>BESS constraints</li><li>Demand response constraint</li></ul>	<ul><li>Simultaneous allocation and operation management of wind-based DG and BESS in distribution system considering DR</li><li>Optimal sizing and siting of wind DG and BESS along with DR participation leads to significant energy savings and improvement of power quality</li><li>When DR participation rate increased, BESS capacity decreased</li></ul>
[120]	<ul><li>Determination of dynamic electric energy retail pricing tariffs</li></ul>	<ul><li>Statistical analysis</li></ul>	<ul><li>Improve the performance of demand response techniques</li></ul>	<ul><li>Minimum power demand</li><li>Load variation</li></ul>	<ul><li>Novel quantitative measure of the load profile that accurately reflected the overall generation expansion planning and utilization costs</li><li>Peak-to-average ratio (PAR) did not reflect the load characteristics</li></ul>
[53]	<ul><li>Price-driven demand response (PDDR) to affect customers' consumption (including critical peak pricing, TOU pricing, real-time pricing</li></ul>		<ul><li>Evaluation of advantages and disadvantages of PDDR</li></ul>		<ul><li>Review of three different PDDR programs at residential sector</li><li>TOU</li><li>CPP</li><li>RTP</li></ul>
[51]	<ul><li>Lack of informed decision from both the supplier and the consumer</li></ul>		<ul><li>DSM to redesign the load profile and to decrease the peak load demand</li></ul>		<ul><li>Review of DSM strategies with both DR and energy efficiency policies</li></ul>
[110]		<ul><li>Generic Algorithm (GA)</li></ul>	<ul><li>Optimal operation strategies</li><li>Validation of economic benefit</li></ul>	<ul><li>Node voltage limit</li><li>Feeder current limit</li><li>Nodal power balance</li></ul>	<ul><li>Optimal operation of BESS can reduce energy loss and increase economic benefits of the DS</li></ul>

Table 4. Cont.

Ref.	Challenges or Issues	Solution Methodology	Research Objectives	Constraints/Objective Function	Paper Contribution
[103]	■ Optimal integration of emerging DERs	■ Bi-level GA using Matlab		■ PV generation limits ■ BESS constraints ■ Feeder thermal limit constraints ■ Power balance constraints	■ DGs were effective in annual energy loss reduction ■ BESSS facilitated higher DG penetration and levelled the load profile ■ DR bridged the gap between peak and valley demands and therefore distresses to the system
[118]		■ Innovative Water Evaporation (IWEO) algorithm	■ Optimal coordination of controllable load scheduling, BESS and uncertain wind power	■ Nodal power balance ■ Feeder thermal limits ■ Controllable load management ■ Network configuration	■ Two-stage framework was developed to coordinate the generation of DGs, scheduling of BESSs, optimal management of controllable load
[105]	■ Fluctuation of grid voltage	■ Power flow simulation			■ Incorporation of BESS can mitigate voltage violation ■ More effective in rural distribution feeder suggesting when the line impedance is high
[80]		■ Non-sorted generic algorithm (NSGA-II) Technique for order of preference by similarity to ideal solution (TOPSIS)	■ Optimal dispatches of BESS ■ Minimize distribution power loss and grid demand cost	■ Nodal voltage limit ■ Power loss minimization ■ Grid demand cost minimization ■ Nodal power balance ■ Feeder thermal limits	■ Optimal coordination of wind power, BESS, SC and TOU-DR significantly reduced the network losses and grid demand consumption cost
[76]	■ To place DG units at more efficient buses rather than end buses of radial links usually used for voltage stability improvement	■ Non-Linear Programming (NLP) ■ Fuzzification applied to objective functions	■ Optimal sizing and location of DG units	■ Number of DGs ■ Power loss minimization ■ Maximize voltage stability margin ■ Branch and voltage limits	■ Modelled all types of DGs ■ Employed adaptive reactive limits rather than fixed limits ■ New technique to formulate the number of DGs without converting the NLP problem into mixed-integer NLP ■ Minimization of the number of DG units led to placement of these units at more efficient buses rather than end buses of radial link

Table 4. Cont.

Ref.	Challenges or Issues	Solution Methodology	Research Objectives	Constraints/Objective Function	Paper Contribution
[102]	■ Simultaneous consideration of cost of energy purchased from the grid, energy losses, emission penalty cost, demand deviation penalty, operation and maintenance cost for NPV benefits	■ Complex mixed-integer, non-linear and non-convex optimization techniques ■ Bi-level optimization problem (BLOP)	■ Multilayer DS and BESS investment planning with coordination of DR ■ The coordination aimed to maximize Net Present Value (NPV) profit	■ PV generation limits ■ BESS capacity limits ■ Power dispatch and SOC limits ■ DR constraints ■ Thermal feeder limits ■ Power balance constraints ■ Cost of energy purchased from the grid ■ Energy losses ■ Emission penalty cost ■ Demand deviation penalty ■ Operation and maintenance cost	■ Impact of DR on investment planning of DG and BESS ■ Simultaneous consideration of cost of energy purchased from the grid, energy losses, emission penalty cost, demand deviation penalty, operation and maintenance cost for NPV benefits ■ Higher NPV benefits ■ Analysed impact of DR on payback period: payback within 9 of 20 years of planning was significant compared to non-DR-based investment planning ■ Other technical benefits
[32]	■ Optimal sizing, siting and configuration of DGs		■ Review on technical benefits of renewable DG ■ Review current status of REN		■ Significant roles that renewable DGs can play in technical, economic and environmental operation
[72]		■ Intelligent Single Particle Optimiser (ISPO) ■ Sequential Monte Carlo simulation method		■ Operation strategy of BESS (power and time periods of charging and discharging) ■ Reliability improvement benefits of BESS ■ Optimal planning model of BESS	■ Comprehensive optimal allocation model of BESS considering operation strategy ■ Numerical method based on expectation for the calculation of system reliability improvement with BESS in planning was proposed ■ Optimal BESS capacity and sizing problems were (simultaneously?) solved by cost–benefit analyses
[62]			■ Optimal planning of DGs ■ Power quality, voltage stability, power loss, reliability and profitability		■ Conventional and metaheuristic techniques ■ Metaheuristics algorithms were popular choice because of their flexibility in multi-objective planning problems

229

Table 4. *Cont.*

Ref.	Challenges or Issues	Solution Methodology	Research Objectives	Constraints/Objective Function	Paper Contribution
[106]	• Determination of optimal location and sizes		• Optimal DG placement and sizing • Models and solutions • Classify current and future research trends in this field	• DG capacity constraints • Operating constraints • Investment constraints	
[63]			• Cope with intermittency and reduce customer electricity purchase cost		• Fluctuation of electricity prices and the uncertainty of RE resources' output did not influence users' economic benefits • Shared energy storage (ES) system among multiple electricity retailers showed more benefit rather than the separately configured ES

5. Further Research Priorities and Conclusions

5.1. Further Research Priorities

From this review study, the following concerns have not been appropriately and exhaustively attended to and therefore still require researchers' attention:

- All-in-One multi-objective DER optimal planning solutions that include the coordination of various variables such as the type of DG technologies, the types of energy storage integration, DSM mechanisms and different DR strategies, for maximum benefits both for the utility and consumers have not yet been sufficiently researched.
- Further investigations are needed in establishing optimization techniques using hybrid techniques that combine analytical, metaheuristic and computational methods to achieve better results.
- The use of optimization algorithms, ensemble methods and weather forecasting to develop models that can predict renewable energy power output considering weather conditions and seasonal variation still need attention and focus from researchers.
- Development of robust models to quantify the impact of uncertainties related to intermittency of renewable DGs. There is a need to gather resources and tools for weather condition predictions.

5.2. Conclusions

The transformation of PS around the world is effective and largely impacted by a rapid growth of various renewable energy grid integration thus affecting the control and operation of contemporary DS which are becoming more and more active network systems. Supporting and remedial actions are required and should be planned accordingly. This paper presents various operational and technical challenges associated with DG integration into DS. It was shown that the challenges of different natures at different levels of the PS are usually addressed individually, prompting that a holistic approach be considered when addressing them. Power quality, voltage stability, PS reliability, loss minimization, cost–benefits and so many other objectives can be achieved with optimal integration and appropriate planning of DGs. The DG grid integration problem is a multi-objective and hence needs advance multi-objective algorithms to address more than one challenge simultaneously. In order to reduce the variability and increase predictability, robust models need to be designed to include accurate forecasting methods, reliable data collection and safe communication to cater to RE technologies' uncertainties and intermittent nature. Further energy storage and demand side management can play a major role in supplying quality and reliable power to the customers and at the same time reduce the burden on DGs and their intricacies such as variability.

Author Contributions: All authors planned the study, contributed to the idea and field of information; Introduction, K.L.; Software, K.L. and S.S.; Analysis, K.L., S.S. N.L., G.S. and P.N.B.; Conclusion, K.L. and G.S.; writing—original draft preparation, K.L., P.N.B. and N.L.; writing—review and editing, K.L. and G.S.; supervision, S.S. and G.S. All authors have read and agreed to the published version of the manuscript.

Funding: This research received no external funding.

Data Availability Statement: Not applicable here.

Conflicts of Interest: The authors declare no conflict of interest.

References

1. Dobson, I.; Green, S.; Rajaraman, R.; DeMarco, C.; Alvarado, F.; Galavic, M.; Zang, J.; Zimmerman, R. *Electric Power Transfer Capability: Concepts, Applications, Sensitivity and Uncertainty*; University of Wisconsin–Madison: Madison, WI, USA, 2001.
2. Machowski, J.; Bialek, J.W.; Bumby, J.R. *Power System Dynamics: Stability and Control*, 2nd ed.; Sons, J.W., Ed.; John Wiley & Sons, Ltd.: Hoboken, NJ, USA, 2008.
3. Gafaro, F.; Portugal, I. Planning the Operability of Power Systems–Overcoming Technical and Operational Bottlenecks. Available online: https://www.irena.org/-/media/Files/IRENA/Agency/Events/2017/Jan/17/IRENA-WFES-Scaling-up-VRE---Operation-Planning---Final.pdf?la=en&hash=DA778DAF2644A2D6DBC70665B109E6B3AACA6673 (accessed on 17 January 2023).
4. Berger, L.T.; Iniewski, K. *Smart Grid: Applications, Communications and Security*; John Wiley & Sons: Hoboken, NJ, USA, 2012.
5. Banerjee, S.; Mesram, A.; Swamy, N.K. Integration of Renewable Energy Sources in Smart Grid. *Int. J. Sci. Res.* **2015**, *4*, 247–250.
6. Freris, L.; Infield, D.G. *Renewable Energy in Power Systems*; John Wiley & Sons Ltd: Hoboken, NJ, USA, 2008.
7. Loji, K.; Davidson, I.E.; Tiako, R. Voltage Profile and Power Losses Analysis on a Modified IEEE 9-Bus System with PV Penetration at the Distributed Ends. In Proceedings of the Southern African Universities Power Engineering Conference/Robotics and Mechatronics/Pattern Recognition Association of South Africa (SAUPEC/RobMech/PRASA), Bloemfontein, South Africa, 27–29 January 2019.
8. Kumar, A.K.; Selvan, M.P.; Rajapandiyan, K. Dynamic Grid Support System for Mitigation of Impact of High Penetration Solar PV into Grid. In Proceedings of the 2017th International Conference on Power Systems (ICPS), Pune, India, 21–23 December 2017.
9. Von Appen, J.; Braun, M.; Stetz, T.; Diwold, K.; Geibel, D. Time in the Sun: The Challenge of High PV Penetration in the German Electric Grid. *IEEE Power Energy Mag.* **2013**, *11*, 55–64. [CrossRef]
10. Dalhues, S.; Zhou, Y.; Pohl, O.; Rewald, F.; Erlemeyer, F.; Schmid, D.; Zwartscholten, J.; Hagemann, Z.; Wagner, C.; Gonzalez, M.D.; et al. Research and Practice of Flexibility in Distribution Systems: A Review. *CSEE J. Power Energy Syst.* **2019**, *5*, 285–294.
11. Ahmad, S.; Ahmad, A.; Naeem, M.; Ejaz, W.; Kim, H.S. A Compedium of Performance Metrics, Pricing Schemes, Optimization objectives and Solution Methodologies of Demand Side Management for the Smart Grid. *Energies* **2018**, *11*, 2801. [CrossRef]
12. US Energy Information Administration. Energy Sources/Activity. Available online: https://www.eia.gov/electricity/reports.php#/T194 (accessed on 22 December 2022).
13. IRENA(2018). *Global Energy Transformation: A Roadmap to 2050*; International Renewable Energy Agency (IRENA): Masdar City, Abu Dhabi, 2018.
14. Uluski, R.; Borlase, S. Next Generation Distribution Management. *Power Ind. 2022 Trends Predict.* **2022**, *2022*.
15. Palmintier, B.; Broderick, R.; Mather, B.; Coddington, M.; Baker, K.; Ding, F.; Reno, M.; Lave, M.; Bharatkumar, A. *On the Path to SunShot: Emerging Issues and Challenges in Integrating Solar with the Distribution System*; NREL/TP-5D00-65331; National Renewable Energy Laboratories: Golden, CO, USA, 2016.
16. Altaf, M.W.; Arif, M.T.; Saha, S.; Islam, N.; Haque, M.E.; Oo, A. Renewable Energy Integration Challenge on Power System Protection and its Mitigation for Reliable Operation. In Proceedings of the IECON 2020 The 46th Annual Conference of the IEEE Industrial Electronics Society, Singapore, 18–21 October 2020; pp. 1917–1922.
17. Manditereza, P.T.; Bansal, R. Renewable distributed generation: The hidden challenges-A review from protection perspective. *Renew. Sustain. Energy Rev.* **2016**, *58*, 1457–1465. [CrossRef]
18. Kucur, G.; Tur, M.R.; Bayindir, R.; Shabinzadeh, H.; Gharehpetian. A review of Emerging Cutting-Edge Energy Storage Technologies for Smart Grids Purposes. In Proceedings of the 2022 9th Iranian Conference on Renewable Energy & Disitributed Generation (ICREDG), Mashhad, Iran, 23–24 February 2022.

19. CIGRE, W.G. 37-23, *Impact of Increasing Contribution of Dispersed Generation on Power System: Final Report*; Electra: Ramat Gan, Israel, 1998.
20. Barker, P.P.; De Mello, R.W. Determining the Impact of Distributed Generation on Power Systems. Part-1. radial Distribution systems. In Proceedings of the IEEE PES Summer Meeting, Seatle, DC, USA, 16–20 July 2000; pp. 1645–1656.
21. Cossent, R.; Gómez, T.; Frías, P. Towards a future with large penetration of distributed generation: Is the current regulation of electricity distribution ready? Regulatory recommendations under a European perspective. *Energy Policy* 2009, 37, 1145–1155. [CrossRef]
22. Feinstein, C.D.; Orans, R.; Chapel, S.W. THE DISTRIBUTED UTILITY: A New Electric Utility Planning and Pricing Paradigm. *Annu. Rev. Energy Environ.* 1997, 22, 155–185. [CrossRef]
23. Kundur, P.; Paserba, J.; Ajjarapu, V.; Andersson, G.; Bose, A.; Canizares, C.; Hatziargyriou, N.; Hill, D.; Stankovic, A.; Taylor, C.; et al. Definition and Classification of Power System Stability-IEEE/CIGRE Joint Task Force on Stability Terms and Definitions. *IEEE Trans. Power Syst.* 2004, 19, 1387–1401.
24. Kundur, P. *Power Ssytem Stability and Control*; McGraw-Hill: New York, NY, USA, 1994.
25. Mangoyana, R.B.; Smith, T.F. Decentralised bioenergy systems: A review of opportunities and threats. *Energy Policy* 2011, 39, 1286–1295. [CrossRef]
26. Reza, M.; Schavemaker, P.H.; Slootweg, J.G.; Van Der Sluis, S.L. Impacts of Distributed Generation Penetration Levels on Power Systems Stability. *IEEE PES Gen. Meet.* 2004, 2, 2150–2155.
27. Slootweg, J.G.; Kling, W.L. Impacts of Distributed Generation on Power system Transient Stability. *IEEE PES Gen. Meet.* 2002, 2, 862–867.
28. Vournas, C.D.; Sauer, P.W.; Pai, M.A. Relashionship Between Voltage and Angle Stability of Power Systems. *Int. J. Electr. Power Energy Syst.* 1996, 18, 493–500. [CrossRef]
29. Basso, T.S. *System Impacts from Interconnection of Distributed Resources: Current Status and Identification of Needs for Further Development*; NREL/TP-550-44727 National Renewable Energy Laboratory NREL: Oak Ridge, TN, USA, 2009.
30. Prakash, V.; Darbari, M. A Review on Security Issues in distributed Systems. *Int. J. Sci. Eng. Res.* 2012, 3, 1–5.
31. Vardakas, J.S.; Zorba, N.; Verikoukis, C.V. Survey on Demand Response Programs in Smart Grids: Pricing Methods and Optimization Algorithms. *IEEE Commun. Surv. Tutor.* 2015, 17, 152–178. [CrossRef]
32. Adefarati, T.; Bansal, R.C. Integration of renewable distributed generators into the distribution system: A review. *IET Renew. Power Gener. (Wiley-Blackwell)* 2016, 10, 873–884. [CrossRef]
33. Bird, L.; Milligan, M.; Lew, D. *Integrating Variable Renewable Energy: Challenges and Solutions*; National Renewable Energy Laboratory: Golden, CO, USA, 2013.
34. Colmenar-Santos, A.; Reino-Rio, C.; Borge-Diez, D.; Collado-Fernández, E. Distributed generation: A review of factors that can contribute most to achieve a scenario of DG units embedded in the new distribution networks. *Renew. Sustain. Energy Rev.* 2016, 59, 1130–1148. [CrossRef]
35. Denholm, P.; Clark, K.; O'Connell, M. *On the Path to SunShot: Emerging Issues and Challenges in Integrating High Levels of Solar into the Electrical Generation and Transmission System*; NREL08563; National Renewable Energy La: Golden, CO, USA, 2016.
36. Erdiwansyah, E.; Mahidin; Husin, H.; Nasaruddin, N.; Zaki, M.; Muhibbuddin. A critical review of the integration of renewable energy sources with various technologies. *Prot. Control Mod. Power Syst.* 2021, 6, 1–18. [CrossRef]
37. Karimi, M.; Mokhlis, H.; Naidu, K.; Uddin, S.; Bakar, A.H.A. Photovoltaic penetration issues and impacts in distribution network–A review. *Renew. Sustain. Energy Rev.* 2016, 53, 594–605. [CrossRef]
38. Shafiullah, M.; Ahmed, S.D.; Al-Sulaiman, F.A. Grid Integration Challenges and Solution Strategies for Solar PV Systems: A Review. *IEEE Access* 2022, 10, 52233–52257. [CrossRef]
39. Öztürk, E.; Rheinberger, K.; Faulwasser, T.; Worthmann, K.; Preißinger, M. Aggregation of Demand-Side Flexibilities: A Comparative Study of Approximation Algorithms. *Energies* 2022, 15, 2501. [CrossRef]
40. Söder, L.; Lund, P.D.; Koduvere, H.; Bolkesjø, T.F.; Rossebø, G.H.; Rosenlund-Soysal, E.; Skytte, K.; Katz, J.; Blumberga, D. A review of demand side flexibility potential in Northern Europe. *Renew. Sustain. Energy Rev.* 2018, 91, 654–664. [CrossRef]
41. Choudhury, S. A comprehensive review on issues, investigations, control and protection trends, technical challenges and future directions for Microgrid technology. *Int. Trans. Electr. Energy Syst.* 2020, 30, 1–16. [CrossRef]
42. Kennedy, J.; Ciufo, P.; Agalgaonkar, A. A review of protection systems for distribution networks embedded with renewable generation. *Renew. Sustain. Energy Rev.* 2016, 58, 1308–1317. [CrossRef]
43. Al-Abri, R. Voltage Stability Analysis with High Distributed Generation (DG) Penetration. University of Waterloo: Waterloo, ON, USA, 2012.
44. Modarresi, J.; Gholipour, E.; Khodabakhshian, A. A comprehensive review of the voltage stability indices. *Renew. Sustain. Energy Rev.* 2016, 63, 1–12. [CrossRef]
45. Razavi, S.-E.; Rahimi, E.; Javadi, M.S.; Nezhad, A.E.; Lotfi, M.; Shafie-khah, M.; Catalão, J.P.S. Impact of distributed generation on protection and voltage regulation of distribution systems: A review. *Renew. Sustain. Energy Rev.* 2019, 105, 157–167. [CrossRef]
46. Wong, J.; Lim, Y.S.; Tang, J.H.; Morris, E. Grid-connected photovoltaic system in Malaysia: A review on voltage issues. *Renew. Sustain. Energy Rev.* 2014, 29, 535–545. [CrossRef]
47. Gensler, A.; Sick, B.; Vogt, S. A review of uncertainty representations and metaverification of uncertainty assessment techniques for renewable energies. *Renew. Sustain. Energy Rev.* 2018, 96, 352–379. [CrossRef]

48. Tian, W.; Heo, Y.; de Wilde, P.; Li, Z.; Yan, D.; Park, C.S.; Feng, X.; Augenbroe, G. A review of uncertainty analysis in building energy assessment. *Renew. Sustain. Energy Rev.* **2018**, *93*, 285–301. [CrossRef]
49. Gyamfi, S.; Krumdieck, S.; Urmee, T. Residential peak electricity demand response—Highlights of some behavioural issues. *Renew. Sustain. Energy Rev.* **2013**, *25*, 71–77. [CrossRef]
50. Menos-Aikateriniadis, C.; Lamprinos, I.; Georgilakis, P.S. Particle Swarm Optimization in Residential Demand-Side Management: A Review on Scheduling and Control Algorithms for Demand Response Provision. *Energies* **2022**, *15*, 2211–2236. [CrossRef]
51. Panda, S.; Rout, P.K.; Sahu, B.K. Residential Sector Demand Side Management: A Review. In Proceedings of the 2021 1st Odisha International Conference on Electrical Power Engineering, Communication and Computing Technology (ODICON), Bhubaneswar, India, 8–9 January 2021; pp. 1–6.
52. Shewale, A.; Mokhade, A.; Funde, N.; Bokde, N.D. A Survey of Efficient Demand-Side Management Techniques for the Residential Appliance Scheduling Problem in Smart Homes. *Energies* **2022**, *15*, 2863–2896. [CrossRef]
53. Yan, X.; Ozturk, Y.; Hu, Z.; Song, Y. A review on price-driven residential demand response. *Renew. Dustainable Energy Rev.* **2018**, *96*, 411–419. [CrossRef]
54. Mohamad, H.; Mokhlis, H.; Bakar, A.H.A.; Ping, H.W. A review on islanding operation and control for distribution network connected with small hydro power plant. *Renew. Sustain. Energy Rev.* **2011**, *15*, 3952–3962. [CrossRef]
55. Bibak, B.; Tekiner-Moğulkoç, H. A comprehensive analysis of Vehicle to Grid (V2G) systems and scholarly literature on the application of such systems. *Renew. Energy Focus* **2021**, *36*, 1–20. [CrossRef]
56. Rahman, S.; Khan, I.A.; Khan, A.A.; Mallik, A.; Nadeem, M.F. Comprehensive review & impact analysis of integrating projected electric vehicle charging load to the existing low voltage distribution system. *Renew. Sustain. Energy Rev.* **2022**, *153*, 111756. [CrossRef]
57. Yong, J.Y.; Ramachandaramurthy, V.K.; Tan, K.M.; Mithulananthan, N. A review on the state-of-the-art technologies of electric vehicle, its impacts and prospects. *Renew. Sustain. Energy Rev.* **2015**, *49*, 365–385. [CrossRef]
58. Khare, V.; Nema, S.; Baredar, P. Solar–wind hybrid renewable energy system: A review. *Renew. Sustain. Energy Rev.* **2016**, *58*, 23–33. [CrossRef]
59. Mohammed, Y.S.; Mustafa, M.W.; Bashir, N. Hybrid renewable energy systems for off-grid electric power: Review of substantial issues. *Renew. Sustain. Energy Rev.* **2014**, *35*, 527–539. [CrossRef]
60. Muyiwa, A. Wind Resources and Future Energy Security: Environmental, Social, and Economic Issues. 2015. Available online: https://search.ebscohost.com/login.aspx?direct=true&db=nlebk&AN=996965&site=eds-live (accessed on 1 January 2015).
61. Safipour, R.; Sadegh, M.O. Optimal Planning of Energy storage Systems in Microgrids for Improvement of operation Indices. *Int. J. Renew. Energy Res.* **2018**, *8*, 1483–1498.
62. Ehsan, A.; Yang, Q. Optimal integration and planning of renewable distributed generation in the power distribution networks: A review of analytical techniques. *Appl. Energy* **2018**, *210*, 44–59. [CrossRef]
63. Liu, J.; Chen, X.; Xiang, Y.; Huo, D.; Liu, J. Optimal palanning and investment benefit analysis of shared energy storage of retailers. *Int. J. Electr. Power Energy Syst.* **2020**, *126*, 106561. [CrossRef]
64. Lin, P.; Peng, Z.; Lai, Y.; Cheng, S.; Chen, Z.; Wu, L. Short-term power prediction for photovoltaic power plants using improved Kmeans-GRA-Elman model based on multivariate meteorological factors and historical power datasets. *Energy Convers. Manag.* **2018**, *177*, 704–717. [CrossRef]
65. Eseye, A.T.; Zhang, J.; Zheng, D. Short-term photovoltaic solar power forecasting using a hybrid Wavelet-PSO-SVM model based on SCADA and meteorological information. *Renew. Energy* **2017**, *118*, 357–367. [CrossRef]
66. Ahmad, M.W.; Mourshed, M.; Rezgui, Y. Tree-based ensemble methods for predicting PV poer generation and their comparison with support vector regression. *Energy* **2018**, *164*, 465–474. [CrossRef]
67. Khan, W.; Walker, S.; Zeiler, W. Improved solar photovoltaic energy generation forecast using deep learnin-based ensemble stacking approach. *Energy* **2022**, *240*, 122812. [CrossRef]
68. Patel, A.; Swathika, O.V.G.; Subramanian, U.; Babu, T.S.; Tripathi, A.; Nag, S.; Karthick, A.; Muhibbullah, M. A practical approach for predicting predicting power in a small-scale off-grid photovoltaic system using machine learning algorithms. *Int. J. Photoenergy* **2022**, *2022*, 9194537. [CrossRef]
69. Adibah, Z.; Kamaruzzaman; Mohamed, A. Dynamic Voltage Stability of a Distribution System with High Penetration of Grid-Connected Photovoltaic Type Solar Generator. *J. Electr. Syst.* **2016**, *12*, 239–248.
70. AlShammari, M.M. Evolution of Issues on Distributed Systems: A Systematic Review. In Proceedings of the 2020 19th International Symposium on Distributed Computing and Applications for Business Engineering and Science (DCABES), Xuzhou, China, 16–19 October 2020; pp. 33–37.
71. Das, C.K.; Bass, O.; Kothapalli, G.; Mahmoud, T.S.; Habibi, D. Overview of energy storage systems in distribution networks: Placement, sizing, operation, and power quality. *Renew. Sustain. Energy Rev.* **2018**, *91*, 1205–1230. [CrossRef]
72. Liu, W.; Niu, S.; Xu, H. Optimal planning of battery energy storage considering reliability benefit and operation strategy in active distribution system. *J. Mod. Power Syst. Clean Energy* **2016**, *5*, 177–186. [CrossRef]
73. Phuangpornpitak, N.; Tia, S. Opportunities and Challenges of Integrating Renewable Energy in Smart Grid System. In *10th Eco-Energy and Materials Science and Engineering Symposium*; Elsevier: Amsterdam, The Netherlands, 2013; pp. 282–290.
74. Strezosky, L.; Sajadi, A.; Prica, M.; Loparo, K.A. Distribution System Operational Challenges Following integration of Renewable Generation. In Proceedings of the IEEE Energy Tech, Cleveland, OH, USA, 5 September 2016.

75. Almas, H.; Shahrouz, A.; Jörn, A. The Development of Renewable Energy Sources and Its Significance for the Environment. 2015. Available online: https://search.ebscohost.com/login.aspx?direct=true&db=nlebk&AN=980370&site=eds-live (accessed on 1 January 2015).
76. Esmaili, M. Placement of minimum distributed generation units observing power losses and voltage stability with network constraints. *IET Gener. Transm. Distrib.* **2013**, *7*, 813–821. [CrossRef]
77. Anderson, P.M.; Fouad, A.A. *Power System Control and Stability*, 2nd ed.; IEEE Press, Inc.: Piscataway, NJ, USA, 2003.
78. Xu, X.; Yan, Z.; Shahidehpour, M.; Wang, H.; Chen, S. Power System Voltage Stability Evaluation Considering Renewable Energy with Correlated Variability. *IEEE Trans. Power Syst.* **2018**, *33*, 3236–3245. [CrossRef]
79. Ayub, M.A.; Khan, H.; Peng, J.; Liu, Y. Consumer-Driven Demand-Side Management Using K-Mean Clustering and Integer Programming in Standalone Renewable Grid. *Energies* **2022**, *15*, 1006. [CrossRef]
80. Sharma, S.; Niazi, K.R.; Verma, K.; Rawat, T. Coordination of different DGs, BESS and demand response for multi-objective optimization of distribution network with special reference to Indian power sector. *Int. J. Electr. Power Energy Syst.* **2020**, *121*, 106074. [CrossRef]
81. Liang, Y.; Deng, T.; Max Shen, Z.-J. Demand-side energy management under time-varying prices. *IISE Trans.* **2019**, *51*, 422–436. [CrossRef]
82. Dey, B.; Basak, S.; Pal, A. Demand-side management based optimal scheduling of distributed generators for clean and economic operation of a microgrid system. *Int. J. Energy Res.* **2022**, *46*, 8817–8837. [CrossRef]
83. Qin, H.; Wu, Z.; Wang, M. Demand-side management for smart grid networks using stochastic linear programming game. *Neural Comput. Appl.* **2020**, *32*, 139–149. [CrossRef]
84. Latifi, M.; Khalili, A.; Rastegarnia, A.; Bazzi, W.M.; Sanel, S. Demand-side management for smart grid via diffusion adaptation. *IThe Inst. Eng. Technol. Smart Grid* **2020**, *3*, 69–82. [CrossRef]
85. Hecht, C.; Sprake, D.; Vagapov, Y.; Anuchin, A. Domestic demand-side management: Analysis of microgrid with renewable energy sources using historical load data. *Electr. Eng.* **2021**, *103*, 1791–1806. [CrossRef]
86. Hossain, A.M.; Chakrabortty, R.K.; Ryan, M.J.; Pota, H.R. Energy Management of Community Energy Storage in Grid-connected Microgrid under Real-time Prices. *Sustain. Cities Soc.* **2020**, *66*, 102658. [CrossRef]
87. Hossain, A.M.; Pota, H.R.; Squartini, S.; Zaman, F.; Guerrero, J.M. Energy Scheduling of Community Microgrid with Battery Cost using Particle Swarm Optimisation. *Appl. Energy* **2019**, *254*, 113723. [CrossRef]
88. Gallo, A.B.; Simões-Moreira, J.R.; Costa, H.K.M.; Santos, M.M.; Moutinho dos Santos, E. Energy storage in the energy transition context: A technology review. *Renew. Sustain. Energy Rev.* **2016**, *65*, 800–822. [CrossRef]
89. Boampong, R. Evaluating the Energy-Saving Effects of a Utility Demand-Side Management Program: A Difference-in-Difference Coarsened Exact Matching Approach. *Energy J.* **2020**, *41*, 185–207. [CrossRef]
90. Rawat, T.; Niazi, K.R.; Gupta, N.; Sharma, S. Impact analysis of demand response on optimal allocation of wind and solar based distributed generations in distribution system. *Energy Sources Part B Econ. Plan. Policy* **2021**, *16*, 75–90. [CrossRef]
91. Mulleriyawage, U.G.K.; Shen, W.X. Impact of demand side management on optimal sizing of residential battery energy storage system. *Renew. Energy Int. J.* **2021**, *172*, 1250–1266. [CrossRef]
92. Rawat, T.; Niazi, K.R.; Gupta, N.; Sharma, S. Joint Allocation and Operational management of DG and BESS in Distribution System in Presence of Demand Response. In Proceedings of the 2019 8th International Conference on Power Systems (ICPS), Jaipur, India, 20–22 December 2019.
93. Ahmad, S.; Alhaisoni, M.; Naeem, M.; Ahmad, A.; Altaf, M. Joint Energy Management and Energy Trading in Residential Microgrid System. *IEEE Access* **2017**, *4*, 123334–123346. [CrossRef]
94. Ochoa, L.F.; Harrison, G.P. Minimizing energy losses: Optimal accommodation and and smart operation of renewable distributed generation. *IEEE Trans. Power Syst.* **2011**, *26*, 198–205. [CrossRef]
95. Hung, D.Q.; Mithulananthan, N. DG allocation in primary distribution systems considering loss reduction. In *Handbook of Renewable Energy Technology*; Zobaa, A.F., Bansal, R.C., Eds.; World Scientific Publishers: Singapore, 2011.
96. Kyritsis, A.; Voglitsis, D.; Papanikolaou, N.; Tselepis, S.; Christodoulou, C.; Gonos, I.; Kalogirou, S.A. Evolution of PV systems in Greece and review of applicable solutions for higher penetration levels. *Renew. Energy Int. J.* **2017**, *109*, 487–499. [CrossRef]
97. Kapoor, K.; Pandey, K.K.; Jain, A.K.; Nandan, A. Evolution of solar energy in India: A review. *Renew. Sustain. Energy Rev.* **2014**, *40*, 475–487. [CrossRef]
98. Castillo-Calzadilla, T.; Cuesta, M.A.; Olivares-Rodriguez, C.; Macarulla, A.M.; Legarda, J.; Borges, C.E. Is it feasible a massive deployment of low voltage direct current microgrids renewable-based? A technical and social sight. *Renew. Sustain. Energy Rev.* **2022**, *161*, 112198. [CrossRef]
99. Centre/BNEF, F.S.-U. Global Trends in Renewable Energy Investment. Available online: http://www.fs-unep-centre.org (accessed on 24 June 2019).
100. Rawat, T.; Niazi, K.R.; Gupta, N.; Sharma, S. A two-stage optimization framework for scheduling of responsive loads in smart distribution system. *Int. J. Electr. Power Energy Syst.* **2021**, *129*, 106859. [CrossRef]
101. Sharma, S.; Niazi, K.R.; Verma, K.; Meena, N.K. Multiple DRPs to maximise the techno-economic benefits of the distribution network. *J. Eng.* **2019**, *2019*, 5240–5244. [CrossRef]
102. Sharma, S.; Niazi, K.R.; Verma, K.; Rawat, T. A bi-level optimization framework for investment planning of distributed generation resources in coordination with demand response. *Energy Sources Part A Recovery Util. Environ. Eff.* **2020**, *42*, 1–18. [CrossRef]

103. Sharma, S.; Niazi, K.R.; Verma, K.; Thokar, R.A. Bilevel optimization framework for impact analysis of DR on optimal accommodation of PV and BESS in distribution system. *Int. J. Electr. Energy Syst.* **2019**, *29*, e12062. [CrossRef]
104. Rawat, T.; Niazi, K.R.; Gupta, N.; Sharma, S. Impact assessment of electric vehicle charging/discharging strategies on the operation management of grid accessible and remote microgrids. *Int. J. Energy Res.* **2019**, *43*, 9034–9048. [CrossRef]
105. Gamage, G.; Withana, N.; Silva, C.; Samarasinghe, R. Battery Energy Storage based Approach for Grid Voltage Regulation in Renewable Rich Distribution Network. In Proceedings of the 2020 2nd IEEE International Conference on Industrial Electronics for Sustainable Energy Systems (IESES), Cagliari, Italy, 1–3 September 2020.
106. Chauhan, A.; Saini, R.P. A review on Integrated Renewable Energy System based power generation for stand-alone applications: Configurations, storage options, sizing methodologies and control. *Renew. Sustain. Energy Rev.* **2014**, *38*, 99–120. [CrossRef]
107. Georgilakis, P.S.; Hatziargyriou, N.D. Optimal Distributed Generation Placement in Power Distribution Networks: Models, Methods and future Research. *IEEE Trans. Power Syst.* **2013**, *28*, 3420–3428. [CrossRef]
108. Riaz, M.; Ahmad, S.; Hussain, I.; Mihet-Popa, L. Probabilistic Optimization Techniques in Smart Power System. *Energies* **2022**, *15*, 825. [CrossRef]
109. Rawat, T.; Niazi, K.R.; Gupta, N.; Sharma, S. A linearized multi-objective Bi-level approach for operation of smart distribution systems encompassing demand response. *Energy* **2022**, *238*, 121991. [CrossRef]
110. Sharma, S.; Niazi, K.R.; Verma, K.; Rawat, T. Impact of Multiple Battery Energy Storage System Strategies on Energy Loss of Active Distribution Network. *Int. J. Renew. Energy Res* **2019**, *9*, 1705–1711.
111. Tan, W.-S.; Hassan, M.Y.; Majid, M.S.; Abdul Rahman, H. Optimal distributed renewable generation planning: A review of different approaches. *Renew. Sustain. Energy Rev.* **2013**, *18*, 626–645. [CrossRef]
112. Thokar, R.A.; Gupta, N.; Swarnkar, A.; Sharma, S.; Meena, N.K. Optimal Integration and Management of Solar Generation and Battery Storage System in Distribution Systems under Uncertain Environment. *Int. J. Renew. Energy Res.* **2020**, *10*, 11–12.
113. Sandhu, K.S.; Mahesh, A. Optimal sizing of PV/wind/battery Hybrid Renewable Energy System Considering Demand Side Management. *Int. J. Electr. Eng. Inform.* **2018**, *10*, 79–93. [CrossRef]
114. Arias, L.A.; Rivas, E.; Santamaria, F.; Hernandez, V. A Review and Analysis of Trends Related to Demand Response. *Energies* **2018**, *11*, 1617. [CrossRef]
115. Hakimi, S.M.; Moghaddas-Tafreshi, S.M. Optimal Planning of a Smart Microgrid Including Demand Response and Intermittent Renewable Energy Resources. *IEEE Trans. Smart Grid* **2014**, *5*, 2889–2900. [CrossRef]
116. Setlhaolo, D.; Xia, X.; Zhang, J. Optimal scheduling of household apppliances for demand response. *Electr. Power Syst. Res.* **2014**, *116*, 24–28. [CrossRef]
117. Hussain, I.; Mohsin, S.; Basit, A.; Khan, Z.A.; Qasim, U.; Javai, N. A review on Demand Response: Pricing, Optimization, and Appliance Scheduling. In *6th International Conference on Ambient Systems, Network and Technologies, ANT 2015*; Elsevier: Amsterdam, The Netherlands, 2015.
118. Sharma, S.; Niazi, K.R.; Verma, K.; Rawat, T. Impact of battery energy storage, controllable load and network reconfiguration on contemporary distribution network under uncertain environment. *IET Gener. Transm. Distrib.* **2020**, *14*, 4719–4727. [CrossRef]
119. Ding, M.; Xu, Z.; Wang, W.; Wang, X.; Song, Y.; Chen, D. A review on China's large-scale PV integration: Progress, challenges and recommendations. *Renew. Sustain. Energy Rev.* **2016**, *53*, 630–652. [CrossRef]
120. Parizy, E.S.; Ardakani, A.J.; Mahammadi, A.; Loparo, K.A. A new quantitative load profile measure for demand response performance evaluation. *Electr. Power Energy Syst.* **2020**, *121*, 106073. [CrossRef]

Disclaimer/Publisher's Note: The statements, opinions and data contained in all publications are solely those of the individual author(s) and contributor(s) and not of MDPI and/or the editor(s). MDPI and/or the editor(s) disclaim responsibility for any injury to people or property resulting from any ideas, methods, instructions or products referred to in the content.

MDPI
St. Alban-Anlage 66
4052 Basel
Switzerland
Tel. –41 61 683 77 34
Fax –41 61 302 89 18
www.mdpi.com

Energies Editorial Office
E-mail: energies@mdpi.com
www.mdpi.com/journal/energies

www.ingramcontent.com/pod-product-compliance
Lightning Source LLC
LaVergne TN
LVHW070438100526
838202LV00014B/1618